STRUCTURAL GEOLOGY

Edited by

Antonella Galetto

TAP

Toronto Academic Press

STRUCTURAL GEOLOGY

Editor: Antonella Galetto

Toronto Academic Press
4164 Lakeshore Road
Burlington ON L7L 1A4
Canada
www.tap-books.com

Email: orders@arclereducation.com

ISBN: 978-1-77956-308-8

ABOUT THE EDITOR

Antonella Galetto, is a PhD geologist with training in various earth sciences disciplines, including renewable energy exploration, geochronology, and geographic information systems. She graduated from the National University of Cordoba in Argentina. Her research projects have primarily focused on the exploration of geothermal energy and studying the evolution of the Andes through collaboration with multiple scientific groups worldwide. Currently, she's working on processing geospatial imagery for various exploration purposes and serve as a tutor of geographic information systems and geodata analysis at the university level. Additionally, she work as an academic editor for a company that offers revision, polishing, and translation services for scientific research papers in the geosciences field.

Table of Contents

List of Figures

List of Tables

PREFACE

For decades, tectonics and structural geology have been pivotal in Earth resource development, shaping the world we live in today. The resource revolutions of the past, such as coal mining during the Industrial Revolution, the extraction of minerals and metals, and the unlocking of hydrocarbons, were largely driven by the expertise of tectonics and structural geologists. However, these activities have significantly contributed to the climate and ecological crises we face today.

Over the past 50 years, structural geology has evolved into an integrative, synthetic science focused on deciphering the history preserved in the rock record and understanding the processes of rock deformation. This field combines historical elements, such as structural inheritance and tectonic history, with theoretical, process-based approaches. Its strength lies in using these approaches simultaneously to determine the three-dimensional architecture, kinematic evolution, and dynamic conditions of lithospheric deformation across a wide range of spatial and temporal scales.

Our approach to using Earth resources is changing out of necessity. We have consumed, produced, and polluted to the point where we must fundamentally change our practices to ensure the survival of the ecosystems that support us. Climate scenarios modeled by the International Panel on Climate Change indicate that we can avoid the most dangerous impacts of climate change if we limit the global temperature rise to 1.5°C above pre-industrial levels. To achieve this, we need to transition to a 'net-zero carbon' world by 2040 and reduce non-CO2 greenhouse gas emissions.

Action is imperative. We need to drastically reduce greenhouse gas emissions and deploy negative-emissions technologies to balance the environmental impact while promoting sustainable global development. This shift will lead to innovative uses of the subsurface, such as fluid cycling through injection and storage, CO2 disposal, and waste management. Resource extraction will diversify and its environmental footprint will shrink. These new uses of the subsurface will increase the demand for structural geological expertise. Consequently, we will need structural geologists who are adept at integrating across disciplines and capable of communicating and collaborating with various stakeholders, including policymakers, the public, and communities, to ensure that our sustainable future benefits everyone.

Contents/Structure of the Book

This book explores the fundamental principles and complex phenomena of geological structures and processes. This book is divided into 10 chapters, each focusing on a critical aspect of structural geology.

Chapter 1 introduces the field of structural geology, outlining its significance in understanding the Earth's crust and its applications in various geological and engineering contexts. It covers the basics of structural analysis, including the identification and description of geological structures, mapping techniques, and the tools used for structural investigations. This chapter sets the foundation for understanding the subsequent topics by establishing the importance of structural geology in interpreting geological history and predicting geological behavior.

Chapter 2 deals with the processes and mechanisms of rock deformation. It explains the conditions under which rocks deform, the types of deformation (elastic, brittle, and ductile), and the factors influencing these processes, such as temperature, pressure, and time. The chapter also covers the deformation mechanisms at the microscopic and macroscopic scales, providing a comprehensive understanding of how rocks respond to tectonic forces.

Focusing on the concept of strain, Chapter 3 discusses how rocks change shape and volume in response to stress. It introduces strain analysis techniques, including graphical methods and mathematical models, to quantify deformation. The chapter also explores different types of strain (homogeneous and heterogeneous) and their significance in interpreting geological structures.

Stress is a fundamental concept in structural geology, and Chapter 4 provides a detailed examination of stress and its role in deformation. It covers the principles of stress, including normal and shear stress, stress tensors, and the relationships between stress and strain. The chapter also discusses the sources of stress in the Earth's crust, such as tectonic forces, gravitational forces, and thermal processes.

Chapter 5 explores the geological settings and processes associated with extensional regimes, where the Earth's crust is being pulled apart. It examines the structures formed in these settings, such as normal faults, rift valleys, and grabens, and the tectonic processes driving extension. The chapter also discusses the implications of extensional tectonics for resource exploration and geohazard assessment.

The focus of Chapter 6 shifts to brittle deformation and the formation of fractures in rocks. The chapter explains the conditions under which brittle deformation occurs, the types of fractures (joints, faults, and veins), and their significance in geological and engineering contexts. It also covers the mechanics of fracturing and the factors influencing fracture propagation.

Faults are significant geological structures. Chapter 7 provides an in-depth analysis of their formation, classification, and behavior. It covers different types of faults (normal, reverse, and strike-slip), fault mechanics, and the role of faults in accommodating crustal deformation. The chapter also explores the implications of faulting for earthquake hazards and resource exploration.

Chapter 8 examines the movement and deformation of rocks in the brittle regime. It introduces kinematic analysis techniques to understand the movement along faults and fractures. The chapter also discusses paleostress analysis, which reconstructs the stress fields responsible for past deformation events, providing insights into the tectonic history of a region.

Salt tectonics is a specialized field within structural geology. Chapter 9 explores the unique behavior of salt and its role in geological structures. It covers the properties of salt, the formation of salt domes and diapirs, and the impact of salt tectonics on sedimentary basins and hydrocarbon reservoirs.

Chapter 10 focuses on the formation and analysis of folds, which are common structures in deformed rocks. It explains the mechanisms of folding, the classification of folds based on geometry and orientation, and the significance of folds in geological interpretation. The chapter also covers the techniques used to analyze folds and their implications for understanding regional tectonics.

Together, these chapters provide a comprehensive understanding of structural geology, equipping readers with the knowledge and skills to analyze and interpret geological structures and processes.

Audience

The audience for this book includes undergraduate and graduate students studying geology, earth sciences, or related fields, as well as professionals and researchers in the geosciences industry. The content is designed to be accessible to those with a basic understanding of geology while also providing in-depth insights and advanced concepts for those seeking a deeper understanding of structural geology principles and applications. The book serves as a comprehensive resource for anyone interested in the deformation of the Earth's crust, tectonics, and the processes that shape the geological features of our planet.

Learning Features

- Start-of-chapter learning objectives
- Key terms in each chapter
- Summary points at the end of each chapter
- Review Questions at the end of each chapter
- Case Studies and Activities in each chapter for solving real problems
- References at the end of each chapter

CHAPTER

1 Structural Geology and Structural Analysis

LEARNING OBJECTIVES

After studying this chapter, you will be able to:

- Understand the basics of structural geology
- Discuss the relationship between structural geology and tectonics
- Describe structural data and field data
- Define remote sensing and geodesy
- Describe seismic data
- Explain structural analysis

KEY TERMS FROM THIS CHAPTER

Brittle deformation	Ductile rocks
Elastic modulus	Geologic map
Geological region	Geotechnical risk
Hydrothermal fluids	Kinematic analysis
Lithosphere	Orogeny
Plate tectonics	Rheologic analysis
Structural geology	Structural mapping

INTRODUCTION

Structural geology is the study of how rocks are distributed in three dimensions and how they have been deformed over time. The main objective of structural geology is to analyze the current shapes of rocks to learn about their past deformation history, ultimately understanding the stress forces that caused these changes. By understanding the stress patterns, scientists can make connections to important geological events in the past, such as mountain building and rifting, due to plate tectonics. Structural geologists analyze micro- and meso-scale structures within rocks to develop tools and methods to identify larger structures that cannot be seen directly, although satellite imagery can now assist in this process.

Because outcrops are frequently rare or unavailable in important areas, direct observation of large structures is not possible, necessitating correlative interpretation. This makes the study of primary and secondary structures a challenging endeavor. Furthermore, because the relief is typically small in comparison to the map area, the geological observation is essentially two dimensional. Thus, even in regions of nearly continuous outcrop, interpretation is still required to create a three-dimensional image.

In addition, structural geologists have to piece together the rocks' deformation history based on the patterns of primary and secondary structures they discover in the field. This goal necessitates fieldwork, i.e., direct examination of rocks in their native settings, such as drill cores, outcrops, and landscapes. Every feature is thought to be a record of an event that the carefully examined outcrop has gone through. Changes in strain or stress axes are thought to be preserved by changes in structural orientations. Understanding the relative ages of various structures, such as folds, faults, and fabrics, through crosscutting or overprinting relationships, is therefore essential to understanding the deformation history.

1.1. BASICS OF STRUCTURAL GEOLOGY

The study of folds, faults, and other lithosphere deformation structures—their appearance, formation process, and motivation—is the focus of structural geology. Structures occur in many different settings and have undergone exciting changes in stress and strain; if we can decipher the code, we can access information ranging from features hundreds of kilometers long to microscopic details.

1.1.1. Use and Importance

In the fields of economic geology, mining geology, and petroleum geology, the study of geologic structures has always been crucial. Often, faulted and folded rock layers create traps that hold and concentrate liquids like natural gas and petroleum. Similar to this, regions that are faulted and structurally complex are noteworthy for being hydrothermally fluid-permeable zones, which leads to concentrated areas of deposits of base and precious metal ore. In structurally complex regions, veins of minerals containing different metals frequently occupy faults and fractures. These faulted and structurally fractured areas frequently coexist with intrusive igneous rocks. They frequently also surround collapse features like old sinkholes and geologic reef complexes. Gold, silver, copper, lead, zinc, and other metal deposits are frequently found in areas with complex structural features (Jones et al., 2009).

An essential component of engineering geology, which studies the mechanical and physical characteristics of natural rocks, is structural geology. Internal flaws in rocks called structural fabrics and defects, such as folds, foliations, joints, and faults, can compromise the stability of man-made constructions like road cuts, dams, open pit mines, underground mines, and tunnels for vehicles.

The only way to examine geotechnical risk, which includes earthquake risk, is to look at both structural geology and geomorphology together. Furthermore, these scientists place special emphasis on karst landscape sections that are situated atop caverns, possible sinkholes, or other collapse features. Moreover, regions with steep slopes may collapse or experience landslides.

The principles of structural geology must be applied by environmental geologists and hydrogeologists in order to comprehend how groundwater penetration and flow are affected by, or are affected by, geologic sites. For example, a hydrogeologist might have to ascertain whether hazardous material seepage from landfills is happening in a residential neighborhood or whether briny water seeping into an aquifer.

The theory of plate tectonics, which was developed in the 1960s, explains how continents move as a result of the separation and collision of crustal plates. It is utilized throughout structural geology as a framework to examine and comprehend features at the local, regional, and global scales. It is essentially structural geology on a planet scale.

1.1.2. Methods

Structural geologists use a variety of methods to (first) measure rock geometries, (second) reconstruct their deformational histories, and (third) estimate the stress field that resulted in that deformation.

1.1.2.1. Geometries

In the field, primary data sets for structural geology are gathered. Planar features such as bedding planes, foliation planes, fold axial planes, fault planes, and joints, as well as linear features like stretching lineations, which are the traces of planar features on other planar surfaces, fold axes, and intersection lineations, which are the minerals that are ductilely extended, are all measured by structural geologists.

1.1.2.1.1. Measurement Conventions

In geology, the strike and dip of a planar structure are used to determine its inclination. The right-hand convention defines the strike as the point where a planar feature intersects a horizontal plane, and the dip as the amount of inclination below the horizontal at a right angle to the strike. For instance, N25E,45SE is the record of a strike that is 25 degrees east of north and dips 45 degrees southeast.

As an alternative, since dip and dip direction are absolute, they can be used. A 360-degree dip is measured, usually starting from the north and moving clockwise. As an illustration, consider a 45-degree dip toward 115-degree azimuth, represented as 45/115. Keep in mind that this is identical to what was said before.

The term "hade," which refers to a plane's deviation from verticality, is used infrequently (90°-dip).

Plunge of the fold axis is measured in dip and dip direction, or more precisely, azimuth of plunge and plunge. A fold axial plane's orientation is expressed in terms of strike and dip or dip and dip direction.

If at all possible, measurements of lines are made in terms of their dip and direction. Lineations are frequently expressed on a planar surface and can be challenging to quantify precisely. In this instance, the lineation can be assessed as a pitch or rake across the surface, measured from the horizontal.

A protractor flat on a planar surface with the flat edge horizontal is used to measure rake. The angle of the lineation is then measured clockwise from horizontal. Using a stereographic projection and the rake and strike-dip data of the plane the lineation was measured from, the lineation's orientation can then be determined.

If movement on the plane, such as slickensides, has created lineations on a fault, this is noted as a lineation with a rake and annotated with the indication of throw on the fault.

Although there is a benefit to using different formats that discriminate between planar and linear data, it is generally easier to record strike and dip information of planar structures in dip/dip direction format as this will match all the other structural information you may be recording about folds, lineations, etc.

1.1.2.1.2. Plane, Fabric, Fold, and Deformation Conventions

The standard procedure for examining structural geology is to locate the linear structures, also known as planar fabrics since this suggests a textural formation, and then analyze the linear structures to determine deformations.

The order of formation of planar structures is indicated by their naming convention, with the original sedimentary layering being designated as S_0 at the lowest level. In instances where it is challenging to identify S_0 in heavily altered rocks, numbering may begin at a random number or a letter may be assigned (such as S_A). If there is bedding-plane foliation resulting from burial metamorphism or diagenesis, it may be denoted as S_{0A}.

The folds, if any, are denoted by the numbers F_1, F_2, etc. Folding typically creates the axial plane foliation or cleavage of a fold, and the number convention should correspond with this. An F_2 fold, for instance, ought to have S_2 axial foliation.

Deformations are labeled in sequential order of formation using the letter D to indicate a deformation event. This means that folds and foliations should correspond to these events. For instance, D_1, D_2, and D_3. Since deformation events form folds and foliations, these events should correlate with each other. A D_2 deformation, for instance, would result in an F_2 fold with an S_2 axial plane foliation.

Multiple deformations may occur during metamorphic events. Identifying them in the same way as the structural characteristics they are accountable for, such as M_2, can be helpful at times. This could be accomplished by geochronology, by identifying metamorphic mineral assemblages formed by distinct events, or by observing porphyroblast formation in cleavages of known deformation age.

Because they result from the intersection of two planar structures, intersection lineations in rocks are named after the two planar structures that gave rise to them. For example, the L_{1-0} intersection lineation (also called the cleavage-bedding lineation) is the intersection lineation of an S_1 cleavage and bedding.

Quantifying stretching lineations can be challenging, particularly in highly stretched ductile rocks with little to no foliation information remaining. They can be recognized similarly to planar surfaces and folds, such as L_1, L_2, when correlated with deformations when at all possible (since few are formed in folds and many are not strictly associated with planar foliations). Although it is usually unnecessary, some geologists find it more convenient to annotate them with a subscript S, such as Ls_1, to distinguish them from intersection lineations.

1.1.2.1.3. Stereographic Projections

A two-dimensional grid projection is used in stereographic projection to plot linear and planar features (structural strike and dip readings, usually obtained with a compass clinometer) passing through an imagined sphere. This allows for a more comprehensive analysis of a set of measurements. Stereographic projection is a technique for analyzing the nature and orientation of deformation stresses, lithological units, and penetrative fabrics. Within the field of structural geology, Richard W. Allmendinger's Stereonet is extensively utilized.

1.1.2.1.4. Rock Macro-Structures

Structural geology is, broadly speaking, the study of the relationships and interactions between stratigraphic units in three dimensions within rock terranes or geological regions.

The orientation, deformation, and relationships of stratigraphy, or bedding, which may have been faulted, folded, or given a foliation by some tectonic event, are the main topics of discussion in this area of structural geology. This is primarily a geometric science, from which one can create three-dimensional block models and cross sections of various rock types, regions, terranes, and portions of the Earth's crust.

Understanding orogeny and plate tectonics, as well as the oil, gas, and mineral exploration industries in particular, require an understanding of regional structure because features like faults, folds, and unconformities are the main factors influencing ore mineralization and oil traps.

Seismic tomography and seismic reflection are being used to study modern regional structure in three dimensions, offering unparalleled views of the Earth's interior, faults, and deep crust. Additional geophysical data, such as airborne magnetics and gravity, can reveal details about the types of rocks that are thought to be in the deep crust (McCaffrey et al., 2005).

1.1.2.1.5. Rock Microstructures

Structural geologists examine the microstructure, or texture, of rocks on a small scale to gain detailed information about metamorphic rocks and certain features of sedimentary rocks, usually regarding folding.

Measurement and characterization of foliations, crenulations, metamorphic minerals, and the timing of the relationships between these structural and mineralogical features are all part of the textural study process. Typically, this entails gathering manual specimens, which can then be sliced into petrographic thin sections for examination under a petrographic microscope. In order to analyze certain rock features that exhibit scale invariance, microstructural analysis is also applied in multi-scale statistical analysis.

1.1.2.2. Kinematics

Rock geometry measurements are used by geologists to learn about the strain history of rocks. Both brittle faulting and ductile folding and shearing are examples of strain. Where temperatures and pressures are higher, ductile deformation occurs in the deeper crust and brittle deformation occurs in the shallow crust.

1.1.2.3. Stress Fields

Geologists can interpret the stress and strain relationships in rocks to understand how rock deformation occurred in the past. The following features are commonly analyzed to determine stress fields from deformation structures.

- In perfectly brittle rocks, faulting occurs at 30° to the greatest compressional stress. (Byerlee's Law)
- The greatest compressive stress is normal to fold axial planes.

1.1.2.4. Modeling

Structural geology modeling is becoming more and more crucial for research and economic geology, including mineral and petroleum development. Anticlines, synclines, fold and thrust belts, and other structural systems can be represented in 2D and 3D models to help better understand how a structure changes over time. Geologists can only understand the surface geological mapping if they do not model or interpret the subsurface. Major economic potential could be lost if the region were solely dependent on its surface geology and ignored its structural and tectonic history.

1.1.3. Characterization of the Mechanical Properties of Rock

Deep beneath the surface of the earth, structures formed by deformation are largely determined by the mechanical characteristics of rock. Different structures that geologists observe above ground in the field will result from the conditions under which a rock is present. The goal of structural geology is to connect the alterations that the rock underwent to arrive at its ultimate structure with the formations that are visible to humans. The history of the rock's deformation can be clarified by understanding the deformational conditions that result in such structures.

The deformation of rock is largely dependent on temperature and pressure. Rocks are ductile at the extremely high temperatures and pressures found beneath the earth's crust. They may break, bend, or fold. The stress and strain fields are other essential elements that help form the structure of rock beneath the surface of the earth.

1.1.3.3. Stress-Strain Curve

As a directed force over area, stress is a form of pressure. A rock undergoes shape changes when it is under stress. The rock might or might not take on its original shape after the tension is released. Strain, or the change in length over the material's original length in one dimension, is used to quantify that change in shape. Strain is brought on by stress, and strain eventually changes the structure.

A reversible deformation is referred to as elastic deformation. Put another way, a rock returns to its original shape when the stress on it is released. Atomic bonds can be stretched, compressed, or distorted in reversible linear elasticity. When the force is released, the material bounces back since there is no bond breaking. A linear relationship, or Hookean relationship, between stress and strain is used to model this kind of deformation.

$$\grave{o} = \frac{\sigma}{E}$$

where E is the material-dependent elastic modulus, σ and ⊠ stand for stress and strain, respectively. In essence, the elastic modulus is a gauge for the strength of atomic bonds.

Non-reversible deformation is referred to as plastic deformation. For permanent deformation, the relationship between strain and stress is nonlinear. Because stress involves the breaking of bonds, the material has undergone a permanent change in shape.

An applied stress can cause dislocations to move, which is one mechanism of plastic deformation. We can think of rocks as poly-crystalline materials because they are essentially aggregates of minerals. An additional or absent half plane of atoms in the periodic array of atoms that make up a crystal lattice is known as a dislocation, and it is one kind of crystallographic defect. Every genuine crystallographic material has dislocations.

1.1.3.4. Hardness

Hardness is hard to measure. It is a measurement of the resistance to permanent deformation, in particular. Hardness as a surface quality—a gauge of a material's abrasiveness or resistance to surface scratches—has a historical background. On the other hand, if the material under test has a uniform composition and structure, its surface is only a few atomic layers thick, and the measurements pertain to the bulk material.

Thus, simple surface measurements yield information about the bulk properties. Ways to measure hardness include:

- Mohs Scale
- Dorry abrasion test
- Deval abrasion test
- Indentation hardness

In metallurgy and materials science, indentation hardness is frequently used to describe a material's resistance to indenter penetration.

1.1.3.5. Toughness

The best way to characterize toughness in a material is its ability to resist cracking. A material absorbs energy until it fractures during plastic deformation. The work necessary to fracture the material is represented by the area under the stress-strain curve.

The toughness modulus is defined as:

$$M_t = \frac{2}{3}\sigma_{UTS}\,\epsilon_f$$

where σ_{UTS} is the ultimate tensile strength, and ϵ_f is the strain at failure. The modulus is the maximum amount of energy per unit volume a material can absorb without fracturing. From the equation for modulus, for large toughness, high strength and high ductility are needed. These two properties are usually mutually exclusive. Brittle materials have low toughness because low plastic deformation decreases the strain (low ductility).

1.1.3.6. Resilience

The amount of elastic energy absorbed by a material under stress is known as its resilience. Put differently, the external force applied to a material during its deformation. The strain energy absorbed per unit volume is found in the area under the elastic portion of the stress-strain curve.

The resilience modulus is defined as:

$$M_R = \frac{(\sigma_y)^2}{2E}$$

where σ_y is the yield strength of the material and E is the elastic modulus of the material. To increase resilience, one needs increased elastic yield strength and decreased modulus of elasticity.

1.2. RELATIONSHIP BETWEEN STRUCTURAL GEOLOGY AND TECTONICS

The word structure is derived from the Latin word *struere*, to build, and we could say:

A geologic structure is a geometric configuration of rocks, and structural geology deals with the geometry, distribution and formation of structures.

It should be noted that primary structures produced by sedimentary or magmatic processes are not covered by structural geology; rather, it focuses solely on structures formed during rock deformation. On the other hand, primary structures can be modified to create deformation structures, as in the case of the folding of bedding in sedimentary rocks. The Greek word *tektos* is the source of the closely related word tectonics. Both structural geology and tectonics deal with the formation and resulting structure of the Earth's lithosphere as well as the movements that sculpt and change the planet's outer regions.

When we say external, we mean outside of the studied rock volume. External processes or causes include, although are not limited to, gravity-driven salt or mud diapirs, flowing glaciers, forceful magma intrusion, and meteor impacts. A tectonic style can be defined by the distinctive structures that each of these "causes" can produce, and the associated tectonics can be given unique names. The large-scale aspect of tectonics that deals directly with the movement and interaction of lithospheric plates is known as plate tectonics. Expressions like subduction tectonics, collision tectonics, and rift tectonics are used more specifically in the context of plate tectonics. The deformation of bedrock, usually sedimentary rocks, at the edge of an advancing ice sheet is known as glaciotectonics. In this instance, the deformation is caused by the ice pushing against the substrate, especially in areas where the glacier's base is cold and frozen.

The deformation brought on by the (mostly) vertical movement of salt through its overburden is the subject of salt tectonics. Although salt tectonics can also be closely linked to plate tectonics, glaciotectonics and salt tectonics are primarily driven by gravity. For instance, tectonic strain can produce fissures that allow salt to enter an object's cover gravitationally. The downward sliding of significant portions of rocks and sediments, particularly continental margin deposits resting on weak salt or overpressured shale layers, is generally referred to as gravity tectonics. In these kinds of settings, raft tectonics—a kind of gravity tectonics—occurs. Some people believe that smaller landslides and their structures are examples of gravity tectonics, while others believe that these surficial processes are not tectonic. The basic compaction of sediments and sedimentary rocks as a result of loading by younger sedimentary strata is a typical example of nontectonic deformation. Neotectonics studies the contemporaneous stress field and recent and continuing crustal motions. Neotectonic structures, which manifest as fault scarps on the surface, are the result of repeated satellite measurements of elevation changes in certain regions and seismic data from earthquakes (Figure 1.1).

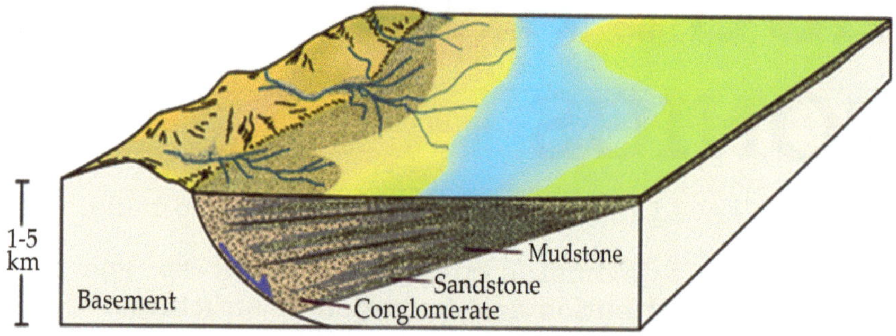

Figure 1.1. Illustration of the close relationship between sedimentary facies, layer thickness variations and syndepositional faulting (growth fault) along the margin of a sedimentary basin.

Source: Haakon Fossen, Structural Geology, 978-0-521-51664-8.

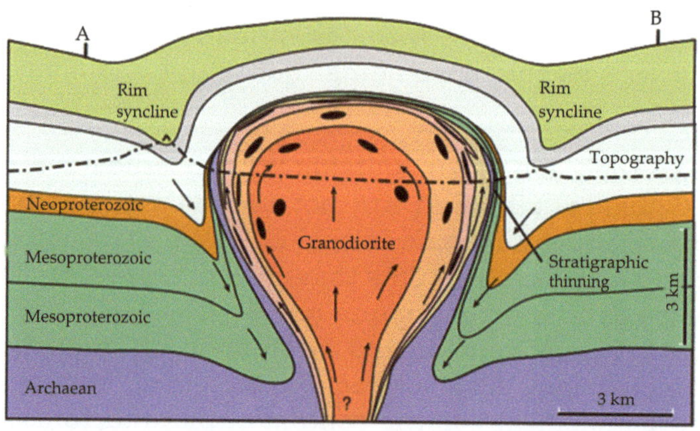

Figure 1.2. Structural geology can be linked to processes and mechanisms other than plate stresses. This map and profile from a granodioritic pluton southwest of Beijing, China, portray close connection between forceful intrusion of magma, strain and folds in the country rock.

Source: Haakon Fossen, Structural Geology, 978-0-521-51664-8.

Folding and shearing of the layers in Figure 1.2 indicate deformation close to the pluton's edge, which is indicative of a strong magma intrusion into the crust. The shape of enclaves, or inclusions, is represented by ellipses in this figure, and it is evident that these elongations increase in proximity to the pluton's edge. Therefore, there has been a history of forceful intrusion that has flattened the outer part of the pluton. Minerals that have undergone metamorphic growth before, during, and after deformation may also contain textures and structures that reflect kinematics and the history of deformation, as well as information about the pressure and temperature conditions that existed during the process. Therefore, the structural geology of a locality or region may be closely related to sedimentary, magmatic, and metamorphic processes.

These examples focus on strain, but structural geologists, particularly those studying brittle structures in the upper crust, also pay close attention to stress. Stress, which is not visible to the naked eye, is a complex and abstract concept for many. However, strain cannot occur without a stress field surpassing the rock's ability to resist deformation. While we can induce stress by applying force to a surface, stress is typically felt uniformly from all directions within the lithosphere. Thus, a comprehensive understanding of stress involves considering its three-dimensional nature. The relationship between stress and strain is always present, but deciphering it from naturally formed deformations can be challenging, unlike in a controlled laboratory setting. Deformation structures that form at or near the Earth's surface, in the chilly upper crust where rocks tend to fracture, in the hotter lower crust where deformation tends to be ductile, and in the underlying mantle are all covered by structural geology. It includes structures that form almost instantly, over tens of millions of years, and at scales ranging from hundreds of kilometers to micro- or atomic-scale structures.

This leads to the existence of a wide range of subdisciplines, strategies, and techniques in the field of structural geology. While the production geologist is concerned about subseismic sealing faults, the oil exploration geologist might be thinking about trap-forming structures created during rifting or salt tectonics. The university professor uses structural mapping, physical modeling, or computer modeling to understand mountain-building processes, while the engineering geologist may take fracture orientations and densities into account in relation to a tunnel project. Numerous techniques and strategies are employed, all of which contribute to our understanding of the tectonic or structural evolution of a given region or to the prediction of its structural pattern. Structural geology is typically based on observations and data that need to be evaluated and interpreted. Thus, a crucial component of structural geology is structural analysis.

Analysis of structural data leads to the development of a tectonic model for a specific area. This model helps explain the observed structures in the context of larger geological processes such as rifting or salt movements. For instance, if a series of normal faults indicating E-W extension are mapped in a mountain range, a suitable model must be identified to account for this extension. This could involve a rift model, extensional collapse during mountain building, or gravity-driven collapse following mountain formation. Age relationships between structures and additional information like radiometric dating, evidence of magmatism, and relative age relations are crucial in selecting the most appropriate model. It is possible that multiple models can explain the same data set, and it is essential to consider and evaluate alternative models. In general, a simpler model is preferred over a more complex one.

1.3. STRUCTURAL DATA SETS

The structures that arise from natural deformation serve as a reflection of the incredibly complex physical system that is the planet Earth, as evidenced by their diverse expressions and histories. Hence, it is necessary to condense and pinpoint the one or a small number of critical elements that best characterize or facilitate the identification of deformation structures that are visible or mappable in naturally deformed rocks. Objective observations and thorough descriptions of naturally deformed rocks are essential to understanding natural deformation. Field observations of deformed rocks and their structures represent the most direct and significant source of information on how rocks deform. In our mapping and description of structures and tectonic deformation, indirect observations of geologic structures through diverse remote sensing techniques, such as satellite data and seismic surveying, are becoming more and more significant. Laboratory experiments provide important insights into the relationship between deformation and different physical conditions, such as temperature, boundary conditions, stress field, and the physical characteristics of the deforming material. We can also control the different parameters and properties that influence deformation using numerical models, which simulate rock deformation on a computer.

In addition to assisting us in comprehending how internal and external physical conditions influence or forecast the formation of deformation structures, experiments and numerical models also shed light on the evolution of deformation structures, or more specifically, the deformation history. On the other hand, naturally deformed rocks are the culmination of histories of natural deformation, and it could be challenging to infer this history from the rocks themselves. Rock properties and boundary conditions can be controlled and their impact on deformation and deformation history investigated using numerical and experimental models. However, the history of deformation is contained in some degree in any deformed rock. Knowing what to look for and how to interpret this data present a challenge. This task can be accomplished with the help of objective and precise field observations, as well as calculations and experimental work.

1.4. FIELD DATA

The value of conventional field observations of distorted rocks and their structures cannot be overstated. More information exists in rocks than we will ever be able to extract, and accurate field observations of rock structures are essential to the success of any physical or numerical model. It is invaluable to come into direct contact with rocks and structures that have not been processed or interpreted by humans or computers.

Unfortunately, our ability to remain unbiased in our observations is limited. Our previous knowledge and experiences strongly influence how we perceive deformed rocks. It is crucial for any student studying deformed rocks to strive for objectivity. Only by doing so can we hope to uncover unexpected findings and develop new interpretations that may advance our understanding of the structural evolution of a region and the broader field of structural geology. Many important structures go unnoticed until someone brings attention to them, at which point they suddenly seem to be everywhere. Shear bands in heavily deformed ductile rocks (mylonites) are a prime example of this. Initially mistaken for cleavage or overlooked altogether, they only gained recognition in the late 1970s following proper description and interpretation. Since then, they have been identified in nearly every major shear zone or mylonite zone worldwide.

Figure 1.3. Direct contact with rocks is important. Make sure you enjoy many days in the field every year where you allow for open discussions and new ideas. Here Donna Whitney and Christian Teyssier are discussing gneissic structures in the Scandinavian Caledonides.

Source: Haakon Fossen, Structural Geology Second Edition, ISBN 978-1-107-05764-7

Using basic instruments like a hammer, measuring device, topomaps, hand lens, and compass, traditional fieldwork gathers data primarily on structural orientations and samples for thin section studies. High-resolution aerial and satellite photos, as well as contemporary global positioning system (GPS) devices, aid in the continued importance of this kind of data collection. A portable laser-scanning device, which records the time at which laser light pulses strike the Earth's surface and return, may be used for more sophisticated and thorough work. Using this data, an intricate topographic or geometrical model of the outcrop can be constructed, and one or more high-resolution field photos can be placed atop it. A flat image cannot convey the benefit of being able to move around virtually in a model, but Figure 1.4 provides an example of one. On a computer, geologic observations can then be made, such as the direction of fold axes or layering.

Figure 1.4. Mediumfjellet, Svalbard, based on LIDAR (Light Detection and Ranging) data (laser scanning from helicopter) and photos. This type of model, which actually is three dimensional, allows for geometric analysis on a computer and provides access to otherwise unreachable exposures.

Source: Haakon Fossen, Structural Geology, 978-0-521-51664-8

1.4.1. Mapping

For the field geologist, a topographic map that can be customized with geologic contacts, locality numbers, and orientation measurements has always been indispensable. In order to map structures and more efficiently gather, plot, and analyze data using smartphones, tablets, and more specialized digital devices with GPS capabilities, the modern field geologist is more likely to make use of digital maps, high-resolution digital satellite images, and aerial photos. Accurate positioning and more effective data collection are the results of this. The fundamental ideas are still the same, though.

1.4.2. Sketching and Photographing

Making meticulous field sketches with the assistance of photos, orientation measurements, and other measurements that may be related to the sketch is frequently the most crucial method of gathering field data. Additionally, sketching compels the field geologist to pay attention to details and features that they might otherwise miss. Even a simple drawing from a good field photo can highlight details that were missed in the original image. Sketches can also be used to highlight important details while downplaying or ignoring unimportant ones. For the purpose of gaining an overview of an outcrop, many of us find that creating an overview field sketch is helpful. Then, if there are any particularly intriguing structures or outcrop sections, a more detailed drawing can be created. Having a backup set of images for your sketches is always a good idea. An external or built-in GPS unit makes this process the most convenient. Drawing in the field is mostly a matter of experience.

1.4.3. Taking Measurements

We won't go into the specifics of using the many manual compasses that are available to measure planar and linear structures here. Additionally, there are a number of highly helpful apps available for smartphones. These apps make it simple to measure, plot, and display measurements either directly or indirectly on maps or images, but they have drawbacks of their own in terms of accuracy, battery life, etc. that needs to be considered. Regretfully, orientation measurements are being performed using disparate notations. Specifically, planes can be represented by their strike/dip values or by dip and dip direction (dip azimuth). Conversion is straightforward, but clear field notes are necessary to prevent misunderstandings.

A plane's strike and dip are often used by geologists to indicate a plane's orientation. The intersection of our plane with the horizontal plane is known as the strike, and the angle formed by the strike and north is known as the strike value.

Another way to characterize a plane is by its direction of dip. The strike value is now substituted for dip azimuth, or the horizontal direction in which our plane is dipping, while the dip value remains the same as before. Thus, our plane's orientation is indicated as 150/45 or 45/150 if it is dipping toward 150 (to the SE). In essence, this increases our strike angle by 90°, making it simple to switch between the two. However, if you obtain data from other geologists, it could also be confusing. Since dip directions are essentially lines, they are plotted as poles in spherical projections.

Linear structures are defined by their trend and plunge, with the trend being the projection of the structure onto the horizontal plane and the plunge being the angle between this line and the lineation.

The difference between true and magnetic north, or magnetic declination, can vary greatly in different locations and needs to be taken into consideration. The measurements can be changed later (in a spreadsheet or by rotating the data in a spherical projection program) or you can adjust your manual compass (just remember to readjust when moving to a different area). Depending on the app, your smartphone may or may not automatically adjust for declination.

1.4.4. Spherical Projections

Plotting of orientation data is done on rose diagrams and spherical projections (equal area or angle), and smartphone compass apps plot data on the screen automatically so that we can assess the results right away and add measurements if needed. Data can be imported or entered into convenient plotting applications from handheld devices. To fully utilize the plots that spherical projections generate, however, one must have a thorough understanding of them.

1.4.5. Scanlines

Scanlines are lines that span structures, like deformation bands or fractures, and are used to record the location of each structure or the quantity of structures per meter. A measuring tape can be used to measure the structures of interest, or cores or borehole image data can be used to gather scanline data.

1.4.6. Geologic Maps

Geologic maps are essential tools for addressing issues related to Earth resources, hazards, and environments. They depict the distribution of various rock types, surface deposits, and geological structures like faults and folds. These maps play a crucial role in land-use planning by guiding the placement of buildings and transportation infrastructure. In addition, geologic maps are vital for New Mexico residents as they help locate groundwater sources, assist in drilling water-supply wells, and identify potential pollution sources like landfills to ensure they are safely situated away from aquifers.

Assessing natural hazards and environmental or socio-economic risk primarily depends on the fourth dimension of time, which geologic maps are actually four-dimensional data systems. Understanding how and when materials and structures formed is just as important as knowing where they are on a geologic map.

Interactive electronic documents called digital geologic maps place problems related to earth science within geographical contexts. They combine data display with the findings of interpretive research to depict the dimensions, forms, depths, and physical and chemical contexts of earth materials. The combination of geologic maps and GIS databases help us address a great variety of complex geologic and hydrologic issues, such as: How does subsurface distribution of porous and impermeable rock affect the flow of water, the potential for contamination, and the volume available for use?

1.4.7. Cross-sections

In order to create a comprehensive geologic map, it is essential to include one or more cross-sections. These sections are constructed by creating a topographic profile using surface data gathered from field observations. Typically, the location of the section is chosen in relation to important lithologic boundaries, foliation, major faults, or fold hinges to capture the tectonic transport direction. The topographic profile is typically generated using digital elevation data and is then extended into the subsurface with the assistance of geophysical data and well information.

1.4.8. Collecting Oriented Samples

Knowing the orientation of a sample can be crucial for purposes such as kinematic analysis or preparing oriented thin sections. To ensure a sample is properly oriented, reposition it, mark a strike and dip symbol on a dipping surface while it is secure, and record these values on the sample itself, as depicted in Figure 1.5. If there is a lineation present, consider marking it with an arrow and indicating its trend/plunge. However, note that a lineation alone may not provide enough information to properly reorient the sample. Additionally, be sure to include sample number and position details, and take a photograph of the sample before placing it in a clearly labeled bag.

Figure 1.5. Collection of an oriented sample requires that the orientation (strike/dip) of one of the faces is measured and marked on the sample before it is permanently removed from the outcrop.

Source: Haakon Fossen, Structural Geology Second Edition, ISBN 978-1-107-05764-7

1.5. REMOTE SENSING AND GEODESY

With the availability of satellite images at ever-higher resolutions, like those displayed in Figure 1.6a and c, cartographers can now effectively map map-scale structures. These kinds of data are becoming more widely available on the Internet, where they can be combined with digital elevation data to generate 3-D models. Orthophotos, or orthorectified aerial photos, can provide additional information with resolutions as low as a few tens of centimeters in certain situations. Aerial and satellite photos can be used to map brittle faults and fractures as well as ductile structures like folds and foliations. (Hollenstein, M. D., et al., 2008)

Figure 1.6. (a) Satellite image of the Canyonlands National Park area, Utah. The image reveals graben systems on the east side of the Colorado River. An orthophoto (b) reveals that the grabens run parallel to fractures, and a high-resolution satellite image (c) shows an example of a graben stepover structure.

Source: http://geologylearn.blogspot.com/2016/11/structural-geology-and-tectonics.html.

Interferometric Synthetic Aperture Radar (InSAR) is a valuable remote sensing method in the field of neotectonics that maps variations in ground surface deformation by combining two or more radar satellite images. Satellites continuously send out beams of radar waves toward Earth, and an image is created based on the data that is returned. The phase of the wave as it hits and becomes reflected is also recorded, but the intensity of the reflected information also reflects the makeup of the ground. By comparing phases, we can keep an eye on millimeter-scale variations in surface elevation and geometry, which could be indicative of active tectonic movements brought on by sliding or earthquakes. This kind of data can also be used to create precise digital elevation models and topographic maps.

In general, GPS data are a valuable resource for obtaining information from GPS satellites in order to quantify plate movements. Such information can also be gathered on the ground using fixed GPS devices that have accuracy as low as millimeters (Figure 11.7).

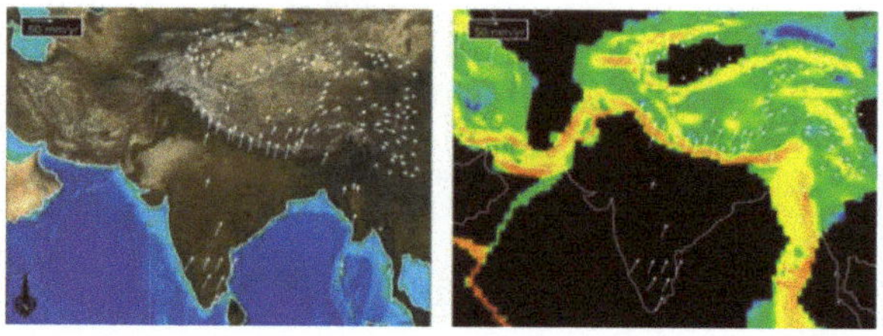

Figure 1.7. Use of GPS data from stationary GPS stations worldwide over time can be used to map relative plate motions and strain rates. (Left) White arrows (velocity vectors) indicating motions relative to Europe.

Source: http://geologylearn.blogspot.com/2016/11/structural-geology-and-tectonics.html.

1.6. DEM, GIS, AND GOOGLE EARTH

Traditional paper maps remain valuable for numerous field mapping applications; however, durable laptops, tablets, and handheld devices are increasingly crucial as they enable the direct digitization of structural features on digital maps and images. With the use of a Geographic Information System (GIS), digital field data can be merged with elevation and other data. Field observations, different geologic maps, aerial photographs, satellite images, gravity and magnetic data, usually in conjunction with a digital elevation model, can all be combined with GIS to carry out a range of statistical and mathematical operations. An Earth's surface is commonly represented by a digital elevation model (DEM), but any geologic surface or interface that can be mapped in three dimensions can also have a DEM created for it. Nowadays, surfaces that have been mapped from seismic data cubes are frequently displayed as DEMs, making it simple to examine their geometry and orientations.

1.6.1. Digital Elevation Model

A 3D computer graphics representation of elevation data used to represent terrain or overlaying objects, usually of a planet, moon, or asteroid, is called a digital elevation model (DEM) or digital surface model (DSM). A discrete global grid is referred to as a global DEM. DEMs are the most widely used foundation for digitally generated relief maps and are widely utilized in geographic information systems (GIS). While a digital terrain model (DTM) primarily depicts the ground, other digital terrain models (DEM and DSM) may depict building roofs or tree top canopy.

1.6.1.1. Types

A triangular irregular network (TIN), which is a vector-based representation of elevation, or a raster—a grid of squares—can be used to depict a digital elevation model (DEM). While the Raster DEM is referred to as a secondary (computed) DEM, the TIN DEM dataset is also known as a primary (measured) DEM. Techniques like photogrammetry, lidar, IfSAR or InSAR, land surveying, etc. could be used to obtain the DEM.

Although land surveying data can also be used to create DEMs, remote sensing techniques are typically used to gather the data.

1.6.1.2. Rendering

The data from a digital elevation model (DEM) is often represented visually to make it easier for humans to understand, even though the DEM itself is just a matrix of numbers. A contoured topographic map could be the format for this visualization, or it could take the form of shading and false color assignment, also known as pseudo-color, to represent elevations as colors. For instance, green could be used for the lowest elevations, shading to red, and white for the highest elevations.

Occasionally, oblique views are also used in visualizations, which recreate a synthetic visual representation of the terrain as it would appear from an angle. Elevations are occasionally scaled using vertical exaggeration in these oblique visualizations to emphasize minute elevation differences. However, some scientists argue that vertical exaggeration misleads the viewer about the actual terrain.

1.6.2. Production

Digital elevation models can be created in a variety of ways by mappers, but often, they rely on remote sensing instead of direct survey data.

In earlier DEM generation methods, digital contour maps that were created through direct land surface surveying are frequently interpolated. In mountainous places where interferometry isn't always reliable, this technique is still in use. Although they might be regarded as digital terrain models, contour line data and any other sampled elevation datasets (by GPS or ground survey) are not DEMs. A DEM suggests that elevation is consistently available at every site within the research region.

1.6.2.1. Satellite Mapping

Interferometric synthetic aperture radar is a potent method for creating digital elevation models. It involves two radar satellite passes (like those of RADARSAT-1, TerraSAR-X, or Cosmo SkyMed) or one pass if the satellite has two antennas (like the SRTM instrumentation). This method gathers enough data to create a digital elevation map that is tens of kilometers on a side with a resolution of about ten meters. The digital image correlation method, which involves taking two optical images at different angles from the same pass of an aircraft or Earth observation satellite (like the HRS instrument of SPOT5 or the VNIR band of ASTER), can be used to create other types of stereoscopic pairs.

The first significant elevation data for a large portion of Earth's landmass was collected by the SPOT 1 satellite in 1986 through two-pass stereoscopic correlation. Subsequent data was obtained by the European Remote-Sensing Satellite (ERS) in 1991, the Shuttle Radar Topography Mission (SRTM) in 2000 using single-pass SAR, and the Advanced Spaceborne Thermal Emission and Reflection Radiometer (ASTER) on the Terra satellite in 2000 using double-pass stereo pairs.

The HRS instrument on SPOT 5 has acquired over 100 million square kilometers of stereo pairs.

Making digital elevation maps of planets using orbital altimetry has proven to be a valuable tool in the field of planetary science. Although radar altimetry is also used, laser altimetry is the main tool for this. The Mars Orbiter Laser Altimeter (MOLA) mapping of Mars, the Lunar Orbital Laser Altimeter (LOLA) and Lunar Altimeter (LALT) mapping of the Moon, and the Mercury Laser Altimeter (MLA) mapping of Mercury are examples of planetary digital elevation maps created using laser altimetry. Every planetary body has a different reference surface when it comes to planetary mapping.

1.6.2.2. Methods for Obtaining Elevation Data Used to Create DEMs

- Lidar
- Radar
- Stereo photogrammetry from aerial surveys
- Structure from motion/Multi-view stereo applied to aerial photography
- Block adjustment from optical satellite imagery
- Interferometry from radar data
- Real Time Kinematic GPS
- Topographic maps
- Theodolite or total station
- Doppler radar
- Focus variation
- Inertial surveys
- Surveying and mapping drones
- Range imaging

1.6.2.3. Accuracy

Evaluating the quality of a Digital Elevation Model (DEM) involves assessing both the absolute accuracy of elevation at each pixel and the relative accuracy of the morphology it portrays. One way to gauge the quality of a DEM is by comparing it to DEMs obtained from various sources.

Several factors play an important role for quality of DEM-derived products:

- terrain roughness;
- sampling density (elevation data collection method);
- grid resolution or pixel size;
- interpolation algorithm;
- vertical resolution;
- terrain analysis algorithm;
- Reference 3D products include quality masks that give information on the coastline, lake, snow, clouds, correlation, etc.

1.6.3. Geographic Information System (GIS)

A geographic information system (GIS) is a combination of computer hardware and software that stores, manages, analyzes, edits, outputs, and visualizes geographic data. While much of this work is typically done within a spatial database, it is not necessary for a system to be considered a GIS. In a broader context, a GIS may also encompass human users, support staff, procedures, workflows, relevant concepts and methods, and institutional organizations.

GIS, the term commonly used for the industry and profession focused on these systems, is often synonymous with geoinformatics. The academic discipline studying these systems and their underlying geographic principles is known as GIScience, a subdiscipline of technical geography. While both GIS and GIScience may refer to this field, GIScience is the more commonly used term.

Geographic information systems are used across a variety of technologies, processes, and methods in engineering, planning, management, transport/logistics, insurance, telecommunications, and business operations. GIS and location intelligence applications are essential for location-enabled services, providing geographic analysis and visualization capabilities.

GIS utilizes location as the "key index variable" to connect previously unrelated information. It records the spacetime coordinates of Earth's locations and extents through date and time, as well as x, y, and z coordinates (representing longitude, latitude, and elevation). By relating spatial-temporal references to each other and ultimately to physical locations, GIS has enabled new avenues of scientific research.

1.6.4. Google Earth

Google Earth is a digital software that creates a 3D model of Earth using satellite images and other data. It allows users to explore cities and landscapes in detail from different perspectives. Users can navigate the globe by entering addresses or coordinates, using a keyboard or mouse, or on a mobile device with touch screen capabilities. They can also add their own data and share it through online platforms. Google Earth covers over 97% of the world and includes 10 million miles of Street View imagery as of 2019.

Through the desktop application, Google Earth offers a number of additional tools in addition to Earth navigation, one of which is a measure distance tool. There are extra globes for the Moon and Mars in addition to a night sky viewing instrument. There's also a game for a flight simulator. Additional features let users see images from different locations that have been added to Panoramio, details about specific locations from Wikipedia, and Street View imagery. Additionally, Google Earth's web version has a feature called Voyager that adds in-program tours on a regular basis. These tours are typically given by scientists and documentarians.

Google Earth has been outlawed in several nations due to concerns raised by some about privacy and national security. Some nations have asked for specific regions—typically those with military installations—to be hidden in Google Earth satellite photos.

1.7. SEISMIC DATA

Seismic data are indispensable for mapping subsurface structures and have transformed our knowledge of fault and fold geometry since the 1960s. While some seismic data are gathered solely for academic research, the majority of seismic data collection is driven by the search for gas and petroleum. Rift basins and continental margins thus provide the majority of seismic data.

Seismic data acquisition is by definition a unique kind of acoustic remote sensing, even though it is always handled differently in the geo-community. By using an air gun as a sound source to create sound waves that pierce the sea floor's crust, boats are used to gather data on marine seismic reflection. On the ocean floor, microphones can also be placed. While more laborious, this technique records seismic P- and S-waves (S-waves do not pass through water). By positioning the sound source and microphones (geophones) on the ground, seismic data can also be gathered onshore. Usually, an explosive device or a vibrating truck would be the onshore sound source, but for very shallow and local targets, a sledgehammer or specially made gun can be used instead.

Where there is an abrupt change in density and/or the velocity at which sound waves travel through the rock, or where there is an increase in acoustic impedance, layer boundaries reflect sound waves. A lengthy row of microphones, known as hydrophones offshore and geophones onshore, records the time that the reflected sound signals surface. These digital data are gathered, and computers process them to create an underground seismic image.

Several methods can be used to process seismic data, depending on the study's objectives. The vertical axis of standard reflection seismic lines is the two-way travel time. For those data to be converted into an ordinary geologic profile, depth conversion is required. A velocity model that takes into account the lithology (sound travels more quickly through sandstone than through shale, and more quickly through limestone) and burial depth (lithification increases velocity) is used to convert depth. Generally speaking, depth conversion occurs in the interpretation. Nevertheless, the seismic sections' vertical axis is depth rather than time if the seismic data itself are depth migrated. This allows for more accurate depictions of faults and layers and accounts for lateral variations in rock velocity that could present geometrical or visual difficulties for the interpreter working with a time-migrated section. On the other hand, the velocity model is what determines how accurate the depth migrated data is.

Where the energy released is high enough to reach deep into the crust and even the upper mantle, deep seismic lines can be gathered. These lines are helpful in examining the lithosphere's large-scale structure. Even though regional and deeply spaced seismic lines are referred to as two-dimensional (2-D) seismic data, an increasing amount of data from commercial (petroleum company) sources is gathered as a three-dimensional (3-D) cube where line spacing is sufficiently close (c. 25 m) where sections through the cube can be made in any direction and where the data can be processed in three dimensions. Lines that run parallel to the direction of collection are sometimes referred to as inlines; lines that run perpendicular to inlines are known as crosslines; and other vertical lines are known as random lines. Time slices, or horizontal sections, are helpful when interpreting faults.

Three-dimensional seismic data offers a unique opportunity for accurately mapping faults and folds in the subsurface. However, the resolution of seismic data is limited, meaning that only layers that are a certain distance apart (usually 5–10 m) can be distinguished, and only faults with a minimum offset can be imaged and interpreted. The quality and precision of 3-D data is typically superior to 2-D lines, as the reflected energy is more accurately restored through 3-D migration. The seismic resolution of high-quality 3-D data is influenced by factors such as depth, acoustic impedance of layer interfaces, data collection methods, and noise, but generally, fault throw can be identified at around 15–20 m resolution.

For 3-D seismic data sets, advanced techniques for data analysis and visualization are now available, which is useful for locating faults and other subterranean structures. Seismic 3-D data sets are typically used in petroleum exploration and exploitation, and geophysicists and structural geologists analyze these sets using computers. Geologic cross-sections and structural contour maps that can be used for structural analysis in a variety of ways, such as structural restoration, are produced by the interpretation.

Structural geologists are also interested in other kinds of seismic data, especially seismic data from earthquakes. This data provides us with crucial information about the tectonic regime, or, to put it simply, whether a region is experiencing strike-slip, extension, or shortening of faults.

1.8. SEISMIC DATA ACQUISITION

The three distinct stages of seismic exploration are seismic interpretation and seismic data processing, with seismic data acquisition coming first. In order to conduct a seismic survey, a seismic source must be used at specific locations. The energy that is released during the propagation of seismic waves from the source is then recorded at specific locations on the surface using devices known as receivers, which can be either geophones or hydrophones.

The process of planning a seismic survey, also known as the survey design, is necessary before seismic data can be obtained. Planning is done for the various survey parameters that are used in this process, such as the type of source, type of receiver, source spacing, number of source shots, number of receivers in a receiver array (i.e., group of receivers), number of receiver channels in a receiver spread, sampling rate, record length (i.e., the duration for which the receiver actively records the seismic signal), and so on. With the designed survey, seismic data can be recorded in the form of seismic traces, also known as seismograms, which directly represent the "response of the elastic wavefield to velocity and density contrasts across interfaces of layers of rock or sediments as energy travels from a source through the subsurface to a receiver or receiver array."

1.8.1. Survey Parameters

1.8.1.1. Source Types for Land Acquisition

Depending on the acquisition settings, various sources may be used for land acquisition.

When it comes to rough terrain, areas with high topographic variability, or environmentally sensitive areas like marshes, farming fields, mountainous regions, etc., explosive sources like dynamite are the preferred seismic sources. In order to minimize safety hazards during detonation and maximize the amount of seismic energy transferred into the subsurface, these kinds of sources must be buried (coupled) into the ground. Explosive sources have the advantage that during seismic processing, the seismic wavelet has an inverse that is stable and causal, which can be used in attempts to remove (deconvolve) the original wavelet. This is because the seismic signal, also known as the seismic wavelet, is minimum phase, meaning that the majority of the wavelet's energy is focused at its onset. The source/seismic wavelet is not precisely known and reproducible, which means that vertical stacking of seismograms or traces from these individual shots may result in less-than-ideal results (i.e., a lower signal-to-noise ratio). This is a major drawback of using explosive sources. Furthermore, it is impossible to precisely isolate the seismic wavelet in order to produce spikes or impulses that correspond to reflections on seismograms (the dirac delta function is the ideal goal). The fact that each explosion at the designated locations modifies the physical characteristics of the subsurface close to the source leads to variations in the seismic wavelet as it passes by these regions, which is one of the factors that contributes to the varying nature of the seismic wavelets corresponding to explosive sources.

Simpler seismic sources called Weight Drop sources, like the hammer source, are usually used in near-surface seismic refraction surveys. Usually, this kind of source only uses one weight source (e.g., hammer) and a plate (with a trigger to start recording on receivers) and is therefore practically possible in the majority of places. Because it produces smaller amplitudes and, consequently, smaller penetration depths than vibratory and explosive sources, it is primarily used in near-surface surveys. Weight drop sources use an unknown source wavelet, similar to explosive sources, which makes deconvolution and optimal vertical stacking challenging.

1.8.2. Source Types for Marine Acquisition

Since the 1970s, air-gun has been the seismic source most frequently used in marine seismic acquisition. The air-gun produces an acoustic pulse (signal) by quickly releasing compressed air that has been subjected to extreme pressure into a chamber. Its widespread use is due in part to the generated pulses' repeatability, controllability, and predictability. Furthermore, it generates the source using air, which is easily accessible and cost-free. Finally, compared to other marine seismic sources, it has a comparatively smaller environmental impact on marine life, which discourages the use of vibratory sources for marine acquisition. Air guns are usually employed in groups or arrays (i.e., several air guns at varying volumes) in order to reduce the appearance of oscillations or bubble pulses on the traces and to optimize the signal-to-noise ratio.

1.8.3. Receiver Type

1.8.3.1. Hydrophone

In marine seismic acquisition, a hydrophone is a type of seismic receiver that is sensitive to pressure changes brought on by acoustic pulses in its surroundings. The piezoelectric transducers used in typical hydrophones sense changes in pressure by producing an electric potential that is a direct indicator of those changes. Similar to air guns, hydrophones are also frequently used in groups or arrays, where a number of hydrophones are connected together to provide the highest possible signal-to-noise ratio.

1.8.3.2. Geophone

When purchasing land, a geophone—a seismic receiver—is frequently used to track the velocity of particles in a particular direction. A geophone can be made of multiple components that are intended to record both shear waves and p-waves (compressional waves) or it can be made of a single component that is intended to record p-waves. For geophones to record the actual ground motion caused by the seismic signal, there must be a strong enough coupling with the ground. This is especially crucial for the higher frequency seismic signal components, as inadequate coupling can cause significant changes to the signals' amplitude and phase.

1.8.4. Sampling Interval and Nyquist Criterion

Because the seismic signal that the receivers must record is by its very nature continuous, it must be discretized. The sampling interval or sampling rate is the rate at which this continuous signal is discretized (for further information, see Sampling (signal processing)). The Nyquist criteria states that the frequency at which the seismic signal must be sampled must be at least twice as high as the signal's maximum frequency component, or $f_{sample} \geq 2f_{max,signal}$. The remaining difficulty is that, in most cases, the highest frequency component is unknown during acquisition, making it impossible to calculate the sampling rate. In order to prevent temporal aliasing, estimates of the highest frequencies that the signal may contain must be made; typically, sampling rates greater than these estimates are preferred.

1.8.5. Record Length

Notwithstanding the term length, the record length describes the amount of time (usually expressed in milliseconds) that the receivers are actively recording and storing the subsurface seismic response. In order to guarantee that the direct waves are detected as the initial arrivals on the near-offset receivers, this recording time should typically begin a little earlier than the source is started. The record length should also be sufficient to guarantee that the most recent anticipated arrivals are noted. Record lengths for deeper exploration surveys are typically adjusted to several seconds (6 seconds is common). A typical exploration time for deep crust is 15 to 20 seconds. The record length is typically preferred to be longer than necessary rather than shorter because the recorded traces can always be clipped for later arrivals during data processing.

1.9. EXPERIMENTAL DATA

Since the early days of structural geology, folding and faulting have been physically modeled. Since the middle of the 20th century, this modeling has become more methodical in nature. A few of the processes and structures that have been simulated in the lab are buckle folding, shear folding, reverse, normal, and strike-slip faulting, fault populations, fault reactivation, porphyroclast rotation, diapirism, and boudinage.

Geologic structures are traditionally modeled by packing a box with clay, sand, plaster, silicone putty, honey, and other materials, then applying deformations like simple shear, extension, contraction, or other. When a lot of shear is needed, a ring shear device is employed. The disk-shaped volume's outer and inner halves are rotated in this configuration. During the deformation history, a large number of models can be captured on camera, photographed, or scanned using computer tomography. The centrifuge is an additional tool that uses centrifugal force to deform material. In these models, the centrifugal force functions similarly to the force of gravity during geologic processes (Figure 1.8).

Figure 1.8. Experimental work carried out by means of plaster and a simple wooden shear frame to study the evolution of faults in a pullapart (transtensional) setting.

Haakon Fossen, Structural Geology Second Edition, ISBN 978-1-107-05764-7.

Our goal is to create a scale model that, in addition to reducing the size and geometry of the natural object or structure it represents, also appropriately scales its physical attributes. Therefore, we require a geometrically comparable model whose lengths match those of the natural example and whose angle equality is maintained. Kinematic similarity is also necessary, with comparability of shape and position changes and proportionality of time. Similar angles of internal friction and proportionate values of cohesion or viscosity contrast are necessary for dynamic similarity.

It is practically impossible to reduce a deformed portion of the Earth's crust to its smallest possible size or set of properties. The natural size of sand grains can reach the size of massive boulders, making it impossible to replicate small-scale structures. Although clay's grain size might be more suitable, we might discover that clay's fine grain size makes it excessively cohesive. Plaster has characteristics that are difficult to precisely characterize since they change throughout the experiment. Physical models are obviously limited, but even with their limitations, observations of progressive deformation under well-defined boundary conditions can yield valuable insights into the workings of natural structures.

To investigate how materials respond to different stress fields and strain rates, experimental deformation of rocks and soils in a deformation rig under the influence of an applied pressure (stress) is used. The samples, which may have a volume of a few tens of cubic centimeters, are subjected to uniaxial compression or tension with a fluid-controlled confining pressure that is related to the crustal depth of interest (uniaxial compression or tension is defined as a force applied in only one direction). Additionally, triaxial tests are carried out, and the resultant deformation may be brittle as well as plastic. We encounter issues with strain rate when dealing with plastic deformation. In order to create plastic structures at laboratory strain rates, we must apply higher temperatures to our laboratory samples because natural plastic strains build up over thousands or millions of years. Thus, the problem of scaling has returned, but this time it concerns temperature, time, and strain rate.

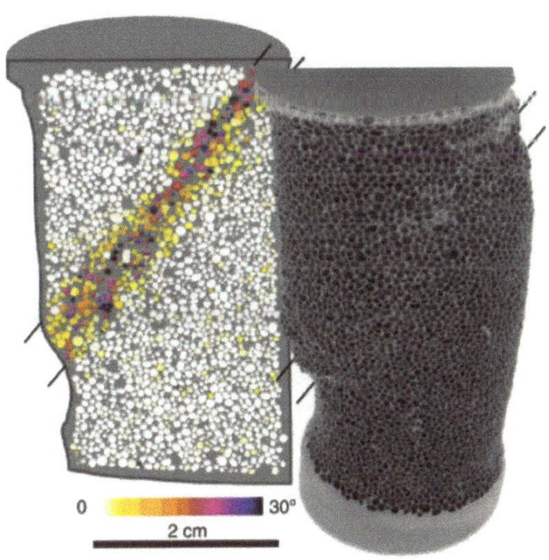

Figure 1.9. Section through a sandstone sample deformed in a triaxial deformation rig, compressed along its vertical axis and confined at a pressure of 100 kPa.

Source: Haakon Fossen, Structural Geology Second Edition, ISBN 978-1-107-05764-7.

1.10. NUMERICAL MODELING

The development of faster and faster computers has made numerical modeling of geologic processes easier. Spreadsheets and Matlab™ are examples of mathematical tools that can be used for simple modeling. More complex modeling, which frequently builds upon finite element and finite difference methods, calls for more expensive and sophisticated software. The models can cover a wide range of topics, from the microscale (e.g., deformation of mineral grains) to the lithosphere as a whole. Such phenomena as changes in the stress field during faulting and fault interaction, the formation of fractures in rocks, the formation of folds under different circumstances, and the microscale diffusion processes involved in plastic deformation can all be modeled. But nature is complicated, and no matter how complex something gets, even the fastest supercomputer eventually runs out of physical resources. Furthermore, current numerical theory is not able to fully describe all aspects of natural deformation. As a result, we must carefully consider the simplifications we make and make use of field and experimental data when organizing the modeling and assessing the outcomes. Thus, geologists with a certain understanding of numerical methodology—including its benefits and drawbacks—and field experience are needed.

1.11. STRUCTURAL ANALYSIS

The majority of structural data characterize the end result of a protracted deformation history, and many structural processes span thousands to millions of years. Data must be carefully analyzed in order to reveal the history itself. It might not be immediately apparent when examining a fold whether it was created by passive bending, shearing, or layer parallel shortening. The same holds true for flaws. The process should always be to analyze the field data and compare with experimental and/or numerical models. It may not always be easy to answer such questions, but the fault formed in what part first? Did it grow from a single point outward, and if so, was this point in the central part of the present fault surface? (Hubbert, 1937)

1.11.1. Geometric Analysis

Geometric analysis is the study and interpretation of structural geometry. This comprises the main structure's (first-order) shape, size, geographic orientation, and geometric relationship to related smaller-scale (second-order) structures. The final point highlights the majority of structures' composite nature and how they manifest at different scales in specific structural associations. As a result, different techniques are required to quantify and characterize structures and structural relationships.

Shape is a spatial description of an open or closed surface, like fault surfaces or folded layer interfaces. While fault curvature may have implications for hanging-wall deformation or may provide information about the slip direction, the shape of folded layers may provide information about the fold-forming process or the mechanical properties of the folded layer.

Possibly the most prevalent kind of structural data are the orientations of linear and planar structures. Vector functions, for example, can be used to describe shapes and geometric features using mathematical functions. Generally speaking, though, natural surfaces are either too uneven to be adequately represented by straightforward vector functions, or it might not be feasible to map folds or faults to the level necessary for a mathematical description. However, it might be required to interpret partially exposed structures geometrically. There will always be some gaps in our data, and when we analyze geologic data, our minds naturally gravitate toward geometric models. For instance, the major fold structures of the Alps were generally thought to be cylindrical when they were first extensively mapped in the twentieth century, which meant that fold hinges were thought to be straight lines. By projecting folds onto cross-sections, this model allowed for the creation of striking sections or geometric models. Later on, it was discovered that the folds were not cylindrical as first thought, and that the hinge lines were curved, necessitating the modification of previous models.

Spherical projection is a very useful tool for representing orientation data in geometric analysis. Stereonet (equal angle) and equal-area projections are commonly used to illustrate or interpret the orientation and geometry of structures in structural geology, where orientation statistics and spatial distributions are taken into account. The most popular method for visualizing and analyzing spatial data is projection, which is also the fastest and most effective. Geometry can generally be shown as maps, profiles, equal-area projections, rose diagrams, or three-dimensional (3-D) models that are derived from field observations, geophysical data, satellite information, or laser scanning equipment. A thorough understanding of spherical projection methods is essential for any serious structural geologist.

1.11.2. Strain and Kinematic Analysis

Strain analysis or quantification may start with geometric description and analysis. Such quantification is helpful in a variety of situations, such as the reconstruction of geologic sections that have been deformed. Finite strain analysis, which examines shape changes from the initial state to the very end result of the deformation, is a common method used in strain analysis. Incremental strain analysis is a useful tool for exploring the deformation history, which is of interest to structural geologists. In this instance, the deformation history is described by a series of increments, and only a portion of it is taken into account.

Strain, by definition, refers to ductile deformation, or deformation in which structures that were initially continuous, like dikes or bedding, continue to be continuous after the deformation. When stress causes rocks to flow (without breaking), this is known as ductile deformation. When rocks fracture or break, they undergo brittle deformation, which is the opposite. Modern geologists, however, do not limit the application of strain to ductile deformation. The discontinuities are disregarded and the term "brittle strain" is applied when there are many fractures occurring on a scale much smaller than the discontinuity each fracture causes. We can do strain analysis on brittle structures, like fault populations, thanks to this simplification.

Kinematic analysis, which examines how rock particles have moved during deformation, is also based on geometric description (the Greek word kinema means movement). Structures useful in kinematic analysis include shear zones, fault surfaces, and the deflection of layering along faults.

1.11.3. Dynamic Analysis

The study of the forces that propel particles into motion is known as dynamic analysis (kinematics). Stress is created by forces operating on a body, and when that stress builds up to a certain point, rocks begin to move. Therefore, dynamics in structural geology refers to the interaction of kinematics and stress. Deformation occurs when certain particles begin to move in relation to other particles; as a result, we may observe shape changes and the emergence of new structures.

Most of the time, dynamic analysis looks at a collection of structures, usually faults and fractures, in an effort to reconstruct the orientation and magnitude of the stress field. When you stress syrup, it behaves differently than when you stress a cold chocolate bar; the syrup will flow and the chocolate bar will shatter. Although the term "rheologic analysis" refers to the portion of dynamics that deals with rock flow, we are still working with dynamic analyses. Mechanical analysis is the study of how things break or fracture, like sugar or rocks. Generally speaking, rocks flow when they are sufficiently warm, which usually translates to deep burial. "Deep enough" refers to temperatures at or near the surface for salt, approximately 300°C for quartz-rich rocks, possibly as high as 550°C for feldspathic rocks, and even higher for olivine-rich rocks. In addition, water content, strain rate, and pressure all matter. It is crucial to understand that while various rocks behave differently in any given situation, they also respond differently to stress in various physical environments. To comprehend how various rocks flow in the lithosphere, rheological testing is carried out in the lab.

1.11.4. Tectonic Analysis

Dynamic, kinematic, and geometric analysis at the level of an orogenic belt or basin are all part of tectonic analysis. Therefore, additional components of sedimentology, paleontology, petrology, geophysics, and other geoscientific subdisciplines may be included in this type of analysis. Tectonicists are another term for structural geologists who study tectonics. Some structural geologists examine the textures and structures that are only visible under a microscope, which is at the other end of the scale. Microstructural analysis, also known as microtectonics, is the study of deformation between and within individual mineral grains. For microstructural analysis, both the optical and scanning electron microscopes (SEM) are helpful instruments.

CASE STUDY

MARINE SEISMIC REFLECTION DATA

Marine Seismic Acquisition Offshore collection of seismic data is done by a vessel that travels at about 5 knots while towing arrays of air guns and streamers containing hydrophones a few meters below the surface of the water. The tail buoy helps the crew locate the end of the streamers. The air guns are activated periodically, such as every 25 m (about every 10 seconds), and the resulting sound wave that travels into the Earth is reflected back by the underlying rock layers to hydrophones on the streamer and then relayed to the recording vessel for further processing.

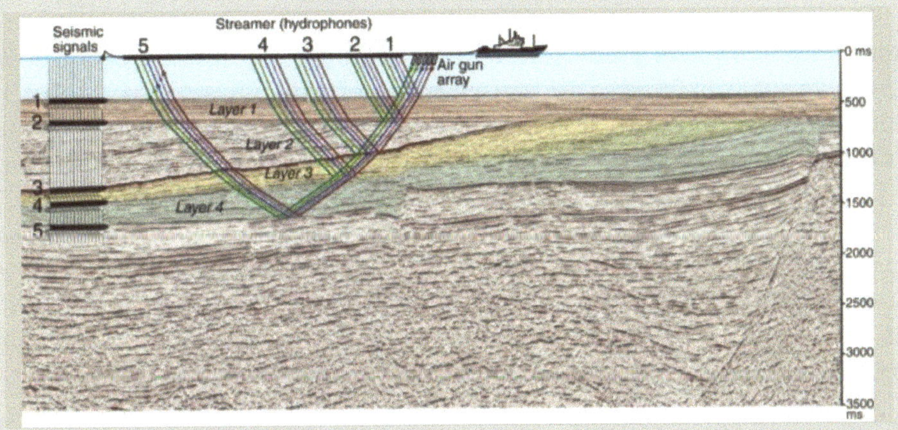

Figure 1. Marine seismic reflection.

Source: Haakon Fossen, Structural Geology Second Edition, ISBN 978-1-107-05764-7

The few sound traces shown on the figure indicate how the sound waves are both refracted across and reflected from the interfaces between the water and Layer 1, between Layer 1 and 2, and between Layer 2 and 3. Reflection occurs if there is an increase in the product between velocity and density from one layer to the next. Such interfaces are called reflectors. Reflectors from a seismic line image the upper stratigraphy of the North Sea Basin (right). Note the upper, horizontal sea bed reflector, horizontal Quaternary reflectors and dipping Tertiary layers. Unconformities like this one typically indicate a tectonic event. Note that most seismic sections have seconds (two-way time) as vertical scale.

The sound waves are reflected from layer boundaries where there is an increase in acoustic impedance, i.e., where there is an abrupt change in density and/or the velocity with which sound waves travel in the rock. A long line of microphones, onshore called geophones and offshore referred to as hydrophones, record the reflected sound signals and the time they appear at the surface. These data are collected in digital form and processed by computers to generate a seismic image of the underground. Seismic data can be processed in a number of ways, depending on the focus of the study. Standard reflection seismic lines are displayed with two-way travel time as the vertical axis. Depth conversion is therefore necessary to create an ordinary geologic profile from those data. Depth conversion is done using a velocity model that depends on the lithology (sound moves faster in sandstone than in shale, and yet faster in limestone) and burial depth (lithification leads to increased velocity). In general it is the interpretation that is depth converted. However, the seismic data themselves can also be depth migrated, in which case the vertical axis of the seismic sections is depth, not time. This provides more realistic displays of faults and layers, and takes into account lateral changes in rock velocity that may create geometrical challenges when interpreting a time-migrated section. The accuracy of the depth-migrated data does however rely on the velocity model.

Deep seismic lines can be collected where the energy emitted is sufficiently high to image the lower crust and even the upper mantle. Such lines are useful for exploring the large-scale structure of the lithosphere. While widely spaced deep seismic lines and regional seismic lines are called two-dimensional (2-D) seismic data, more and more commercial (petroleum company) data are collected as a 3-D cube where line spacing is close enough (ca. 25 m) that the data can be processed in three dimensions, and where sections through the cube can be made in any direction. The lines parallel to the direction of collection are sometimes called inlines, those orthogonal to inlines are referred to as crosslines, while other vertical lines are random lines. Horizontal sections are called time slices, and can be useful during fault interpretation.

Three-dimensional seismic data provide unique opportunities for 3-D mapping of faults and folds in the subsurface. However, seismic data are restricted by seismic resolution, which means that one can only distinguish layers that are a certain distance apart (typically around 5–10 m for good quality oil industry data), and only faults with a certain minimum offset can be imaged and interpreted. The quality and resolution of 3-D data are generally better than those of 2-D lines because the reflected energy is restored more precisely through 3-D migration. The seismic resolution of high-quality 3-D data depends on depth, acoustic impedance of the layer interfaces, data collection and noise, but would typically be at around 15–20 m for identification of fault throw.

Sophisticated methods of data analysis and visualization are now available for 3-D seismic data sets, helpful for identifying faults and other structures that are underground. Petroleum exploration and exploitation usually rely on seismic 3-D data sets interpreted on computers by geophysicists and structural geologists. The interpretation makes it possible to generate structural contour maps and geologic cross-sections that can be analyzed structurally in various ways, e.g., by structural restoration. Other types of seismic data are also of interest to structural geologists, particularly seismic information from earthquakes. This information gives us important information about current fault motions and tectonic regime, which in simple terms means whether an area is undergoing shortening, extension or strike-slip deformation.

ACTIVITY

For the following map view diagram, determine the correct map symbol that should appear in the blue oval. Note: the purple, green, and red lines represent beds, the wavy blue line is a stream, and the dotted lines are elevation contours.

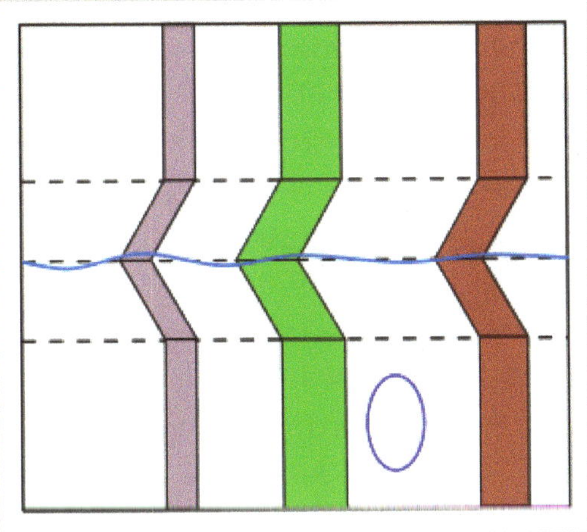

SUMMARY

- Structural geology is the study of the three-dimensional distribution of rock units with respect to their deformational histories.
- The study of geologic structures has been of prime importance in economic geology, both petroleum geology and mining geology.
- Structural geology is a critical part of engineering geology, which is concerned with the physical and mechanical properties of natural rocks.
- Structural geologists use a variety of methods to (first) measure rock geometries, (second) reconstruct their deformational histories, and (third) estimate the stress field that resulted in that deformation.
- For economic geology such as petroleum and mineral development, as well as research, modeling of structural geology is becoming increasingly important.
- The mechanical properties of rock play a vital role in the structures that form during deformation deep below the earth's crust.
- A geologic structure is a geometric configuration of rocks, and structural geology deals with the geometry, distribution and formation of structures.
- Glaciotectonics is the deformation of sediments and bedrock (generally sedimentary rocks) at the toe of an advancing ice sheet. In this case it is the pushing of the ice that creates the deformation, particularly where the base of the glacier is cold (frozen to the substrate).
- Planet Earth represents an incredibly complex physical system, and the structures that result from natural deformation reflect this fact through their multitude of expressions and histories.
- A topographic map onto which geologic contacts, locality numbers and orientation measurements can be added has always been essential to the field geologist.
- Scanlines are lines across structures such as fractures or deformation bands, where the location of each structure or number of structures per meter is recorded.
- In the field of neotectonics, InSAR (Interferometric Synthetic Aperture Radar) is a useful remote sensing technique that combines two or more radar satellite images to map variations in ground surface deformation.
- In the mapping of subsurface structures, seismic data are invaluable and since the 1960s have revolutionized our understanding of fault and fold geometry.
- Seismic data acquisition is the first of the three distinct stages of seismic exploration, the other two being seismic data processing and seismic interpretation.
- Physical modeling of folding and faulting have been performed since the earliest days of structural geology, and since the middle part of the twentieth century such modeling has been carried out in a more systematic way.

REVIEW QUESTIONS

1. What is structural geology all about?
2. Name the four principal ways a structural geologist can learn about structural geology and rock deformation. How would you rank them?
3. How can we collect structural data sets? Name important data types that can be used for structural analysis.
4. What are the advantages and disadvantages of seismic reflection data sets?
5. What is a scale model?
6. What is kinematic analysis?

REFERENCES

1. Boyer, S. E. (1992). Geometric evidence for synchronous thrusting in the southern Alberta and northwest Montana thrusts belts. In K. R. McClay (Ed.), *Thrust Tectonics* (pp. 377–390). Chapman and Hall.
2. Geng, M., Welford, J. K., Farquharson, C. G., & Peace, A. L. (2018). 3-D inversion of airborne gravity gradiometric data for the Budgell Harbor Stock: A case history of using a probabilistic approach. In *Society of Exploration Geophysicists, SEG International Exposition and 88th Annual Meeting, Anaheim, CA, USA*. https://doi.org/10.1190/segam2018-2978510.1
3. Hollenstein, M. D., Müller, A., & Geiger, H. G. K. (2008). Crustal motion and deformation in Greece from a decade of GPS measurements, 1993–2003. *Tectonophysics, 449*, 17–40.
4. Hubbert, M. K. (1937). Theory of scale models as applied to the study of geologic structures. *Bulletin of the Geological Society of America, 48*, 1459–1520.
5. Huismans, R. S., & Beaumont, C. (2003). Symmetric and asymmetric lithospheric extension: Relative effects of frictional-plastic and viscous strain softening. *Journal of Geophysical Research, 108*. https://doi.org/10.1029/2002JB002026.
6. Jones, R. R., McCaffrey, K. J. W., Clegg, P., Wilson, R. W., & Holliman, N. S. (2009). Integration of regional to outcrop digital data: 3D visualization of multi-scale geological models. *Computers and Geosciences, 35*, 4–18.
7. Lapointe, P. L. (1979). Paleomagnetism of the Notre dame bay lamprophyre dikes, Newfoundland, and the opening of the North Atlantic Ocean. *Canadian Journal of Earth Sciences, 16*, 1823–1831.
8. Maerten, L., & Maerten, F. (2006). Chronologic modeling of faulted and fractured reservoirs using geomechanically based restoration: Technique and industry applications. *American Association of Petroleum Geologists Bulletin, 90*, 1201–1226.
9. Mataragio, J., & Kieley, J. (2009). Application of full tensor gradient invariants in detection of intrusion-hosted sulphide mineralization: Implications for deposition mechanisms. *First Break, 27*, 95–98.

10. McCaffrey, K., Jones, R. R., Holdsworth, R. E., Wilson, R. W., Clegg, P., Imber, J., Hollinman, N., & Trinks, I. (2005). Unlocking the spatial dimension: Digital technologies and the future of geoscience fieldwork. *Journal of the Geological Society, 162*, 927–938.
11. Peace, A. L., Welford, J. K., Geng, M., Sandeman, H., Gaetz, B. D., & Ryan, S. S. (2018). Rift-related magmatism on magma-poor margins: Structural and potential-field analyses of the Mesozoic Notre Dame Bay intrusions, Newfoundland, Canada and their link to North Atlantic opening. *Tectonophysics, 745*, 24–45.
12. Savage, H. M., & Cooke, M. L. (2003). Can flat-ramp-flat fault geometry be inferred from fold shape? A comparison of kinematic and mechanical folds. *Journal of Structural Geology, 25*, 2023–2034.
13. Wanless, R. K., Stevens, R. D., Lachance, G. R., & Rimsaite, R. Y. H. (1965). Age Determinations and Geological Studies. Part 1–Isotopic Ages. *Geological Survey of Canada, Report 64–17.*
14. Yin, A., & Nie, S. (1993). An indentation model for North and South China collision and the development of the Tanlu and Honam fault systems, eastern Asia. *Tectonics, 12*, 801–813.

CHAPTER

2 Deformation

LEARNING OBJECTIVES

After studying this chapter, you will be able to:

- Understand basic concepts
- Discuss homogeneous and heterogeneous deformation
- Explain deformation mechanism
- Focus on deformation and strain

KEY TERMS FROM THIS CHAPTER

Brittle deformation
Compressional stress
Dislocation creep
Dynamic recrystallization
Foliation
Geometry
Grain size
Cataclasis
Decomposition
Ductile deformation
Earth
Fracturing
Grain boundary sliding
Heterogeneous strain

INTRODUCTION

On our planet, deformation is a natural occurrence. From the perspective of earth science, we can observe that it is a process that modifies a rock's size and shape. We frequently notice that different rocks have different kinds of lines and swirls that do not repeat in any other rock. Have you ever looked at the rocks and wondered how each is so unique? These patterns are the result of deformation. It describes the process that modifies the volume, size, or shape of a portion of the earth's crust. The kind of deformation differs depending on the kind of stress and kind of rock present in a particular region of the earth's crust. It mostly occurs as a result of stress, or more accurately, a force applied to the affected area. Moreover, this process occurs for a number of reasons. For example, temperature changes, plate tectonics, sediment accumulation, and other factors can readily induce it.

2.1. BASIC CONCEPTS

The process by which rocks alter in size or shape in reaction to stress is known as rock deformation. Stress is defined as the force exerted per unit area and is a result of many different processes, such as plate tectonics, gravity, and human activity.

2.1.1. Principles of Progressive Deformation

The net change in volume and/or geometry between an initial and a final state achieved over a predetermined period of time is known as finite strain. The rock goes through several shifting geometrical states (incremental strain) during this period of time. The deformation history or path is defined by these shifting states of a deformation process. A given finite deformation can arise from an infinite number of geometrically possible deformation paths, but only one of them has ever really happened.

2.1.2. Components of Deformation

Depending on the goal of the decomposition, the displacement field can be divided into a number of different parts. The traditional method of dissecting it involves dividing the change in shape and volume from the rigid body deformation, which takes the form of rigid translation and rotation.

2.1.2.1. Translation

Every particle in the rock is moved by translation in the same direction and over the same distance, and the parallel vectors in its displacement field are all the same length. Translations can be significant, as in the case of thrust nappes—detachable rock slices—that are moved hundreds or tens of kilometers. Jotun Nappe (A sample from the Scandinavian Caledonides is given in Figure 2.1). Here, rigid translation accounts for the majority of the deformation. Field observations show that the strain, or change in shape, is mostly confined to the lower parts. However, we are unable to estimate the rigid rotation (see below) because we do not know the precise orientation of this nappe prior to the onset of deformation. Thus, a large translation component, an unidentified but potentially tiny rigid rotation component, and a strain component localized to the nappe's base make up the total deformation.

Mineral grains, fault blocks, and layers in rocks can all be translated along planar faults or slip planes on a smaller scale without undergoing an internal structural change.

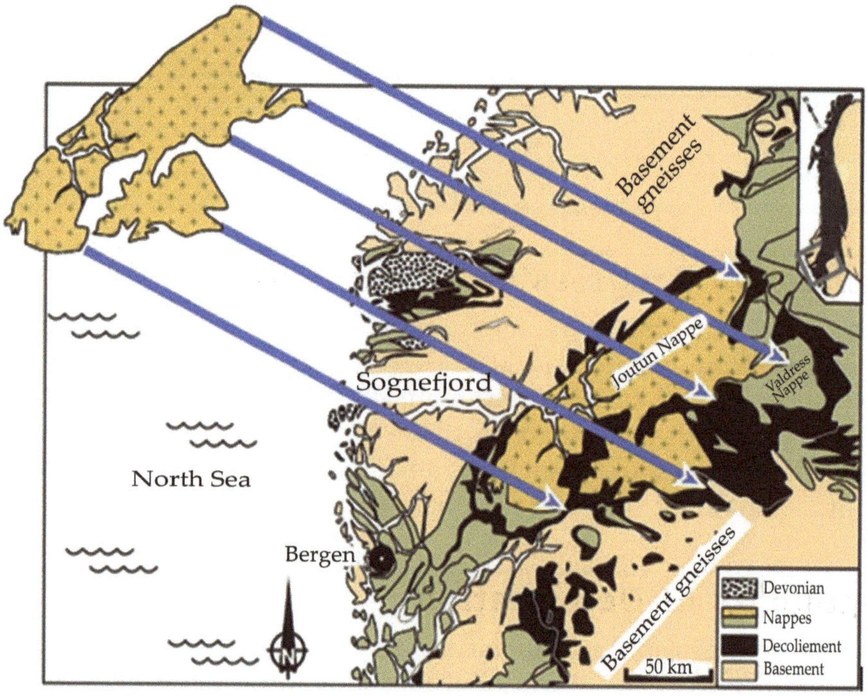

Figure 2.1. The Jotun Nappe in the Scandinavian Caledonides seems to have been transported more than 300 km to the southeast, based on restoration and the orientation of lineations. The displacement vectors are indicated, but the amount of rigid rotation around the vertical axis is unknown. The amount of strain is generally concentrated to the base.

Source: https://geologylearn.blogspot.com/2016/11/what-is-deformation.html.

2.1.2.2. Rotation

Here, "rotation" refers to the complete deformed rock volume under study being rotated rigidly. It is not the same as the rotation of the strain ellipse's (imaginary) axes during progressive deformation. A rock volume, such as a shear zone, rotates uniformly in a physical sense in relation to an external coordinate system when it is in rigid rotation. A major thrust nappe's or tectonic plate's large-scale rotations usually revolve around vertical axes. On the other hand, small-scale rotations can happen about any axis, and fault blocks in extensional settings can rotate around horizontal axes.

2.1.2.3. Strain

Non-rigid deformation, also known as strain or distortion, is comparatively easy to define.

Strain is the term used to describe any change in shape, with or without a change in volume, which suggests that the positions of the particles within a rock have shifted.

Even though a rock volume can be rigidly translated and rotated in any direction, we will never be able to tell from simply looking at the rock itself. In the field or in samples, all that is visible is strain, and possibly the manner in which that strain has built up. Think about your lunchbox. You can't see this deformation directly, but you can bring it to work or school, where there will be a lot of rotation and translation. If you compare the appearance of your lunch bag before you left home, you can determine whether it was squeezed on the way to school. You would use your understanding of the proper shape for a lunch bag to determine the amount of strain (or change in shape) if someone else made your meal and placed it in your bag.

With very few exceptions, we have never seen the deformed rock in its undisturbed state, so the last point is extremely pertinent. The next step is to apply our understanding of the typical appearance of these rocks in an unstrained state. For instance, we might anticipate that strained ooliths or reduction spots would have been spherical (circular in cross-section) in their undeformed state if we find them in the rock.

2.1.2.4. Volume Change

A rock volume may have contracted or expanded even though its shape hasn't changed. Therefore, in order to fully describe deformation, we must also include volume change, or area change in two dimensions. Dilation, another name for volume change, is frequently understood to be a particular kind of strain known as volumetric strain. If at all possible, it is helpful to keep this kind of deformation apart.

2.1.3. Types of Deformation

There are two main ways that rocks can deform: brittlely and ductilely. When rocks are under stress, they fracture or break, causing brittle deformation. Ductile deformation is the bending or flowing of rocks under stress. Numerous variables, such as the kind of rock, the pressure and temperature levels, and the rate of stress application, affect the type of deformation that takes place.

2.1.3.1. Brittle Deformation

Because of the relatively low temperature and pressure in the upper crust of the Earth, brittle deformation is most common there. Several characteristics can be caused by brittle deformation, such as:

- Faults: Rock fractures known as faults are caused by displacement. Numerous stresses, such as compression, tension, and shear, can result in faults.
- Joints: In rocks, joints are fractures that do not exhibit displacement. Rocks that have cooled or dried out frequently develop joints.
- Breccia: Rocks called breccia are composed of sharp pieces of other rocks. Rocks that are brittlely deformed frequently form breccia.

2.1.3.2. Ductile Deformation

Because of the relatively high temperatures and pressures found in the Earth's lower crust and mantle, ductile deformation is most prevalent there. Among the features that can be produced by ductile deformation are the following.

- Folds: In rocks, folds are bends. Numerous stresses, such as tension and compression, can result in folds.
- Foliation: A layering known as foliation is present in certain metamorphic rocks. The

arrangement of the minerals in the rock during ductile deformation is what leads to foliation.

- Lineation: A linear characteristic found in certain metamorphic rocks is called lineation. When ductile deformation occurs, elongated grains or minerals in the rock align, leading to lineation.

2.1.4. System of References

Selecting a reference or coordinate system is necessary for deformation studies. It can appear as though the dock is moving rather than the ship when you are standing on a dock and watching a large ship arrive or depart. The rest of the world moves by translation in relation to the ship when the reference system is unconsciously fixed to it. Although intriguing, this is not a very practical source of information. In order to minimize complexity, rock deformation must also be taken into account within the context of a reference coordinate system, which must be carefully chosen.

When discussing displacements and kinematics, a reference frame is always necessary.

Orienting the coordinate system along significant geologic structures is frequently beneficial. This might be a local shear zone, the base of a thrust nappe, or a plate boundary. We want to get rid of rigid rotation and translation in many situations. When it comes to shear zones, we typically align two axes parallel to the zone and one perpendicular to it. The origin could be fixed to the shear zone's edge if the deformation in the zone as a whole is of interest to us. By "gluing" the origin to a particle inside the zone (which is still parallel to or perpendicular to the shear zone boundaries), we can learn more about the environment surrounding any particular particle in the zone. Because the coordinate system rotates and translates in tandem with the shear zone, translation and rigid rotation of the shear zone are eliminated in both scenarios. Although an oblique coordinate system with respect to the shear zone boundaries is not inherently problematic, it can complicate matters both mathematically and aesthetically.

2.1.5. Deformation: Detached from History

Deformation is simply the discrepancy between the deformed and original states, providing no insight into the specific events that occurred during the deformation process.

There are an infinite number of ways in which a strain can have accumulated.

Envision a fatigued pupil, or instructor, for that matter, dozing off in a boat while cruising a lake or the ocean. When the student wakes up, he or she quickly remembers where they were before falling asleep, but they are unsure of the precise route that the boat has taken due to winds and currents. The student can only assess the strain (or, ideally, zero, change in shape) of the boat based on its position before and after the nap. The deformation can be mapped, but the deformation history cannot.

Let us also consider particle flow: Just like students moving between lecture halls, there are endless paths they can take, each with a different length of time. However, this movement does not involve time itself. The lecturer, in a rush between classes, only sees that the students have transitioned from one lecture hall to the other. The lecturer is unaware of the students' journey (though they may have some ideas based on coffee cups, etc.). Similarly, rock particles can also move along various paths from their original state to a deformed state. One distinction between rock particles and students is that while students can move independently, rock particles, like mineral grains in a rock, are stuck together in a solid continuum and cannot move freely.

2.1.6. Factors Affecting Deformation

The kind of deformation that rocks undergo when subjected to stress is caused by numerous factors. The first is the kind of rock that is changing. The rate of strain comes next, followed by pressure and temperature. For example, increased pressure and temperature can cause ductile deformation. For this reason, rocks close to the earth's surface have greater ductility.

In a similar vein, rocks undergo continuous pressure, which causes them to alter over time in terms of shape. Stress alone, a force applied across an area, is the cause of this change. It is evident from this that confining stress—applied uniformly so as to prevent the rock from changing—is possible.

Then, when force is applied unevenly, we have differential stress. This directed stress is further separated into three categories, which are as follows:

- Tensional Stress: It spreads out the rocks. This resulted in the formation of valleys over time. When the crust is pulled in opposing directions, tensional stress results.
- Compressional Stress: Tall mountain ranges like the Smokey Mountains and the Himalayas are the result of this kind of stress. When two substantial portions of the Earth's crust collide, like when two continental plates collide, compressional stress results.
- Shear Stress: Shear stress leads to the formation of fault lines. When two plates rub against one another in opposing directions, shear stress is created.

It is created when two plates collide and result in a fault line similar to the San Andreas Fault.

2.2. HOMOGENEOUS AND HETEROGENEOUS DEFORMATION

Usually, homogeneous strains or distortions are assumed in mathematical treatments of strain, as opposed to heterogeneous ones. Any heterogeneously strained rock body, nevertheless, can be split into tiny sections that display homogeneous strain's traits. Sometimes referred to as non-homogeneous strain, heterogeneous strain affects non-rigid bodies in an irregular, non-uniform manner (Figure 2.2). In a heterogeneous strain, complex forms are formed by the distortion of circles, squares, or their three-dimensional counterparts, cubes and spheres, into parallel lines that are not parallel after the strain. Non-rigid rock bodies are uniformly and regularly affected by homogeneous strain (Figure 2.3). Homogeneous strain causes spheres and circles to become ellipsoids and ellipses, respectively, while cubes and squares are deformed into prisms and parallelograms, respectively, because parallel lines that were parallel before the strain continue to be parallel after it. The strain must be uniform and systematic throughout the deformed body for these generalizations to be valid.

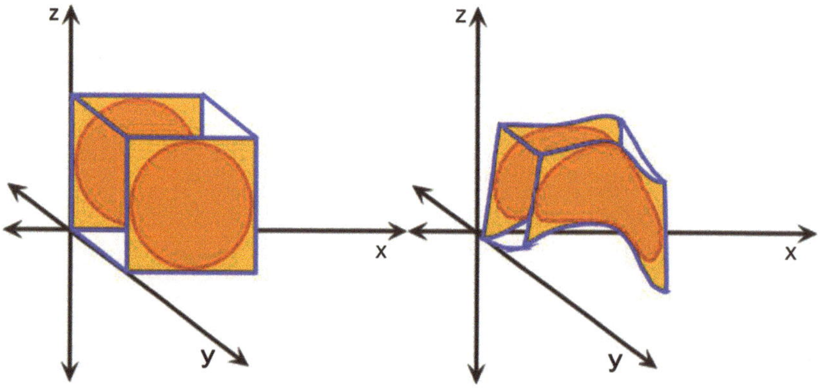

Figure 2.2. Heterogeneous or non-homogenous strain results in distorted complex forms.

Source: https://earth.sdsu.edu/visualstructure/vss/htm_hlp/hom_het.htm.

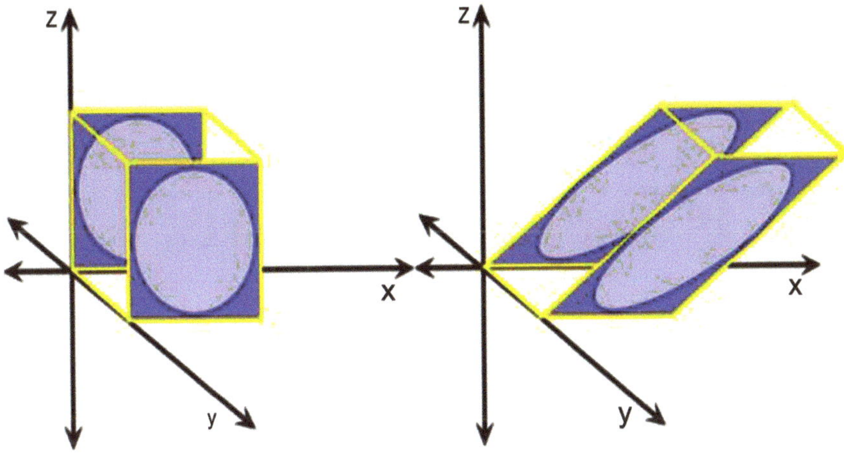

Figure 2.3. Homogenous strain affects bodies in a regular and uniform manner.

Source: https://earth.sdsu.edu/visualstructure/vss/htm_hlp/hom_het.htm.

2.3. DEFORMATION MECHANISM

Common terms used to describe deformation mechanisms include brittle, ductile, and brittle-ductile. The interplay of internal (such as composition, grain size, and lattice-preferred orientation) and external (such as temperature and fluid pressure) factors is the driving mechanism. The conditions, rheology, dynamics, and motions of tectonic events are constrained by a variety of micro-structures that are produced by these mechanisms and studied in rocks. In a given set of circumstances, multiple mechanisms may be active, and some mechanisms may develop on their own. For certain materials, the circumstances and timing under which distinct deformation mechanisms predominate can be determined through in-depth microstructure analysis. Typical processes for deformation mechanisms include:

- Fracturing
- Cataclastic flow
- Grain boundary sliding
- Diffusive mass transfer
- Dislocation creep
- Dynamic recrystallization (recovery)

2.3.1. Fracturing

Fracture is a brittle deformation process that leaves materials with permanent linear breaks devoid of internal material displacement (Karato, 2011). Fractures may be individual or connected, with fracturing requiring the material's ultimate strength to be surpassed causing rupture. Rupture is facilitated by high differential stress accumulation, leading fractures to potentially evolve into faults. The term "fault" is reserved for fractures that include movement. Fractures can occur at various scales, from micro to macro, in rock formations (Figure 2.4).

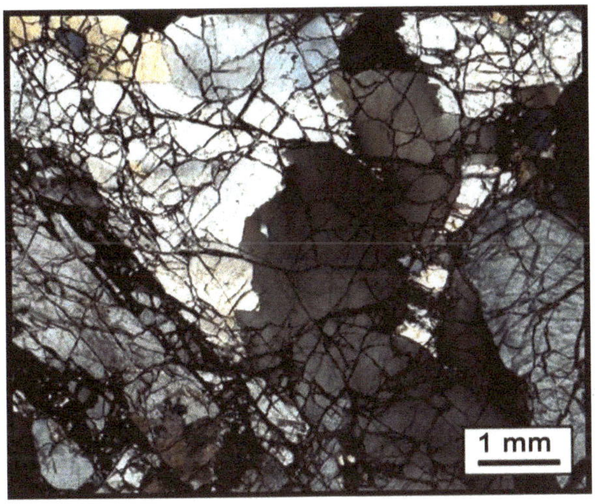

Figure 2.4. Cross polarized image of a high concentration of variably oriented joints within a granitic rock from the San Andreas Fault, California. No obvious displacement along fractures.

Source: By Cstudnic – Own work, CC BY-SA 4.0, https://commons.wikimedia.org/w/index.php?curid=84217839.

2.3.2. Cataclastic Flow

A non-elastic brittle mechanism known as "cataclasis," or "comminution," occurs at low to moderate homologous temperatures, low confining pressures, and relatively high strain rates. It only happens above a specific differential stress threshold that is influenced by temperature and fluid pressure. Grain size reduction from breaking and crushing, rigid body grain rotation, and frictional sliding on grain boundaries are all made possible by cataclasis. Where severe grain size reduction takes place, such as in thin zones along slip or fault surfaces, intense cataclasis takes place. Cataclasis in rocks creates cataclasite, a cohesive, fine-grained fault rock. Cataclastic flow is the result of rock deformation during shearing, specifically microfracturing and frictional sliding where small fractures, or microcracks, and the corresponding rock fragments slide past one another. Typically, cataclastic flow happens in low-grade to diagenetic metamorphic settings. That being said, this is contingent upon the material's mineralogy and the degree of pore fluid pressure. In general, cataclastic flow is unstable and ends when deformation localizes into slip on fault planes (Figure 2.5).

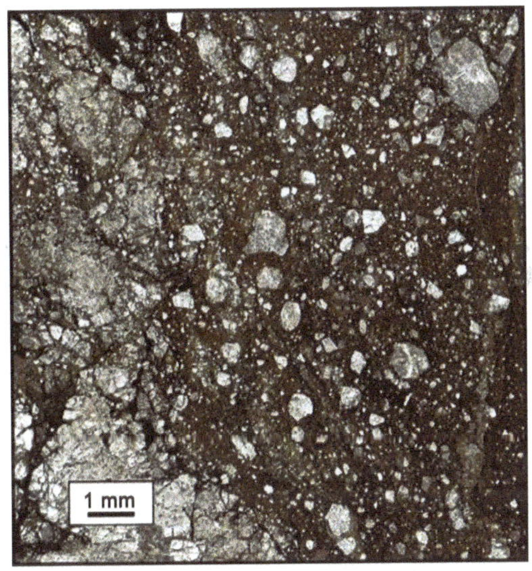

Figure 2.5. Rounded to sub-rounded grains within very fine grained matrix. Fracture processes "grind"/"roll"/"slide" grains past each other creating the rounded appearance of the individual grains.

Source: By Cstudnic – Own work, CC BY-SA 4.0, https://commons.wikimedia.org/w/index.php?curid=84217457.

2.3.3. Grain Boundary Sliding

Grain boundary sliding is a plastic deformation mechanism in which crystals can pass one another through diffusion without causing noticeable voids or friction. Granular flow is the name given to the deformation mechanism connected to this mechanism. There are no voids because of locally enhanced crystal plastic deformation, solid-state diffusive mass transfer, or the solution and precipitation of a grain boundary fluid. Neighbor switching results in a low strain rate at which this mechanism functions. Both temperature and grain size affect grain boundary sliding. High temperatures and the presence of extremely fine-grained aggregates with short diffusion paths are favorable conditions for it. Except for at the grain boundary to allow for grain sliding, large strains operating in this mechanism do not cause the development of a lattice preferred orientation or any discernible internal deformation of the grains; this process is known as superplastic deformation.

2.3.4. Diffusive Mass Transfer

The migration of vacancies in the crystallographic lattice accommodates the strain in this class of mechanisms. As a result, there is a shift in the crystal's shape due to mass diffusion. These migrations, which condition a crystallographic shape fabric or strain, are directed towards sites of maximum stress and constrained by the grain boundaries. A more perfect crystal is the end product. The migration of lattice defects from regions of low to high compressive stress accommodates this grain-size-sensitive process, which happens at low strain rates or very high temperatures. Pressure solution, Coble creep, and Nabarro-Herring creep are the three primary mechanisms of diffusive mass transfer. Nabarro-herring creep, also known as volume diffusion, is dependent on grain size and operates at high homologous temperatures. The strain rate is inversely proportional to the square of the grain size, meaning that as the grain size increases, the creep rate decreases. Grain elongation along the stress axis is a result of the diffusion of vacancies through the crystal lattice (microtectonics) during Nabarro-Herring creep. The Nabarro-Herring creep exhibits a mild dependence on stress. Grain-boundary diffusion, also known as coble creep, is the process by which vacancies diffuse along grain boundaries, elongating the grains along the stress axis. Compared to Nabarro-Herring creep, coble creep occurs at lower temperatures and exhibits a stronger grain-size dependence. It is more significant in the deformation of the plastic crust and plays a bigger role than Nabarro-Herring creep.

2.3.5. Dislocation Creep

A non-linear (plastic) deformation mechanism known as "dislocation creep" occurs when crystal vacancies move past sites of obstruction in the crystal lattice and ascend. Elevated differential stress causes these migrations within the crystal lattice, which can happen in one or more directions. The temperature at which it happens is lower than that of diffusion creep. Slip is the mechanical process that dislocation creep presents. A combination of weak crystallographic orientations from vacancies and flaws in the atomic structure and slip planes define the main direction in which dislocation occurs. A portion of the crystal shifts relative to the rest of the crystal by one lattice point along the slip plane with each dislocation. The lengths of displacement vary depending on the distances between atoms or ions in the crystal lattice of each crystalline material. The Burgers vector is the vector that describes the displacement's length and orientation. Given that dislocations only move in particular lattice planes, the emergence of a strong lattice preferred orientation can be seen as proof of dislocation creep.

Because of the effects of strain-hardening, a dislocation "tangle" can impede the movement of other dislocations, which pile up behind the blocked ones and make the crystal difficult to deform, dislocation glide is unable to act alone to produce large strains. Simultaneous dislocation creep and diffusion are possible. The mechanism that produces the least viscosity will determine the effective viscosity of a stressed material under specific temperature, pressure, and strain rate conditions (Sibson, 2002). There must also be an active recovery process, such as a dislocation climb or grain-border migration. The crystal achieves a more stable state when the dislocation slips because the underlying imperfection is eliminated. Compared to brittle fracture, it requires a lot less differential stress. This process doesn't weaken crystals' internal structure or harm the mineral.

2.3.6. Dynamic Recrystallization

Dynamic recrystallization is the process of eliminating internal strain in grains caused by deformation. This involves reorganizing a material by changing grain size, shape, and orientation within the same mineral. When recrystallization occurs after deformation stops, especially at high temperatures, it is known as static recrystallization or annealing. Dynamic recrystallization leads to smaller grains, while static recrystallization results in the formation of larger equant grains. A broad variety of metamorphic conditions can lead to dynamic recrystallization, which has a significant impact on the mechanical characteristics of the material undergoing deformation. Two end-member processes lead to dynamic recrystallization: (1) Grain-boundary migration (migration recrystallization) and (2) the formation and rotation of subgrains (rotation recrystallization).

1. Rotation recrystallization (subgrain rotation) is the increase in the crystallographic mismatch across the boundary caused by a subgrain's progressive misorientation as more dislocations enter the dislocation wall, a zone of dislocations caused by climb, cross-slip, and glide. Eventually, the misorientation across the boundary (typically 10–15° misorientation) becomes large enough to identify individual grains. With numerous subgrains and an elongated or ribbon-like shape, grains are known for their gradual transition from low-angle subgrains to high-angle boundaries.
2. Migration recrystallization (grain-boundary migration) is the method through which a grain expands at the expense of nearby grains. Low-temperature grain boundary migration, also known as bulging recrystallization, is the process by which a grain boundary bulges into a nearby grain with a high dislocation density to form new, smaller, independent crystals. This mobility of the grain boundary may be local at low temperatures. The bulges generated can migrate the grain boundary or form subgrain (low-angle) boundaries that can eventually become grain boundaries, causing the original grain to split off and form new grains. Bulging recrystallization frequently happens at triple junctions, the edges of old grains. High-temperature grain-boundary migration crystallization causes the grain boundary to sweep through nearby grains to remove dislocations because the growing grain has a lower dislocation density than the grain or grains consumed at high temperatures. New grains are often larger than preexisting subgrains, and grain boundaries are lobate with variable grain sizes. Grains can be almost strain-free at very high temperatures, but they are highly lobate or ameboid.

2.4. DEFORMATION AND STRAIN

Deformation occurs when a geological body is impacted by stress. Below is a summary of the various forms of deformation that will be covered in this section in Figure 2.6. Body deformation can be broadly classified into two categories: non-rigid and rigid. Rock may be translated (all points within a body move parallel to each other) or rotated during rigid body deformation, but its size or shape remain unchanged. On the other hand, alterations in size or form do happen during non-rigid body deformation, also referred to as strain. Whereas dilation refers to a change in size, distortion is a change in shape. Whether the size grows or decreases, the term dilation still holds true.

When all three of the principal stresses are equal, pure dilation—a change in size without a change in shape—occurs. These equal stresses can be seen, for instance, at the bottom of a swimming pool, where the water's weight acts uniformly in all directions (hydrostatic pressure); they can also be seen in objects on the ground, where the weight of the atmosphere acts uniformly in all directions (air pressure); and in situations where an object is buried under a load of rock that covers it (lithostatic pressure). Though they come from different sources, air pressure, hydrostatic pressure, and lithostatic pressure all have a compressive effect on material in all directions. As a result, if the pressure is greater than the affected material's compressive strength, they act to decrease the volume of objects. An increase in volume can happen when there is a drop in pressure, like when a balloon is raised from the bottom of a swimming pool to the surface.

In order for stress to cause a body to change shape, it is necessary for at least one of the principal stresses to differ from the other two, and for this difference in stress to exceed the material's strength. An actual modeling clay sphere can help you see this effect. The air pressure acting on the sphere could be broken down into three equal-magnitude principal stresses if the sphere were placed on a table. You could probably add more pressure than just air pressure to the sphere by carefully placing your hand on top of it without compressing it. Although $\sigma_1 > \sigma_2 = \sigma_3$ in this instance, the difference $\sigma_1 - \sigma_3$ is still less than the clay's material strength. As long as the material's strength is not surpassed, the shape remains unchanged. Should you exert sufficient pressure on the sphere to cause it to become the shape of a cookie, you will have increased σ1 beyond the limits of σ_2 and σ_3, as well as beyond the clay's material strength. The vertical maximum principal stress (σ_1) would be present in the stress ellipsoid that depicts the stress applied to the clay. A stress ellipsoid with horizontal σ_1 is more common in geology, and it is created when horizontal stresses from plate collisions shorten rocks in the earth's crust. Depending on scale, these deformational processes have different names. That is, depending on the size or volume of the rock of interest, we may refer to a deformation using a different term. Imagine a section of an orogen that is several kilometers wide and has developed a fault. One example of distortion on the orogen scale is the fault's horizontal shortening. The fault has translated the rock bodies that it moved. The dislodged rocks themselves might not have changed shape, though. On a smaller scale, translation has happened, but distortion has not happened. Therefore, even though strain describes non-rigid body deformation, rigid body deformation on a smaller scale can also cause strain on one scale of observation.

Any scale can experience translation, rotation, dilation, and distortion; however, the mechanism causing the deformation will vary according to the scale. Large fault blocks moving across one another, grains sliding past one another, or sub-microscopic displacements of the crystal structure of mineral grains are a few examples of how translation happens. Note: the entire mineral grain is distorted due to translations occurring at the atomic level. Another illustration of scale-dependent labeling and observation is this one. A whole orogen may be bent by rotation, rock beds may rotate to create folds, or individual grains may rotate within a fault zone.

When a rock dilates, its volume increases if the entire rock body experiences microscopically small fractures that all open minutely. On the other hand, if the rock gets compacted, the pore spaces, also known as voids, get smaller and the volume of the rock lowers. In either scenario, dilation will happen to the rock overall but not to the individual grains that comprise that rock. The grains will dilate with the rock if there is any degree of grain dissolution. Geosynclines can become orogens due to distortion, which can also change the shape of individual grains in a variety of ways and cause single plutons to revert to their original intrusive forms.

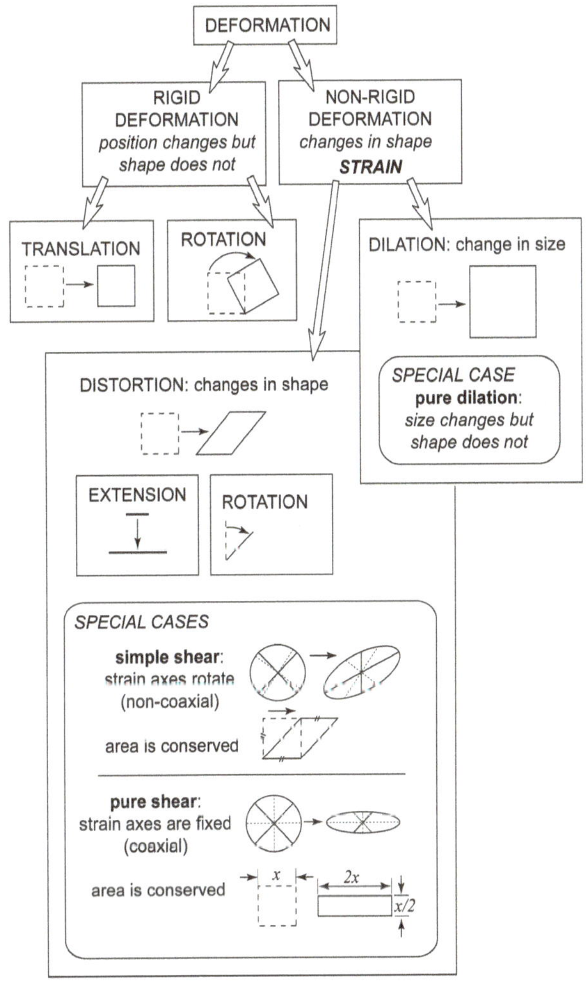

Figure 2.6. Types of deformation.

Source: https://geol319r4.athabascau.ca/unit03.htm#lesson2.

2.4.1. Measuring Strain

2.4.1.1. Homogeneous and Heterogeneous Strain

Objects (such as grains, fossils, or plutons) may shorten or stretch as a result of a rock body's deformation, and their orientation may also alter. A deformation is considered homogeneous if these changes take place uniformly, meaning that each area experiences the same amount of deformation. The deformation is referred to as heterogeneous (or inhomogeneous) if it is locally variable. An ellipse would result from homogeneous deformation of a round object, such as a grain or pluton. The ellipse will become flatter the more deformation there is. Conversely, inhomogeneous deformation could cause the object to take on the shape of a dumbbell or a teardrop, depending on where the greatest strain was applied.

A sufficiently large mass of rock will show that every strain is not uniform. However, there are nearly always a few isolated regions where the strain has been roughly uniform. For instance, the strain brought on by a pluton's intrusion will be higher in the country rock next to it than in the country rock farther away. Nonetheless, strain can be almost uniform on the scale of a single outcrop, say 10 meters across. Thus, by segmenting inhomogeneous strain into regions where homogeneous strain has occurred, it can be described.

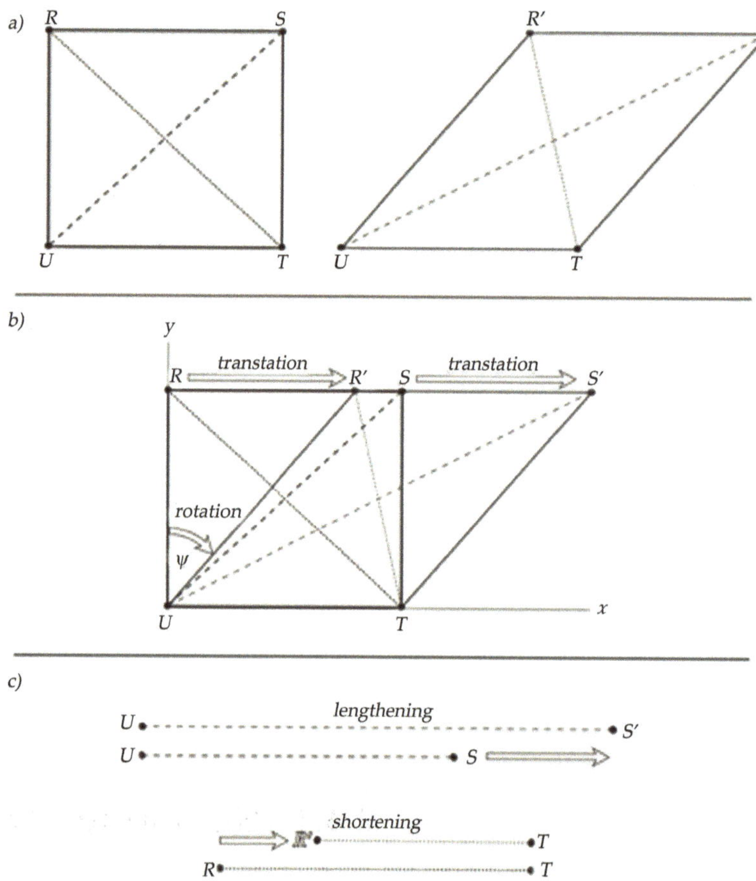

Figure 2.7. Analysis of distortion.

Source: https://geol319r4.athabascau.ca/unit03.htm#lesson2.

An examination of a rock body's distortion and dilatation allows for the description of homogeneous strain. Unfortunately, because we usually don't know the size of the body before dilation occurs, it's usually impossible to determine dilation. For this reason, the remainder of this conversation will only address distortion.

When the box RSTU is homogeneously deformed into the parallelogram R′S′TU, Figure 2.7 illustrates the geometric result of homogeneous strain in two dimensions. The object's height and the lengths of the segments RS (R′S′) and TU remain unchanged in this example. All other lines' lengths, however, have either grown (e.g., US′ > US) or fell (e.g., R′T is less than RT). In addition, all lines other than those that are parallel to the diagram's x-axis have had their orientations changed by an angle of ψ.

2.4.1.2. Measuring Changes in Line Length

The change in line length due to strain can be measured in a variety of ways. Here, we'll concentrate on the technique of calculating the extension, e. A line's length changes as follows: $e > 0$ indicates stretching; $e < 0$ indicates shortening; and $e = 0$ indicates no change in length.

2.4.1.3. Measuring Changes in Angles Between Lines

There are two measurements to take into account when measuring changes in the angles between lines. A line that was initially perpendicular to a line of interest can deviate from being perpendicular; this is described by the first shear, angular shear (ψ). The angle between line B's initial position (dashed line) and its position following deformation (solid line) in this example represents the angular shear along line A. When the rotation is counterclockwise, the angular shear takes on a negative sign; when it is clockwise, it takes on a positive sign.

This definition of angular shear gives the impression that we are discussing line B's rotation, but what really interests us is what transpired to line A. Interpreting "angular shear along the line A" as "the rotation of line B relative to the axes defined by line A" is one way to make sense of this."

We will now look at shear strain, which is represented by the Greek letter gamma (γ). Shear strain is the term used to describe how a line's points shift in position when it rotates to understand the meaning of shear strain intuitively. Observe that $\gamma = 0.5$, which indicates that point A has been translated to the right by half of the object's total width. Point A is translated to the right by an amount equal to the object's width when $\gamma = 1$. Put differently, given w being the object's width, the translation distance of a point on its undeformed side is equal to $\gamma \times w$.

2.4.2. The Strain Ellipsoid

A helpful idea for illustrating the amount of distortional strain experienced by a body of deformed rock is the strain ellipsoid (Figure 2.8). An initially spherical body that has undergone a homogeneous shape change assumes the shape of a strain ellipsoid.

Since perfectly spherical objects are very uncommon in rocks, the strain ellipsoid that results and the original sphere are conceptual idealizations. However, the strain ellipsoid enables us to study the morphological changes of geological bodies of various sizes and initial forms, ranging from tiny grains to pebbles, plutons, and entire orogens.

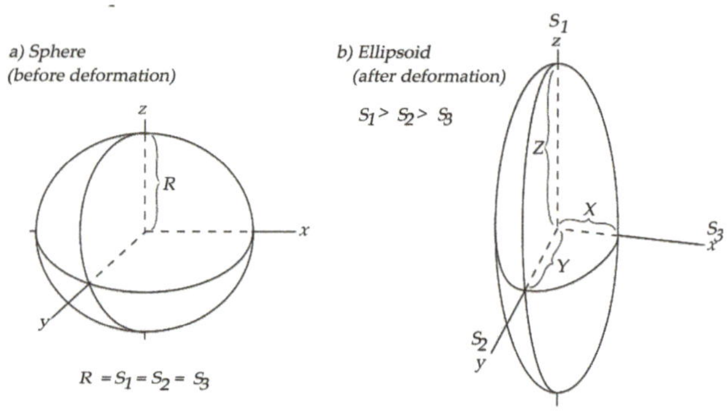

Figure 2.8. Strain ellipsoid.

Source: https://geol319r4.athabascau.ca/unit03.htm#lesson2.

The strain ellipsoid has a maximum principal axis of strain S_1, an intermediate principal axis of strain S_2, and a minor principal axis of strain S_3 ($S_1 > S_2 > S_3$). The original sphere in Figure 2.8a has radius Rdot. The ellipsoid of strain in Figure 2.8b would result from both x- and y-direction shortening (i.e., as well as elongation in the z-direction (that is, along the S_2 and S_3 principal axes). on the principal axis S_1. With the aid of Equation, we can determine the extensions made in the x-, y-, and z-directions. The extensions in the x-, y-, and z-directions are e_x, e_y, and e_z, respectively, and Eq. (2.1), where $l_o = R$. The lengths X, Y, and Z, respectively, are the lf in these directions. Changing these variables in the formula. 2.1, the following is obtained:

$ex=X-R/R$ Eq. 2.1(a)

$ey=Y-R/R$ Eq. 2.1(b)

$ez=Z-R/R$ Eq. 2.1(c)

Observe that the extension is positive where the ellipsoid is larger and negative where it is smaller than the sphere's original radius. By multiplying the extension by 100%, the extension can also be expressed as a change percentage. It is necessary to specify if the percent change represents an increase or decrease; a decrease can be denoted by a negative sign.

Even in cases where dilation has taken place, we can determine the strain experienced by an initially spherical object by calculating the axial ratio of the strain ellipsoid, which spares us from dealing with the original diameter. The triple fraction X|Y|Z is used to express the axial ratio.

Notes about the axial ratio:

- It is intended for the axial ratio to consist of three parts. Don't boil it down to just one number. It is only helpful when we are able to visualize the ellipsoid that would be formed by all three values.
- Dividing all values by the smallest value simplifies the axial ratio. For instance, divide by 10 to obtain an axial ratio of 4|1|2.5 if X = 40, Y = 10, and Z = 25. It is acceptable for two of the numbers in the axial ratio to be non-integers (i.e., in order for numbers to come after the decimal place. There is only one non-integer value in our example, 2.5.
- The units (e.g., mm, s, g, etc.) are typically just as significant in a computation as the numbers. Generally speaking, they ought to be incorporated at the outset and maintained throughout the computation. However, we omit the units in an axial ratio. This is so because the axial ratio makes sense even in the absence of the units, which do not cancel in the three-part ratio. The only restriction on units is that all measures must be expressed in the same unit, such as mm or m, in order for axial lengths to be initially described by that value. The axial ratio would

incorrectly suggest that the ellipsoid was an elliptical pancake lying on its side if calculations were made using X in millimeters and Y and Z in millimeters.

Although there are three unique situations, strain ellipsoids can have any number of axial ratios:

1. Pure flattening ($S_1 = S_2 > S_3$): Equal lengths of extension in S1 and S2 accommodate shortening along S3. The strain ellipsoid will take on a pancake-like shape if the strain is high (S1 = S2 >> S3).
2. Pure constriction ($S_1 > S_2 = S_3$): Roughly equal amounts of shortening along S2 and S3 accommodate stretching along S1. A cigar-shaped strain ellipsoid will form if the strain is high (S1 >> S2 = S3).
3. Plane strain ($S_1 > S_2 > S_3$, but $l_f = S_2$): Shortening only along S3 allows for stretching along S1. Along S2, there is no distortion.

This implies that even though an object changes shape while under strain, its volume will remain unchanged. Because the shapes of the constituent materials, such as grains, pebbles, and phenocrysts, change when rocks are deformed in accordance with these special cases, the result is a layered appearance. Clefts are a type of deformational structure that are characterized by the alignment of flattened rock components, or pure flattening. The cleavage direction is the easiest for many rocks to split in.

Rocks primarily deformed by plane strain and pure constriction exhibit an elongated appearance akin to a bundle of cigars or pencils. We refer to this deformational structure as lineation. Rocks deformed due to fault displacement or a combination of constriction and flattening will exhibit both cleavage and lineation with axial ratios close to pure constriction.

2.4.3. Shear

Shear is another crucial idea that deals with whether or not the strain ellipsoid rotates under applied stress. Pure shear does not involve the strain ellipsoid's rotation, whereas simple shear does. There are always two dimensions to simple shear. Because length does not change in the S2 direction, it is a type of plane strain. Conversely, pure shear can result in either a lengthening or a shortening of the S2 direction or in no change at all (i.e., plane strain).

CASE STUDY

ARLANDABANAN TUNNEL – SHUTTLE STATION 2

The Arlandabanan is a railroad connecting the Arlanda airport and the city of Stockholm and it start with a tunnel beneath Arlanda airport. The Arlandabanan tunnel consists of three underground stations: (i) two stations, "Shuttle Stations." for the direct line between Stockholm Central and Arlanda, and (ii) one station, "Intercity Station." for the distance train. The Shuttle station 2 was constructed directly beneath the existing Terminal 5. The design work of the Arlandabanan tunnel was an "active design." i.e., the design work was done simultaneously with the rock excavation. Geological mapping was carried out for every blast by the shift working geologist. Deformations and drilling rates were measured and joint mapping was conducted with weekly follow-up. The follow-ups were then used along with weekly visits at the excavation site to determine the amount of reinforcement that was needed.

General information

- Excavation site: Shuttle station 2 and track 31 (tunnels U2/N2, sections 39/140–39/450) of a railroad tunnel under Terminal 5 of Arlanda airport.
- Excavation method: Two smaller pilot tunnels (U2 and N2), followed by pillar removal.
- Tunnel dimension: 165 m long, 23 m span and 8–13 m of overburden.
- Production technique: Drilling-and-blasting.
- Type of rock reinforcement: Shotcrete and bolts.

Rock Properties – Shuttle Station 2

Rock mass type: Mica schist and mica gneiss.

- Rock mass properties (see Table 1).
- Joint set properties (see Figure 1):
    - Two larger structures: A weaker zone consisting of a pegmatite dyke and clay gouge.
    - Orientations: Strike 10° to 20° to the tunnel axis and dip about 70°.

Table 1: Rock properties of Shuttle station 2.

Property	UCS* Intact rock (MPa)	GSI	E (GPa)	c (MPa)	ϕ (°)	σ_t (MPa)
According to Chang and Hellstadius (1998):						
Characteristics of schistosity plane:	-	-	7 – 50	0.25	25	-
Characteristics of any possible failure/weak zone:	-	-	2.5	0.20	20	-
Values used by Töyrä, 2006 for numerical analysis:						
Rock mass	50-75	58	17.6	0.80 – 1.10	55	0.21-0.32
Clay filled structures:	-	-	-	0.05	25	0
Ubiquities joints:	-	-	-	0.30	40	0

* It was estimated using R-scale of Brown (1981) according to Töyrä (2006)

Source: https://www.diva-portal.org/smash/get/diva2:997206/FULLTEXT01.pdf.

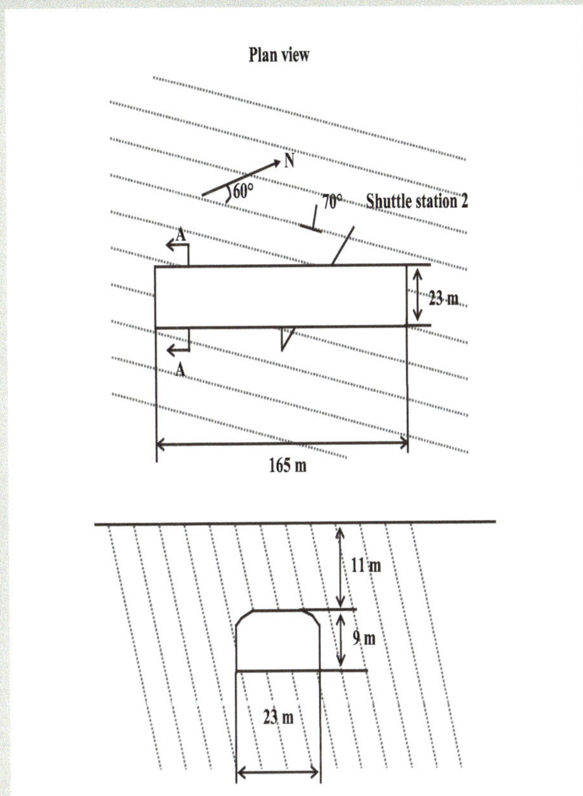

Figure 1. Orientation of the shuttle station 2 in relation to the strike and dip of the mica schist.

Source: https://www.diva-portal.org/smash/get/diva2:997206/FULLTEXT01.pdf.

Stress States – Shuttle Station 2

- Type of stress measurement: Three-dimensional overcoring measurement.
- Depth-equation of the stresses: –
  - Major horizontal stress: σ_H = z/5.27. MPa.
  - Minor horizontal stress: σ_h = z/10.0. MPa.
  - Vertical stress: σ_v = z/ρgz MPa
- Orientations of the major horizontal stress: 120° – 150° from North, hence σ_H is oriented more-or-less perpendicular to the tunnel axis.

Failure and Deformation Measurement

No major instability problems occurred during the excavations of Shuttle station 2 (Töyrä, 2006). Figure 2 shows measurements with 7 extensometers stations and 11 convergence installations between sections 39/227 and 39/457. Tables 2 and 3 give deformation values obtained through extensometer and convergence measurements.

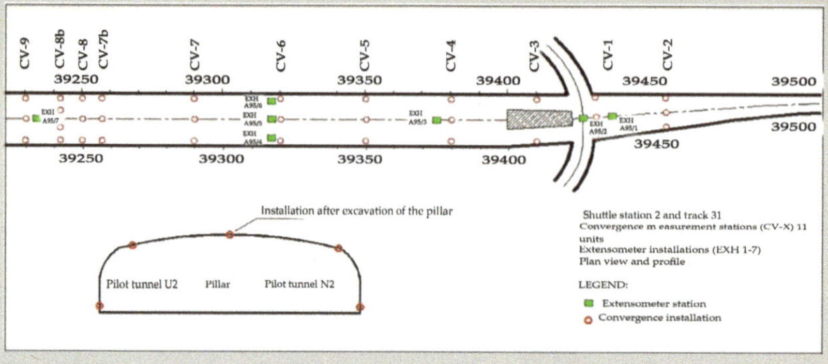

Figure 2. Plan view of the shuttle station 2. Note: from the profile there are five convergence measurement points for each convergence installation (points 1 and 2 at U2; point 3 at pillar; points 4 and 5 at N2).

Source: https://www.diva-portal.org/smash/get/diva2:997206/FULLTEXT01.pdf.

Table 2. Deformation Values Using Extensometers (Track 31)

Exh.	Section	Movement (mm)	Plot period	Observation regarding stability
Exh. 1	39/437	3.5	95-12-01 to 97-04-14	Stable at track 31
Exh. 2	39/425	1.0	95-12-01 to 97-05-14	Stable at track 31
Exh. 3	39/375	1.3	96-03-20 to 97-05-20	Stable shuttle station 2 north
Exh. 4	39/318	0.5	96-03-20 to 97-06-30	Stable at shaft 22
Exh. 5	39/318	-	96-03-20 to 96-12-10	Damaged by blasting of shaft No. 22
Exh. 6	39/318	-0.3	96-03-20 to 97-06-30	Stable at shaft 22
Exh. 7	39/232	-0.2	96-03-20 to 97-02-21	Damaged by blasting of elevator No. 21

Source: https://www.diva-portal.org/smash/get/diva2:997206/FULLTEXT01.pdf.

Table 3. Deformation Values Using Convergences (Track 31)

CV-X	Section	Location	Maximum settlement[7] (mm)	Plot period	Observation regarding stability
CV-1	39/430 track 31	Point 2	-10.5	Feb. 96 to May 97	Completed & Stable
CV-2	39/457 track 31	Point 2	-2.7	Mar. 96 to May 97	Completed & Stable
CV-3	39/412	Point 4	-2.5	Mar 96 to Jun 97	Completed & Stable
CV-4	39/378 Shaft 22	Point 3	-2.3	April 96 to June 97	Completed & Stable
CV-5	39/350	Point 2	-2.0	April 96 to July 97	Completed & Stable
CV-6	39/320	Point 4	2.0	May 96 to July 97	Completed & Stable
CV-7	39/290	Point 2	- 2.5	May 96 to June 97	Completed & Stable
CV-7b	39/257	Point 4	-1.0	April 97 to July 97	Completed & Stable
CV 8	39/250 Shaft 21	Point 2	-7.8	June 96 to June 97	Completed & Stable
CV 8b	39/243 Shaft 21	Point 4	-1.0	April 97 to July 97	Completed & Stable
CV-9	39/227	Point 4	-3.6	July 96 to	Completed & Stable

Note: 3 plots: settlement, H. displacement and L. displacement.

Source: https://www.diva-portal.org/smash/get/diva2:997206/FULLTEXT01.pdf.

CLASS ACTIVITY

CLASS MUSEUM DISPLAY

Children should be asked to search for and bring in samples of geology; these could include items they have purchased or discovered, such as pebbles or fossils (if you're lucky) from the garden or beach.

Show these with hand lenses so that you can perform close observation. Remind the kids that they are curators and that it is their responsibility to make sure others can learn from their artifacts. Encourage the kids to create and finish their own display cards with pertinent details about the specimen. Name, brief description, place of origin (instruct them to research the region/nation instead of writing "a shop"), and a paragraph explaining why the item interests them, among other things, could all be included. Put the collector's name (or their name) on the card. You could invite other classes to visit your mini-museum if you have a really nice display.

Use this as a starting point to learn about collecting, museums, and object care.

SUMMARY

- Rock deformation is the process by which rocks change shape or size in response to stress. Stress is a force applied per unit area, and it can be caused by a variety of factors, including plate tectonics, gravity, and human activity.
- The displacement field can be decomposed into various components, depending on the purpose of the decomposition. The classic way of decomposing it is by separating rigid body deformation in the form of rigid translation and rotation from change in shape and volume.
- Rocks can deform in two main ways: brittlely or ductilely. Brittle deformation occurs when rocks break or fracture under stress. Ductile deformation occurs when rocks bend or flow under stress. The type of deformation that occurs depends on a number of factors, including the type of rock, the temperature and pressure conditions, and the rate at which the stress is applied.
- For studies of deformation, a reference or coordinate system must be chosen. Standing on a dock watching a big ship entering or departing can give the impression that the dock, not the ship, is moving.
- Mathematical treatments of strain commonly assume homogeneous rather than heterogeneous distortions or strains. However, any heterogeneously strained rock body can be subdivided into small areas that exhibit the characteristics of homogeneous strain.
- Deformation mechanisms are commonly characterized as brittle, ductile, and brittle-ductile. The driving mechanism responsible is an interplay between internal (e.g., composition, grain size and lattice-preferred orientation) and external (e.g., temperature and fluid pressure) factors. These mechanisms produce a range of micro-structures studied in rocks to constrain the conditions, rheology, dynamics, and motions of tectonic events.
- Grain boundary sliding is a plastic deformation mechanism where crystals can slide past each other without friction and without creating significant voids as a result of diffusion. The deformation process associated with this mechanism is referred to as granular flow.
- Dislocation creep is a non-linear (plastic) deformation mechanism in which vacancies in the crystal glide and climb past obstruction sites within the crystal lattice. These migrations within the crystal lattice can occur in one or more directions and are triggered by the effects of increased differential stress. It occurs at lower temperatures relative to diffusion creep.
- During rigid body deformation, rock may be translated (all points within a body move parallel to each other) or rotated, but there is no change in its size or shape. In contrast, during non-rigid body deformation, also called strain, changes in size or shape do occur. Distortion is a change in shape, while dilation is a change in size. The term dilation applies whether the size increases or decreases.

REVIEW QUESTIONS

1. What is deformation and its types?
2. Give an example of a deformation.
3. Are earthquakes brittle deformation?
4. How is brittle deformation different from ductile deformation?
5. Does brittle deformation cause folding?
6. What causes deformation of the crust?
7. What is the difference between homogeneous and heterogenous strain?

REFERENCES

1. Argles, T., Prince, C., Foster, G., & Vance, D. (1999). New garnets for old? Cautionary tales from young mountain belts. Earth and Planetary Science Letters, 172(3–4), 301–309. https://doi.org/10.1016/S0012-821X(99)00209-5.
2. Karato, S. (2011). Deformation of Earth Materials: An Introduction to The Rheology of Solid Earth. Cambridge University Press. ISBN 978-1107406056. OCLC 1101360962.
3. Sibson, R. H. (2002). Geology of the crustal earthquake source. In W. H. K. Lee, H. Kanamori, P. C. Jennings, & C. Kisslinger (Eds.), International Handbook of Earthquake and Engineering Seismology (Vol. 81, pp. 455–473). Elsevier. https://doi.org/10.1016/s0074-6142(02)80232-7.
4. Tanner, P., & Shackleton, R. (1979). Structure and stratigraphy of the Dalradian rocks of the Bennabeola area, Connemara, Eire. Geological Society, London, Special Publications, 8(1), 243–256. https://doi.org/10.1144/GSL.SP.1979.008.01.25.
5. Töyrä, J. (2006). Behavior and Stability of Shallow Underground Constructions (Licentiate Thesis 2006:76). Luleå University of Technology. 135 p. ISSN: 1402–1757.

CHAPTER

3 Strain in Rocks

LEARNING OBJECTIVES

After studying this chapter, you will be able to:

- Discuss on rock geology
- Learn about the various properties
- Examine the strain effect
- Define the strain energy

KEY TERMS FROM THIS CHAPTER

Civil engineering	Crystallization
Diagenesis	Homogeneity
Intensity	Lithification
Living organisms	Metamorphism
Mineral grains	Natural cement
Porosity	Pyroclastic rocks
Pyrometamorphism	Recrystallization

INTRODUCTION

The term "strain" in rocks describes the deformation or shape change brought on by applied stress over time. Rocks can change shape in reaction to a variety of external forces, including shear, tension, and compression. This deformation can be plastic, meaning the rock keeps the deformation even after the stress is released, or elastic, meaning the rock returns to its original shape after the stress is released. Squeezing rocks together causes them to compress, shortening in one dimension and thickening in another. This frequently happens at tectonic plate boundaries where tectonic plates collide, causing the compression and folding of rocks to form mountain ranges. Rocks under tension, on the other hand, elongate in one dimension and thin out in another when they are pulled apart. This frequently occurs at divergent plate boundaries, where the movement of tectonic plates causes the creation of mid-ocean ridges and rift valleys.

When parallel forces acting in opposite directions are applied to rocks, shear strain results, causing the rocks to deform as they slide past one another. This kind of strain is frequently seen at transform plate boundaries, which are the places where tectonic plates slide past one another horizontally, causing faults and earthquakes. The intensity and duration of the applied stress, as well as the makeup and structure of the rocks themselves, can all affect how much strain the rocks experience. In contrast to denser, more tightly bound rocks, rocks with high porosity or weak mineral grains are more likely to deform easily under stress.

3.1. ROCK GEOLOGY

In geology, a rock is a coherent, naturally occurring aggregate of one or more minerals. These aggregates usually form identifiable and mappable volumes and are the fundamental building block of the solid Earth. In general, rocks are categorized into three main classes based on the processes that led to their formation. These three classes of rocks are: (1) igneous rocks, which are formed from molten material known as magma; (2) sedimentary rocks, which are made up of materials precipitated from solutions or rock fragments derived from preexisting rocks; and (3) metamorphic rocks, which are formed from sedimentary or igneous rocks under conditions that have changed their mineralogical composition, texture, and internal structure. The most significant factors that determine the subdivided groups and types within these three classes are chemical, mineralogical, and textural attributes.

3.1.1. Rock Types

Rocks classified as igneous solidify from magma, which is a molten mixture of minerals that form rocks and typically volatile substances like steam and gases. Igneous rocks are formed at high temperatures because the constituent minerals of the rock crystallize from molten material. They come from processes that occur deep within the Earth, usually in the upper mantle or the mid-to lower crust, at depths of between 50 and 200 kilometers (30 and 120 miles). There are two types of igneous rocks: those that are intrusive (found in the crust) and extrusive (found on the surface of the land or ocean floor), in which case the molten material that is cooling is referred to as lava. Rocks classified as sedimentary are those that have been deposited at the Earth's surface and have been lithified—that is, compacted and cemented together by ice, wind, water, or living things. The majority are deposited in lakes, rivers, and oceans from the surface of the land. In general, sedimentary rocks are stratified—i.e., there is layering. Layers can be identified by variations in cement type, color, size of the particles, or internal organization.

Metamorphic rocks are created through the transformation of existing rocks due to high temperature, pressure, and chemically active solutions. These changes can be both chemical and physical in nature. This process typically occurs deep within the Earth, resulting in the formation of new minerals, textures, and crystal structures. The recrystallization occurs in a solid state, rather than through complete melting, and can be facilitated by ductile deformation and the presence of fluids such as water. Metamorphism often leads to the development of apparent layering or banding, as minerals segregate into distinct bands. Metamorphic processes can also occur at the Earth's surface due to meteorite impacts and pyrometamorphism near burning coal seams ignited by lightning strikes.

3.1.2. Rock Cycle

Different types of geologic materials, such as mineral crystals and their host rock types, undergo a cycle where they are transformed into various forms. This transformation is influenced by factors like temperature, pressure, time, and changes in environmental conditions within the Earth's crust and on its surface. The rock cycle, shown in Figure 3.1, demonstrates the interrelationships between igneous, metamorphic, and sedimentary rocks. Erosion, which involves both physical and chemical weathering of minerals, contributes to the transportation of these materials to deposition sites. Diagenesis, on the other hand, involves the formation of sedimentary rock through processes like compaction, natural cementation, crystallization from water or solutions, and recrystallization. The process of transforming sediment into rock is known as lithification.

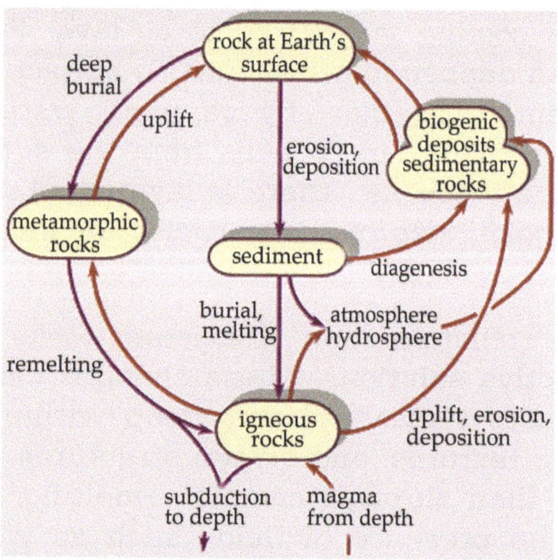

Figure 3.1. The rock cycle.

Source: https://cdn.britannica.com/83/2683-004-486C380B/rock-cycle.jpg.

3.1.3. Texture

The size, shape, and arrangement of the crystals (in the case of igneous and metamorphic rocks) or grains (in sedimentary rocks) determine a rock's texture. Furthermore significant are the degree of homogeneity in the rock (i.e., consistency of composition across the board) and the level of isotropy. The latter is the degree to which the bulk composition and structure of the rock are uniform in all directions. The source material of the rock, the circumstances and environment of deposition (for sedimentary rock) or crystallization and recrystallization (for igneous and metamorphic rock, respectively), and the ensuing geologic history and change can all be inferred from an analysis of the texture of the rock.

3.1.3.1. Classification by Grain or Crystal Size

The common terms used to describe the textures of rocks according to the size of their crystals or grains. The Udden-Wentworth scale, which was created for sediment, is the source of the particle-size categories. The terms are typically used as modifiers for igneous and metamorphic rocks—e.g., granite with medium grains. For smaller crystals, the term "aphanitic" is used, while "phaneritic" is used for larger ones. Pegmatitic refers to very coarse crystals, which are larger than 3 centimeters (1 and a half inches). The three main classifications of sediment size for sedimentary rocks are fine (less than 1/16 millimeter), medium (between 2 and 2 millimeters), and coarse (greater than 2 millimeters, or 0.08 inch). The latter includes silt and clay, which are also referred to as dust and both have sizes that are indiscernible to the naked eye. The majority of shales, which are lithified forms of clay, include some silt. The rocks known as pyroclastic are made of material ejected from volcanoes and are named after the Greek word for broken. Bombs are ejected with molten material, whereas blocks are broken pieces of solid rock.

3.1.3.2. Porosity

The bulk volume of the substance, which includes any grains or crystals and any contained void space, is referred to as "rock." Porosity is the volumetric fraction of bulk rock that is not occupied by crystals, grains, or naturally occurring cement. Put another way, porosity is defined as the grain-to-bulk volume ratio (void space plus void volume). This empty space includes crack space as well as the pore space between crystals or grains. The degree of pore space in sedimentary rocks is determined by several factors, including the degree of cementation, the degree of sorting, the packing arrangement and shape of the grains, and the degree of compaction, which typically increases with burial depth. Minerals containing siliceous, calcareous, carbonate, or iron are common ingredients in cements. The tendency of grains in sedimentary rocks to have similar sizes is known as sorting, to offer a limited selection of sizes. Sediment that has been improperly sorted has a large range of grain sizes and therefore less porosity. A fairly uniform distribution of grain sizes is indicative of well-sorting. Porosity can be significant depending on the kind of close-packing of the grains. Noteworthy is the fact that in engineering usage—e.g., geotechnical or civil engineering—the term "grading" is used instead of the other way around. A poorly graded sediment is a well-sorted one, and a well-graded sediment is a (geologically) poorly sorted one.

Total porosity encompasses all the void space, including those pores that are interconnected to the surface of the sample as well as those that are sealed off by natural cement or other obstructions. Thus the total porosity (ϕ_T) is

$$\aleph_T \quad 100\left(1 \quad \frac{ü_{\ G}}{ü_{\ B}}\right)\%$$

where Vol_G is the volume of grains (and cement, if any) and Vol_B is the total bulk volume. Alternatively, one can calculate ϕ_T from the measured densities of the bulk rock and of the (mono)mineralic constituent. Thus,

$$\aleph_T \quad 100\left(1 \quad \frac{\rho_B}{\rho_G}\right)\%$$

where ρ_B is the density of the bulk rock and ρ_G is the density of the grains (i.e., the mineral, if the composition is monomineralogic and homogeneous). For example, if a sandstone has a ρ_B of 2.38 grams per cubic centimetre (g/cm^3) and is composed of quartz (SiO_2) grains having ρ_G of 2.65 g/cm^3, the total porosity is

$$\varphi_T = 100\left(1 - \frac{2.38}{2.65}\right) = 10.2\%$$

Apparent (effective, or net) porosity is the proportion of void space that excludes the sealed-off pores. It thus measures the pore volume that is effectively interconnected and accessible to the surface of the sample, which is important when considering the storage and movement of subsurface fluids such as petroleum, groundwater, or contaminated fluids.

3.1.4. Stress and Strain

Applying a stress σ (force per unit area) to a material like rock causes the material to change in volume, shape, or dimension. Strain (ε) is the name for this alteration, or deformation. An axial stress can occur—e.g., either simple compression or directed tension, or shear (tangential), or all-sided (e.g., hydrostatic compression). Although the terms pressure and stress are occasionally used synonymously, pressure (P) refers to hydrostatic compression and stress (S) usually refers to directional stress or shear stress. Small stresses result in elastic strains, which are linearly proportional to applied stress and recoverable upon removal of the stress. The strain may be permanent or inelastic under certain circumstances and for higher stresses.

3.1.4.1. Elastic Constants

In elastic deformation, there are various constants that relate the magnitude of the strain response to the applied stress. These elastic constants include the following:

(1) Young's modulus (E) is the ratio of the applied stress to the fractional extension (or shortening) of the sample length parallel to the tension (or compression). The strain is the linear change in dimension divided by the original length (2) Shear modulus (μ) is the ratio of the applied stress to the distortion (rotation) of a plane originally perpendicular to the applied shear stress; it is also termed the modulus of rigidity (3) Bulk modulus (k) is the ratio of the confining pressure to the fractional reduction of volume in response to the applied hydrostatic pressure. The volume strain is the change in volume of the sample divided by the original volume. Bulk modulus is also termed the modulus of incompressibility (4) Poisson's ratio (σ_p) is the ratio of lateral strain (perpendicular to an applied stress) to the longitudinal strain (parallel to applied stress).

For elastic and isotropic materials, the elastic constants are interrelated. For example,

$$\sigma_p = \frac{E}{2\mu} - 1;$$

$$E = 3k(1 - 2\sigma_p)$$

and

$$\mu = \frac{3}{2}k\left(\frac{1-2\sigma_p}{1+\sigma_p}\right)$$

The following are the common units of stress:

1 bar=10^6 dynes per square centimeter

= 10^5 newtons per square metre, or pascals (Pa)

=0.1 megapascal (i.e., 0.1×10^6 Pa)

Thus 10 kilobars = 1 gigapascal (i.e., 10^9 Pa).

3.1.4.2. Rock Mechanics

Rock mechanics is the study of deformation brought on by the strain that rocks experience in response to stresses. The field of study is known as geotectonics when the scale of the deformation is extended to large geologic structures in the Earth's crust. Through laboratory experiments, theoretical model development based on material properties, and field study of deformed rocks and structures, the mechanisms and nature of the deformation of rocks and Earth materials can be explored. Numerous conditions can be directly simulated in the lab, or they can be appropriately scaled through the manipulation of experimental parameters. There are two different kinds of pressure that can be simulated: internal (pore) pressure from pore fluids in void spaces in the rock, and confining (hydrostatic) pressure from burial beneath rock overburden. Studies are conducted on directed applied stress, including compression, tension, and shear, as well as the consequences of rising temperatures brought about by depressurization of the Earth's crust. The results of the length of time and the speed at which stress is applied (i.e., loading) in relation to time are investigated. Additionally, the function of fluids is examined, especially if they have chemical activity.

A directed (uniaxial) compression is applied while a confining pressure is applied (by pressurized fluid) around the cylindrical specimen in certain basic rock-deformation apparatuses. This mimics the deformation that occurs deep inside the Earth. It is also possible to apply an independent internal pore-fluid pressure. It is possible to cover the rock specimen with a thin, impermeable sleeve (e.g., copper or rubber) to isolate the internal pore fluids (if any) from the external pressure medium. The specimen's dimensions are usually a few centimeters. Akira Sawaoka, Naoto Kawai, and Robert Carmichael created another device in 1968 that could apply high pressure to a sample and provide hydrostatic confining pressures of up to 12 kilobars (1.2 gigapascal), additional directed stress, and temperatures of several hundred degrees Celsius. The specimen is placed on the baseplate, and a hydraulic press is used to drive in pistons to apply pressure. To hold the pressure for time experiments and to make the device portable, the end caps can be locked down.

Researchers such as Peter M. Bell and Ho-Kwang Mao have pioneered the development of apparatuses, often using multianvil designs, that can extend the range of static experimental conditions to pressures up to 1,700 kilobars and temperatures of about 2,000°C for small specimens and limited times. This work has been conducted at the Geophysical Laboratory of the Carnegie Institution in Washington, D.C. Additionally, with dynamic techniques utilizing shock from explosive impact generated by gun-type designs, even higher pressures up to 7,000 kilobars (700 gigapascal) can be achieved for very short times. A. Sawaoka at the Tokyo Institute of Technology is a leading figure in this field of ultrapressure research.

Depending on the most recent crustal geologic history, the hydrostatic pressure and temperature in the Earth's upper crust rise at roughly 320 bars and 40°C/kilometer, respectively. Up to two kilobars can be produced by additional directed stress, which is a result of large-scale crustal deformation (tectonism). This is roughly equivalent to the maximum strength of solid crystalline rock at surface pressure and temperature (before fracture) (see below). A single large earthquake, or shift on a fault plane, releases between 50 and 150 bars of stress.

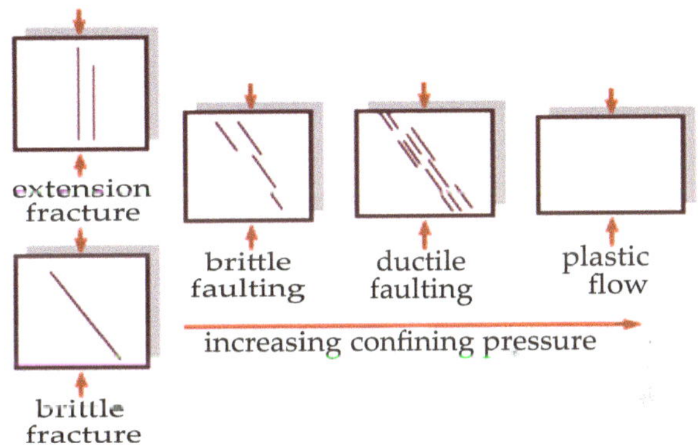

Figure 3.2. Deformation as affected by increased confining pressure.

Source: https://cdn.britannica.com/91/2691-004-29131866/Deformation-pressure.jpg.

One can begin researching the deformation of rocks by assuming ideal behavior, such as elastic strain and homogeneous, isotropic stress and strain. In actuality, igneous and metamorphic rocks contain a fabric of crystals, and sediments contain grains and pores. Large-scale variations in structure and composition can be seen in rock bodies. In addition, factors like prolonged duration, confining pressure, and subsurface fluids influence the rates of deformation change. The generalized shift in response to applied compressional stress and the increasing confining pressure from brittle fracture through faulting to plastic-flow deformation is depicted in Figure 3.2.

3.1.5. Stress-Strain Relationships

The behavior of materials under stress is determined by their stress-strain relationships. Elastic materials exhibit strain that is directly proportional to the applied load, with immediate and reversible deformation up to the yield point stress. Beyond this point, permanent deformation occurs. Viscous materials display slow, smooth flow due to internal friction, known as viscosity, which changes with stress, strain rate, and temperature. Plastic materials show continuous strain after reaching the yield point stress, with some permanent deformation. Elasticoviscous materials combine elastic and viscous behavior, yielding continuously under a constant load. Creep is an example of this behavior, involving slow and continuous deformation over time. Firmoviscous materials behave like solids but exhibit delayed strain release. Plasticoviscous materials behave elastically at first, transitioning to viscous flow after the yield point stress is exceeded.

Some representative values of elastic constants and properties are listed in Table 3.1. The coefficient of viscosity (η) is the ratio of applied stress to the rate of straining (change of strain with time). It is measured in units of poise; one poise equals one dyne-second per square centimeter.

Table 3.1. Some Typical Values of Elastic Constants and Properties

	Elastic constants (at room temperature and pressure)	
Material	**Young's modulus (in 10^6 bars)**	**Shear modulus (in 10^6 bars)**
Ice	0.1	0.03
Shale	0.2–0.3	0.15
Limestone	0.4–0.7	0.22–0.26
Granite	0.3–0.6	0.2
Basalt	0.7–0.9	0.3
Steel	2.1	0.83
Material	Temperature (degrees Celsius)	Coefficient of viscosity (poises)
Lava (Mount Vesuvius)	1,100 1,400	28,300 250
Lava (Oshima, Japan)	1,038 1,125	230,000 5,600
Andesite lava	1,400	150–1,500
Material	Compressive strength (at room temperature and pressure, in kilobars)	
Shale	0.8–1.8	

Sandstone	0.5–2
Limestone	1–2
Granite	1.7–2.5
Basalt	1–3.4

Rheology is the study of the flow deformation of materials. The concept of rheidity refers to the capacity of a material to flow, arbitrarily defined as the time required with a shear stress applied for the viscous strain to be 1,000 times greater than the elastic strain. It is thus a measure of the threshold of fluidlike behavior. Although such behavior depends on temperature, relative comparisons can be made. Some representative values of rheidity times are given in Table 3.2.

Table 3.2. Rheidity Threshold of Fluidlike Deformation

Material	Approximate time
Ice (e.g., glacier)	2 weeks
Gypsum	1 year
Rock salt (e.g., saltdome)	10–20 years
Serpentine (a mafic silicate mineral)	10,000 years

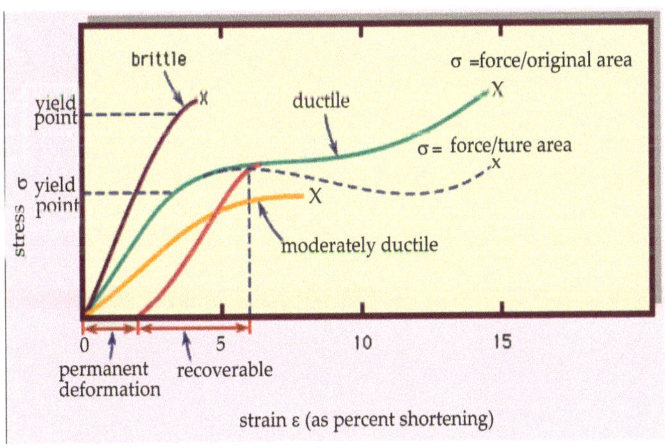

Figure 3.3. Typical stress-strain curves for rock materials. Each X represents the point of fracture for the corresponding material.

Source: https://cdn.britannica.com/92/2692-004-2A85F1FA/stress-strain-curves-rock-materials-X-point-material.jpg.

Figure 3.3 displays typical rock material stress-strain (deformation) curves. Force per unit area is the stress σ, or compression in the figure. The strain, represented in percent, is the specimen's fractional shortening parallel to the applied compression; it is called ε. Before breaking, the ductile (plastically deformable) material exhibits a range of plastic deformation, while the brittle material behaves nearly elastically up to the point of fracture (denoted X). For something to be ductile, it must be able to undergo significant permanent deformation before breaking. Within crystal grains, intracrystalline motion (slip and twinning), intercrystalline motion (crushing and fracture; cataclasis), and recrystallization through solutioning or solid diffusion are the flow mechanisms involved in plastic deformation. A ductile material that is in the plastic range will recover part of its strain elastically if the applied stress is removed, but permanent deformation will result. The ultimate strength is found at fracture, which is the total loss of cohesiveness. It is the highest point (stress) on a stress-strain curve. A material's strength is determined by how resistant it is to failure (structural destruction) through flow or fracture; it is a measurement of the stress needed to cause a body to deform.

3.2. PROPERTIES

Numerous professions, such as geology, petrophysics, geophysics, materials science, geochemistry, and geotechnical engineering, are interested in and use the physical properties of rocks. The scope of the research spans from crystalline and molecular processes to terrestrial explorations of Earth and other planetary bodies. Seismologists use premonitory physical or chemical changes to predict future earthquakes; crystallographers study the synthesis of minerals with unique optical or physical properties; exploration geophysicists investigate the variation of physical properties of subsurface rocks to make possible detection of natural resources like oil and gas, geothermal energy, and metal ores; geotechnical engineers examine the nature and behavior of materials on, in, or of which structures such as buildings, dams, tunnels, bridges, and underground storage vaults are to be constructed; solid-state physicists study the mechanical, electrical, and magnetic aspects of materials.

Since rocks are collections of individual mineral grains or crystals, the characteristics of the different constituent minerals greatly influence the characteristics of the rocks themselves. These overall characteristics of a rock are calculated by averaging the relative characteristics, and occasionally the orientations, of the different grains or crystals. Thus, certain characteristics that are anisotropic (i.e., vary with direction) on a crystalline or submicroscopic scale are largely isotropic for a sizable portion of the bulk rock. A number of other factors include the type and quantity of contained fluids (e.g., natural cements in sedimentary rocks), temperature and pressure, the amount and distribution of void space, the size, shape, and packing arrangement of grains or crystals, and many others (e.g., gases, water, and petroleum). A statistical variation is frequently used to assign representative values for a given property because many rocks show a significant range in these factors. Certain properties can differ significantly based on whether they are measured under simulated laboratory conditions or in situ, or in the subsurface. For example, the fluid content of the rock in place and the temperature at that specific depth have a significant impact on electrical resistivity.

3.2.1. Density

Due to variations in mineralogy and porosity, density varies greatly amongst different types of rock. Interpreting the subsurface geologic structure and rock type can be made easier with an understanding of the distribution of underground rock densities. Density is commonly understood to be the weight of a sample in air per unit volume at a given temperature, although its precise definition is the mass of a substance per unit volume. Mass, a measurement of the matter in a body, is a fundamental property that remains constant no matter where it is, but weight, the force that gravity applies to a body, varies with location. Because the difference between mass and weight would cause less error in the computed density than experimental errors introduced in the volume measurement, sample weights in routine density measurements of rocks are thought to be equivalent to their masses. As a result, weight is frequently used to calculate density rather than mass. Although it should be expressed correctly in kilograms per cubic meter (kg/m^3), density is frequently expressed in grams per cubic centimeter (g/cm^3).

Another property closely related to density is specific gravity. It is defined, as noted above, as the ratio of the weight or mass in air of a unit volume of material at a stated temperature to the weight or mass in air of a unit volume of distilled water at the same temperature. Specific gravity is dimensionless (i.e., has no units).

The bulk density of a rock is $\rho_B = W_G/V_B$, where W_G is the weight of grains (sedimentary rocks) or crystals (igneous and metamorphic rocks) and natural cements, if any, and V_B is the total volume of the grains or crystals plus the void (pore) space. The density can be dry if the pore space is empty, or it can be saturated if the pores are filled with fluid (e.g., water), which is more typical of the subsurface (in situ) situation. If there is pore fluid present,

$$(\text{saturated})\rho_B = \frac{W_G + W_{fl}}{V_B}$$

where W_{fl} is the weight of pore fluid. In terms of total porosity, saturated density is

$$(\text{saturated})\rho_B = \rho_G(1-\varphi_T) + \rho_{fl}\varphi_T$$

and thus

$$\varphi_T = \frac{\rho_G - \rho_B}{\rho_G - \rho_{fl}}$$

where ρ_{fl} is the density of the pore fluid. Density measurements for a given specimen involve the determination of any two of the following quantities: pore volume, bulk volume, or grain volume, along with the weight.

Plotting the statistical range of a set of data as a histogram is a helpful method for determining the density of rocks. The average value is called the mean, and the most common value is called the mode. These expressions can be used to express the representative value and its variation, i.e., the distribution curve's peak. Median is the value of the data set's middle sample (i.e., the value at which half of the samples fall below and half above); and standard deviation is a statistical indicator of the data's distribution that comprises roughly two-thirds of the data when it is plus or minus one standard deviation from the mean value.

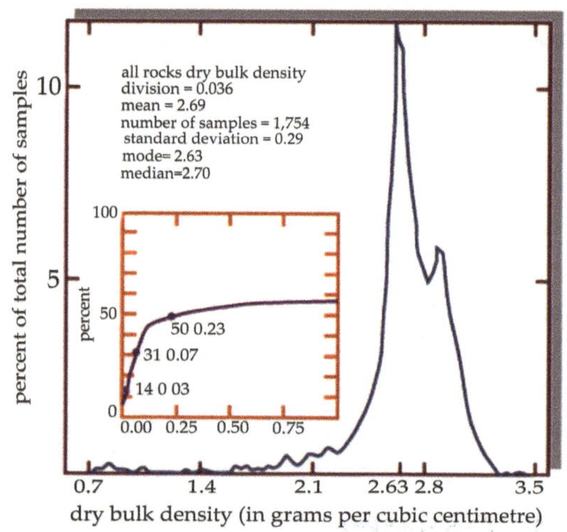

Figure 3.4. Dry bulk densities (distribution with density).

Source: https://cdn.britannica.com/86/2686-004-27B8754C/bulk-densities-rocks-table.jpg.

Table 3.3 contains a list of the dry bulk densities for the different types of rocks that make up the Earth's upper crust. Figure

3.4 displays a histogram plot of these data that shows the percentage of samples as a function of density. Among the parameters provided are (1) sample division and the density range in a single data column, e.g., for Figure 3.4, 0.036 g/cm^3, (2) sample count, and (3) standard deviation. The percentage of samples (on the vertical axis) that fall between the intervals "mode – x" and "mode + x," where x is the horizontal axis, is represented by the small inset plot.

Table 3.3. Dry Bulk Densities for Various Rock Types

Rock type	Number of samples	Mean (grams per cubic cm)	Standard deviation	Mode (grams per cubic cm)	Median (grams per cubic cm)
All rocks	1,647	2.73	0.26	2.65	2.86
Andesite	197	2.65	0.13	2.58	2.66
Basalt	323	2.74	0.47	2.88	2.87
Diorite	68	2.86	0.12	2.89	2.87
Dolerite (diabase)	224	2.89	0.13	2.96	2.90
Gabbro	98	2.95	0.14	2.99	2.97
Granite	334	2.66	0.06	2.66	2.66
quartz porphyry	76	2.62	0.06	2.60	2.62
rhyolite	94	2.51	0.13	2.60	2.49
syenite	93	2.70	0.10	2.67	2.68
trachyte	71	2.57	0.10	2.62	2.57
sandstone	107	2.22	0.23	2.22	2.22

Source: https://www.britannica.com/science/rock-geology/Physical-properties. After data from H.S. Washington (1917) and R.J. Piersol, L.E. Workman, and M.C. Watson (1940) as compiled by Gary R. Olhoeft and Gordon R. Johnson in Robert S. Carmichael (ed.), Handbook of Physical Properties of Rocks, vol. III, CRC Press, Inc. (1984).

The distribution's most frequent (modal) value in Figure 3.4 is 2.63 g/cm^3, or about the density of quartz, a common mineral that forms rocks. For these upper crustal rocks, density values rarely exceed 3.03 g/cm^3. A small number consistently fall below the mode, sometimes even below 1 g/cm3. Granite is an intrusive igneous rock with a narrow density range, low porosity, and a well-defined chemical (mineral) composition. Since entrained gases leave behind voids known as vesicles, basalt is typically an extrusive igneous rock that exhibits a wide range in porosity; as a result, some extremely porous samples may have low densities. The degree of sorting, compaction, grain packing arrangement, and cementation can all affect the porosity of sandstone, a clastic sedimentary rock. As a result, the bulk density changes.

Table 3.4. Typical Density Ranges for Some Other Rock Types

Rock type	Density (grams per cubic cm)
Amphibolite	2.79–3.14
Andesite glass	2.40–2.57
Anhydrite	2.82–2.93
Anorthosite	2.64–2.92
Basalt glass	2.70–2.85
Chalk	2.23
Dolomite	2.72–2.84
Dunite	2.98–3.76
Eclogite	3.32–3.45
Gneiss	2.59–2.84
Granodiorite	2.67–2.78
Limestone	1.55–2.75
Marble	2.67–2.75
Norite	2.72–3.02
Peridotite	3.15–3.28
Quartzite	2.65
Rock salt	2.10–2.20
Schist	2.73–3.19
Shale	2.06–2.67
Slate	2.72–2.84

Source: After data from R.A. Daly, G.E. Manger, and S.P. Clark, Jr. (1966); A.F. Birch (1966); F. Press (1966); and R.N. Schock, B.P. Bonner, and H. Louis (1974) in Robert S. Carmichael (ed.), Handbook of Physical Properties of Rocks, vol. III, CRC Press, Inc. (1984). https://www.britannica.com/science/rock-geology/Physical-properties.

The density of clastic sedimentary rocks increases as the rocks are progressively buried. This is because of the increase of overburden pressure, which causes compaction, and the progressive cementation with age. Both compaction and cementation decrease the porosity.

3.2.2. Electrical Properties

Conductivity, which is the inverse of resistivity, and dielectric constant, along with coefficients indicating how these change with temperature, frequency of measurement, and other factors, are what define a material's electrical nature. The electrical property values can differ significantly for rocks with a wide range of chemical compositions and physical characteristics, such as fluid content and porosity. The definition of resistance

(R) is one ohm, which is obtained by applying a voltage (V) of one volt magnitude across a specimen to produce a current (i) of one ampere ($V = Ri$). One of the material's inherent characteristics is its electrical resistivity (ρ). Stated differently, it is inherent and not reliant on the sample size or the current course. Resistance is related to it by $R = \rho L/A$, where L is the specimen's length, A is its cross-sectional area, and ρ is measured in ohm-centimeters; one ohm-centimeter is equal to 0.01 ohm-meters. The conductivity (σ) can be expressed as $1/\rho$ $ohm^{-1} \cdot centimeter^{-1}$, or mhos/cm. It is stated in either mhos/meter or siemens/meter when using SI units.

Table 3.5. Typical Resistivities

Material		**Resistivity (ohm-centimeter)**
Seawater (18°C)		21
Uncontaminated surface water		$2(10^4)$
Distilled water		$0.2–1(10^6)$
Water (4°C)		$9(10^6)$
Ice		$3(10^8)$
Rocks in situ		
Sedimentary	Clay, soft shale	$100–5(10^3)$
	Hard shale	$7–50(10^3)$
	Sand	$5–40(10^3)$
	Sandstone	$(10^4)–(10^5)$
	Glacial moraine	$1–500(10^3)$
	Porous limestone	$1–30(10^4)$
	Dense limestone	$>(10^6)$
	Rock salt	$(10^8)–(10^9)$
Igneous		$5(10^4)–(10^8)$
Metamorphic		$5(10^4)–5(10^9)$
Rocks in laboratory		
Dry granite		10^{12}
Minerals		
Copper (18°C)		$1.7(10^{-6})$
Graphite		$5–500(10^{-4})$
Pyrrhotite		0.1–0.6
Magnetite crystals		0.6–0.8
Pyrite ore		$1–(10^5)$
Magnetite ore		$(10^2)–5(10^5)$
Chromite ore		$>10^6$
Quartz (18°C)		$(10^{14})–(10^{16})$

Source: https://www.britannica.com/science/rock-geology/physical-properties.

Table 3.5 contains a list of some typical electrical resistivity values for rocks and other materials. Materials with resistivities of 10^{-5}–10 ohm-centimeter (10^{-7}–10^{-1} ohm-meter) and conductivities of 10–10^{7} mhos/meter are typically regarded as "good" conductors. The resistivity and conductivity of those categorized as intermediate conductors are 10^{0}–10^{9} ohm-centimeter (1–10^{7} ohm-meter) and 10^{-7}–1 mhos/meter, respectively. "Poor" conductors, sometimes referred to as insulators, have a conductivity of 10^{-15}–10^{-8} and a resistivity of 10^{10}–10^{17} ohm–centimeter (10^{8}–10^{15} ohm–meter). Seawater has far superior conductivity. because it contains more dissolved salts than fresh water, it has less resistivity; dry rock, on the other hand, has a high resistivity. Fluids usually fill pores in the subsurface to some extent. Materials vary widely in their resistivity; copper and quartz, for instance, differ by 22 orders of magnitude.

The electrical response of a rock to high-frequency alternating currents is partially determined by the dielectric constant, ε. This represents the rock's ability to retain electric charge and serves as a gauge of its polarizability in an electric field. In a vacuum, the dielectric constant is 1.0 in cgs units. The SI unit of measurement is farads per meter, which can also be expressed as the ratio of the material's specific capacity to the vacuum's specific capacity (8.85×10^{-12} farads per meter). For frequencies that are significantly higher than 100 hertz (cycles per second), the dielectric constant depends on both temperature and frequency. Rocks conduct electricity through (1) fluid conduction, i.e., electrolytic conduction through ionic transfer in briny pore water—and (2) semiconductor and metallic (e.g., electron conduction in some sulfide ores). The fluid usually dominates the conductivity response if there is any porosity in the rock and it contains fluid. The conductivity of the fluid (as well as its chemical makeup), temperature, porosity and permeability, and the degree of fluid saturation all affect how conductive the rock is. Rocks usually become more resistive when they lose water, as occurs when clastic sedimentary rocks compact at depth.

3.2.3. Magnetic Properties

The magnetic characteristics of the individual mineral grains and crystals that make up rocks are what give them their magnetic qualities. Usually, magnetic minerals make up a very small portion of the rock. The magnetic properties and overall magnetization of a given rock are determined by this small fraction of grains. This has two implications: (1) chemical inhomogeneities, depositional or crystallization conditions, and post-formation rock behavior can all have a significant impact on a given rock's magnetic properties; and (2) rocks with the same lithology (type and name) do not always have the same magnetic properties. Lithologic classifications are typically based on the predominance of silicate minerals, but the small fraction of magnetic mineral grains, like iron oxides, determines magnetization. The two main magnetic minerals that form rocks are iron oxides and sulfides. Despite the fact that different rocks within the same classification may have different magnetic properties, a rock's general magnetic properties are typically determined by its type and overall composition. If one has specific knowledge about the magnetic properties of crystalline materials and minerals, as well as how those properties are influenced by variables like temperature, pressure, chemical composition, and grain size, one can comprehend the magnetic properties of a given rock fairly well. Knowledge of how typical rock properties vary under different conditions and are dependent on the geologic environment helps to further our understanding.

3.2.3.1. Applications of the Study of Rock Magnetization

Geology, materials science, and prospecting are three fields in which knowledge of rock magnetization is crucial. Mapping the buried rocks' composition, size, type, and depth is a goal of magnetic prospecting. Prospecting is a crucial initial step in the exploration of buried geologic structures. It can be conducted from the surface of the earth, from a ship or an aircraft, and it can help identify potential locations for economic mineral deposits, natural gas, and oil. Geology has historically placed a high value on rock magnetization. By identifying the types of remanent magnetization (see below), paleomagnetic research aims to ascertain the nature of the Earth's field during the formation of specific rocks. The findings of this kind of study have significant implications for age dating, stratigraphic correlation, and reconstructing the crust's historical movements. The notions of continental drift and seafloor spreading, both of which are essential to the theory of plate tectonics, were in fact corroborated by magnetic surveys of the oceanic crust, which for the first time provided the quantitative evidence required to persuasively demonstrate that segments of the crust had undergone large-scale lateral displacements over geologic time (see plate tectonics).

The understanding of magnetization is increasingly important in materials science as well. The design and manufacture of efficient memory cores, magnetic tapes, and permanent magnets increasingly rely on the ability to create materials having desired magnetic properties.

3.2.3.2. Basic Types of Magnetization

There are six basic types of magnetization: (1) diamagnetism, (2) paramagnetism, (3) ferromagnetism, (4) antiferromagnetism, (5) ferrimagnetism, and (6) superparamagnetism.

The orbiting electrons that surround each atomic nucleus are the source of diamagnetism. The atoms create their own magnetic field in opposition to the applied field when an external magnetic field is applied because of the orbital shifts. Stated differently, the external field is opposed by the induced diamagnetic field. All materials contain diamagnetism, although it is weak and only manifests itself in the presence of an applied field. Susceptibility (k) is the tendency of a material to become magnetized in an external field; it is defined as J/H, where J is the magnetization (intensity) per unit volume and H is the applied field strength. Diamagnetic susceptibility has a negative sign because the applied field and the induced field are always opposed. A diamagnetic material has a susceptibility of roughly -10^{-6} electromagnetic units per cubic centimeter (emu/cm^3). Susceptibility expressed as κ per unit mass of material is sometimes used.

Unpaired electrons' electron spin causes paramagnetism. When a collection of electrons is exposed to a magnetic field, their magnetic dipole moments have a tendency to align with the field because each electron has a magnetic dipole moment, which causes it to act like a tiny bar magnet. In the direction of the applied field, the effect increases the net magnetization. Similar to diamagnetism, paramagnetism is weak and only manifests when an applied field is present; however, because the effect amplifies the applied field, paramagnetic susceptibility is always positive. A paramagnetic substance has a susceptibility of approximately 10^{-4} to 10^{-6} emu/cm^3.

The existence of ferromagnetism is also due to the electron's magnetic characteristics. On the other hand, ferromagnetism can exist in the absence of an applied external field, unlike paramagnetism. Because it is advantageous for them to do so energetically, the atoms' magnetic dipole moments align themselves spontaneously. One can maintain a remanent magnetization. The temperature at which the dipole moments would align completely would be at zero kelvin, or −273.15° C. Thermal motions start to disorder the magnetic moments above absolute zero. The ferromagnetic properties of a substance vanish when thermally induced disorder surpasses alignment at a temperature known as the Curie temperature, which varies depending on the material. Ferromagnetic materials have a high and positive susceptibility. It is between 10 and 10^4 emu/cm^3. Iron, cobalt, and nickel are among the few materials that possess a strong residual magnetization and are strictly speaking ferromagnetic. Any material that is noticeably magnetic is commonly referred to as ferromagnetic in general usage, especially in engineering.

When a material's atoms' dipole moments take on an antiparallel configuration without an applied field, it is said to exhibit antiferromagnetism. As a result, there is no net magnetization in the sample. The susceptibility strength is similar to that of paramagnetic materials. Thermal motions destroy the antiparallel arrangement above a temperature known as the Néel temperature, which causes the material to become paramagnetic. A unique state known as spin-canted (anti)ferromagnetism is produced when antiparallel magnetic moments are redirected out of the antiferromagnetic plane, leaving a weak net magnetism in their wake. Hematite (α-Fe_2O_3) is one of these substances. Due to unequal moments of the magnetic sublattices, ferrimagnetism is the antiparallel alignment of atomic dipole moments that does produce a noticeable net magnetization. There is remanent magnetization visible. A substance turns paramagnetic above the Curie temperature. Ferrimagnetic materials include magnetite (Fe_3O_4), the most magnetic common mineral. Superparamagnetism arises in materials whose grains are so tiny (roughly 100 angstroms) that thermal energy overcomes any cooperative dipole moment alignment.

3.3. STRAIN EFFECT

The term "strain" describes the deformation that a material undergoes in reaction to loads or applied forces. It is an essential idea in material science and engineering that sheds light on how various situations affect the behavior of materials. Strain is the measurement of the shape or size change that a material experiences when it is subjected to an external force. Shear strain and normal strain are two of the various types of strain. A material undergoes normal strain when an axial load causes a change in length or volume along a particular axis. It is represented as a ratio of the material's original length to its altered length. On the other hand, shear strain happens when forces are applied parallel to the material's surface, causing it to deform in a way that resembles sliding or shearing. The mechanical characteristics of a material, in particular its modulus of elasticity, affect how strain affects the material's behavior. The material's elastic deformation in response to stress is indicated by this modulus. Hooke's Law states that stress and strain are directly proportionate within the elastic range. This indicates that when the applied load is released, the material reverts to its initial form.

However, plastic deformation happens when the applied forces surpass a particular threshold, which is referred to as the material's yield point. This indicates that even after the load is released, the material experiences a permanent change in shape. The magnitude of the applied load and the characteristics of the material determine the degree of plastic deformation. In engineering applications, knowing the point at which a material changes from elastic to plastic behavior is essential because it establishes the highest load that a structure can bear without permanently deforming. Excessive strain can cause a variety of material failures. For instance, high strain can lead to dislocations in the crystal structure of metals, which can weaken the material and possibly cause fracture. Strain in polymers can cause molecular reorganization and chain slippage, which can impact the material's overall mechanical properties. When building structures or components, it is crucial to take into account a material's strain limits to make sure the elements can safely support anticipated loads. A material's deformation as a result of an outside force is measured by its strain. This indicates the degree of deformation that a material experiences when stressed. Strain can be expressed as a percentage or as a unitless quantity, and it is defined as the ratio of a material's changed length (or other dimensions) to its initial length (or dimension). The following is the equation for the change in a specimen's length.

Slide 1 of 1

Strain formula:

$$\varepsilon = \frac{L - L0}{L0}$$

where

- Strain (ε) is the fractional or percentage change in length or another dimension.
- L is the length of the material after an external load is applied.
- L0 is its original length measured in the same units as "L."

Tensile, shear, compressive, volumetric, and thermal strains are the names for the different types of stress that are applied, and these names correspond to the strains that are observed. When a material is stretched or elongated, it experiences tensile strain; when it is compressed or shortened, it experiences compressive strain. Stress applied in a direction parallel to the cross-section of interest results in a shear strain. Volumetric strain is the term used to describe changes in a material's volume as a result of alterations in its size or shape. Finally, a material's change in length or volume as a result of temperature changes is referred to as thermal strain.

3.3.1. Strain in Materials Work

When an external force is applied to a material, it causes deformation, which is measured as strain. The deformation behavior of the material under stress depends on various factors such as the type of material, its composition, and the magnitude, direction, and nature of the applied stresses. In general, materials can deform in three main ways under stress:

1. Elastic Deformation: Occurs when the material deforms under stress but returns to its original shape when the stress is removed. In other words, the material can withstand a certain amount of strain without undergoing permanent deformation.
2. Plastic Deformation: Occurs when the material is subjected to stress beyond its elastic limit, causing permanent deformation. The material undergoes a change in shape that is not recoverable even after the stress is removed. The amount of plastic deformation that a material can undergo before it breaks is known as its ductility.
3. Fracture: Occurs when the applied stress exceeds the strength of the material, causing it to break or fail. The strength of a material indicates its ability to withstand stress without breaking or undergoing plastic deformation.

3.3.2. Mechanics of Materials: Strain

We have been concentrating on the stress inside structural elements thus far. An object deforms when it is under stress. Consider a rubber band. When you pull on it, it stretches and becomes longer. Strain is the ratio of an object's deformation to its original length, while deformation is a measurement of how much an object is stretched. Consider strain as the percentage of elongation, or the amount that the object grows or contracts when it is loaded. Similar to stress, a structure may encounter two different kinds of strain: (1) Standard Strain and (2) Shear Stress. A force exerts a normal stress when it acts perpendicular, or normal, to an object's surface. Shear stress is produced when a force acts parallel to an object's surface. Consider a rod that is subject to uniaxial tension. Under this tension, the rod elongates to a new length; the normal strain is the product of this tiny deformation and the rod's initial length.

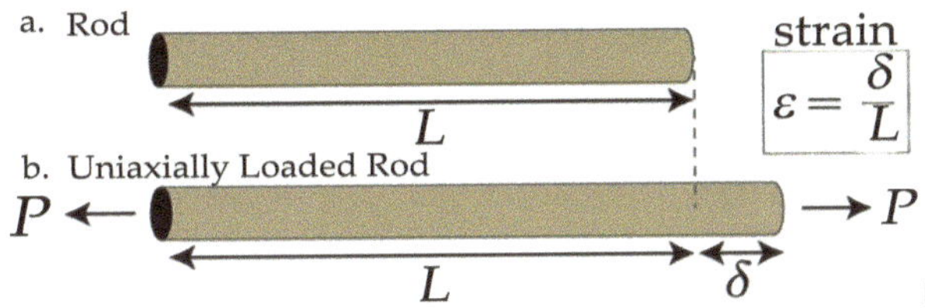

Figure 3.5. Rod under uniaxial tension

Source: https://www.bu.edu/moss/files/2015/01/strain.jpg.

The amount that an object expands or contracts in response to an applied load is measured unitlessly as strain. Normal strain happens when an object elongates in reaction to a typical stress (i.e., perpendicular to a surface), and the Greek letter epsilon is used to represent it. A compressive value is represented by a negative value, and a tensile strain by a positive value. Shear strain happens when an object deforms in reaction to a shear stress (i.e., parallel to a surface), and the Greek letter gamma is used to represent it.

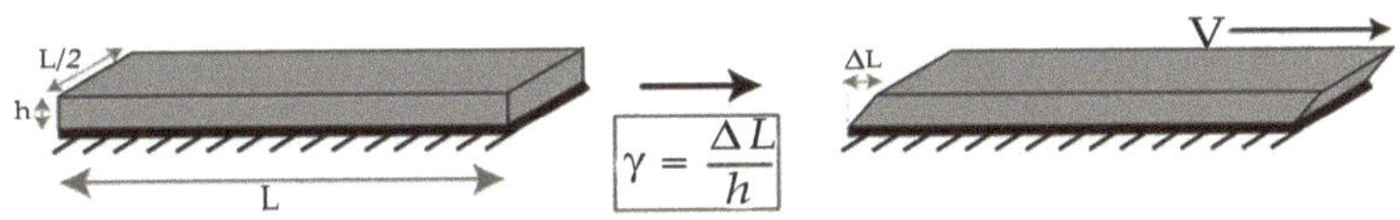

Figure 3.6. Shear strain in materials.

Source: https://www.bu.edu/moss/files/2015/01/shearstrain.jpg.

3.3.2.1. Mechanical Behavior of Materials

Stress and strain are undoubtedly related. A constitutive law links stress and strain, and we can measure the relationship between the two experimentally by figuring out how much stress is needed to stretch a material. A tensile test can be used to accomplish this measurement. In the simplest scenario, an object will deform more the more you pull on it; for small strain values, this relationship is linear. Hooke's Law is the name given to this elastic, linear relationship between stress and strain. Plotting stress against strain will result in a linear graph for small strains; the slope of the line represents the material's Young's elastic modulus. From 1 kPa for Jello to 100 GPa for steel, this value can vary significantly. The linear region of the stress-strain diagram only appears for very small strains (<0.1%) for the majority of engineering materials. We will only be studying materials that are linearly elastic in this course. i.e., they are isotropic (behave in the same way regardless of the direction of pull) and obey Hooke's law.

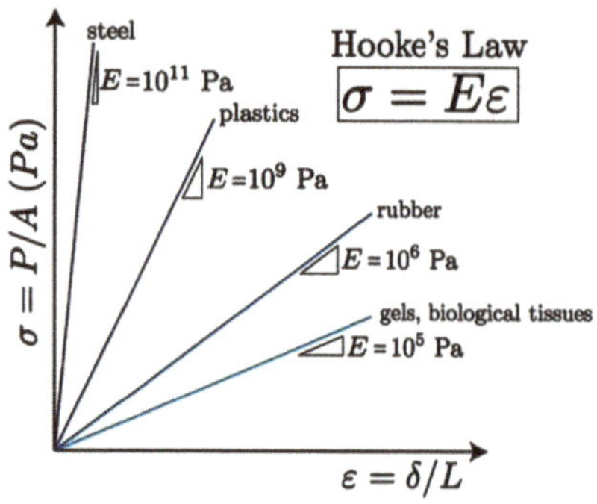

Figure 3.7. Mechanical behavior of materials.

Source: https://www.bu.edu/moss/files/2015/01/stressstraincurve.jpg.

From Hooke's law and our definitions of stress and strain, we can easily get a simple relationship for the deformation of a material.

$$\sigma = E\varepsilon$$
$$\frac{P}{A} = E\frac{\delta}{L} \rightarrow \boxed{\delta = \frac{PL}{EA}}$$

Intuitively, this exam makes a bit of sense: apply more load, get a larger deformation; apply the same load to a stiffer or thicker material, get less deformation. If the structure changes shape, or material, or is loaded differently at various points, then we can split up these multiple loadings using the principle of superposition.

$$\delta_{total} = \sum_i \frac{P_i L_i}{E_i A_i}$$

3.3.2.2. Generalized Hooke's Law

First things first, even just pulling (or pushing) on most materials in one direction actually causes deformation in all three orthogonal directions. Let's go back to that first illustration of strain. This time, we will account for the fact that pulling on an object axially causes it to compress laterally in the transverse directions:

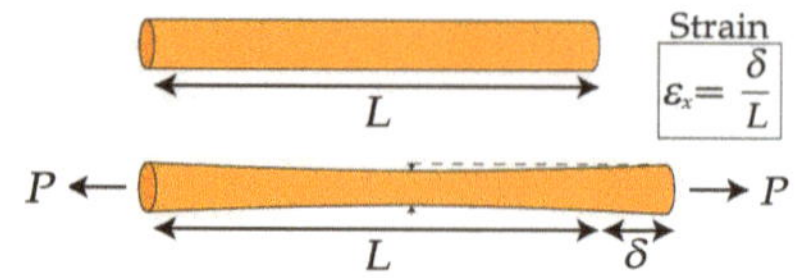

Figure 3.8. Deformation in orthogonal directions.

Source: https://www.bu.edu/moss/files/2015/01/strain_poissons.jpg.

So, pulling on it in the x-direction causes it to shrink in the y & z directions. This property of a material is known as Poisson's ratio, and it is denoted by the Greek letter nu, and is defined as:

$$\nu = -\frac{\text{lateral strain}}{\text{axial strain}}$$

Or, more mathematically, using the axial load shown in the above image, we can write this out as an equation:

$$\nu = -\frac{\varepsilon_y}{\varepsilon_x} = -\frac{\varepsilon_z}{\varepsilon_x}$$

Since Poisson's ratio is a ratio of two strains, and strain is dimensionless, Poisson's ratio is also unitless. Poisson's ratio is a material property. Poisson's ratio can range from a value of −1 to 0.5. For most engineering materials, for example steel or aluminum have a Poisson's ratio around 0.3, and rubbers have a Poisson's ratio around 0.5, which are referred to as "incompressible." Incompressible simply means that any amount you compress it in one direction, it will expand the same amount in its other directions – hence, its volume will not change. There has been some very interesting research in the last decade in creating structured materials that utilize geometry and elastic instabilities to create auxetic materials – materials with a negative Poisson's ratio. Physically, this means that when you pull on the material in one direction it expands in all directions (and vice versa):

This principle can be applied in 3D to make expandable/collapsible shells as well:

Through Poisson's ratio, we now have an equation that relates strain in the y or z direction to strain in the z direction. We can in turn relate this back to stress through Hooke's law. This is an important note: pulling on an object in one direction causes stress in only that direction, and causes strain in all three directions. So, $sigma_y = sigma_z = 0$. Let's write out the strains in the y and z direction in terms of the stress in the x direction.

$$\varepsilon_x = \frac{\sigma_x}{E}$$
$$\varepsilon_y = \varepsilon_z = -\nu\frac{\sigma_x}{E}$$

Remember, up until this point, we've only considered uniaxial deformation. In reality, structures can be simultaneously loaded in multiple directions, causing stress in those directions. A helpful way to understand this is to imagine a very tiny "cube" of material within an object. That cube can have stresses that are normal to each surface, like this:

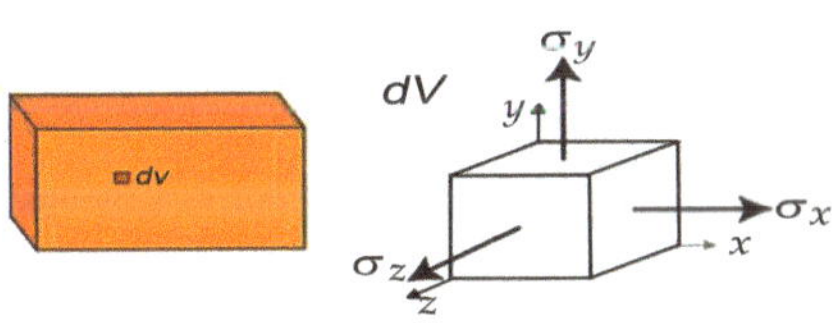

Figure 3.9. Considered uniaxial deformation.

Source: https://www.bu.edu/moss/files/2015/01/cube_normalstresses.jpg.

So, applying a load in the x direction causes a normal stress in that direction, and the same is true for normal stresses in the y and z directions. And, as we now know, stress in one direction causes strain in all three directions. So now we incorporate this idea into Hooke's law, and write down equations for the strain in each direction as:

$$\varepsilon_x = \frac{\sigma_x}{E} - \nu\frac{\sigma_y}{E} - \nu\frac{\sigma_z}{E}$$
$$\varepsilon_y = -\nu\frac{\sigma_x}{E} + \frac{\sigma_y}{E} - \nu\frac{\sigma_z}{E}$$
$$\varepsilon_z = -\nu\frac{\sigma_x}{E} - \nu\frac{\sigma_y}{E} + \frac{\sigma_z}{E}$$

These equations look harder than they really are: strain in each direction (or, each component of strain) depends on the normal stress in that direction, and the Poisson's ratio times the strain in the other two directions. Now we have equations for how an object will change shape in three orthogonal directions. Well, if an object changes shape in all three directions, that means it will change its volume. A simple measure for this volume change can be found by adding up the three normal components of strain:

$$e = \varepsilon_x + \varepsilon_y + \varepsilon_z$$

Now that we have an equation for volume change, or dilation, in terms of normal strains, we can rewrite it in terms of normal stresses.

$$e = \frac{\sigma_x + \sigma_y + \sigma_z}{E} - \frac{2\nu(\sigma_x + \sigma_y + \sigma_z)}{E}$$
$$e = \frac{1-2\nu}{E}(\sigma_x + \sigma_y + \sigma_z)$$

A very common type of stress that causes dilation is known as hydrostatic stress. This is just simply a pressure that acts equally on the entire material. Since it is acting equally, that means:

$$p = \sigma_x = \sigma_y = \sigma_z$$

So, in the case of hydrostatic pressure we can reduce our final equation for dilation to the following:

$$e = \frac{3(1-2\nu)}{E}p$$

This final relationship is important, because it is a constitutive relationship for how a material's volume changes under hydrostatic pressure. The prefactor to p can be rewritten as a material's bulk modulus, K.

$$K = \frac{E}{3(1-2\nu)}$$

Finally, let's get back to the idea of "incompressible" materials. What happens to K – the measure of how a material changes volume under a given pressure – if Poisson's ratio for the material is 0.5?

3.3.2.3. Hooke's Law in Shear

In addition to external forces causing stresses that are normal to each surface of the cube, the forces can causes stresses that are parallel to each cube face. And, as we know, stresses parallel to a cross section are shear stresses.

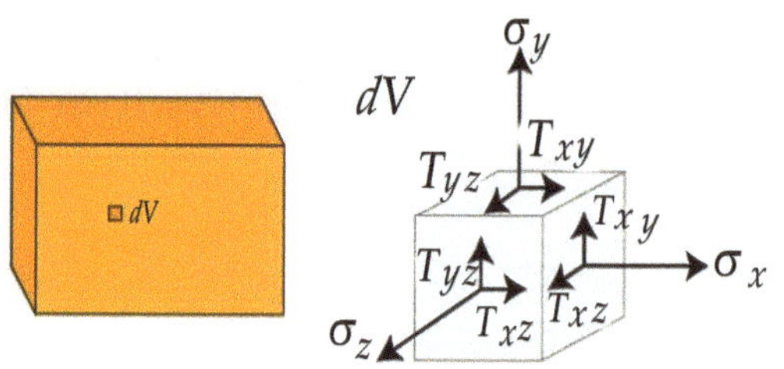

Figure 3.10. A cross section are shear stresses.

Source: https://www.bu.edu/moss/files/2015/01/cube_normalshearstresses.jpg.

Now that cube of material looks a lot more complicated, but it's really not too bad. On each surface there are two shear stresses, and the subscripts tell you which direction they point in and which surface they are parallel to. For instance, take the right face of the cube. Stresses normal to this face are normal stresses in the x direction. There are two stresses parallel to this surface, one pointing in the y direction (denoted tau$_{xy}$) and one pointing in the z direction (denoted tau$_{xz}$). In order for the cube to be in equilibrium, tau$_{xy}$ = tau$_{yx}$ (otherwise, the cube would rotate). Therefore, there are now six stresses (sigma$_{x}$, sigmay, sigmaz, tau$_{xy,}$ tau$_{yz,}$ tau$_{xz}$) that characterize the state of stress within a homogenous, isotropic, elastic material.

So, how do these shear stresses relate to shear strains? Hooke's law in shear looks very similar to the equation we saw for normal stress and strain:

$$\tau = G\gamma$$

In this equation, the proportionality between shear stress and shear strain is known as the shear modulus of a material. That's the equation in its general form, but we can rewrite it more explicitly in terms of its components of x,y, and z. Doing so will give us the generalized Hooke's law for homogenous, isotropic, elastic materials.

$$\varepsilon_x = \frac{\sigma_x}{E} - \nu\frac{\sigma_y}{E} - \nu\frac{\sigma_z}{E}$$
$$\varepsilon_y = -\nu\frac{\sigma_x}{E} + \frac{\sigma_y}{E} - \nu\frac{\sigma_z}{E}$$
$$\varepsilon_z = -\nu\frac{\sigma_x}{E} - \nu\frac{\sigma_y}{E} + \frac{\sigma_z}{E}$$
$$\gamma_{xy} = \frac{\tau_{xy}}{G} \quad \gamma_{yz} = \frac{\tau_{yz}}{G} \quad \gamma_{zx} = \frac{\tau_{zx}}{G}$$

In the generalized Hooke's law we have our six components of stress and strain, and three material properties. A natural question to as is how do these three material properties relate to each other? That relationship is given by the following equation:

$$G = \frac{E}{2(1+\nu)}$$

3.4. STRAIN ENERGY

Potential energy in the form of strain energy is created when a solid undergoes deformation. The energy required to deform a solid object is stored as it is deformed. When the distorted item reverts to its original form, this energy is released. Strain energy is a term used to characterize the impact of this stored energy on the solid. This energy belongs to a bigger category of energy known as potential energy, which is composed of energy that has been stored and is just waiting to be used. An object is strained when it is made to perform an action against its will. When an object is subjected to strain energy, it is compelled to maintain a shape or position that is contrary to its desires. This can be as easy as bending a tree branch or as difficult as a fighter jet landing's systems deforming. In most cases, the energy applied to the object that caused the strain is equal to the strain energy within the object. This essentially indicates that an applied energy will be absorbed and stored by a solid object. Generally, the amount of energy held equals the amount released. Because of this, energy can be applied to an object, stored until needed, and then released in its entirety and without loss back into the system.

A loss will only happen if the strain energy gets close to or beyond the solid's point of plastic deformation. The solid will now permanently change into a different shape. The change is catalyzed by the energy contained in the object. The strain energy will gradually be lost as a result of this deformation and end up as heat or the transformation. A lesser amount of energy will release normally if the object is released before the deformation is finished. A basic spring is the most effective way to visualize strain energy. A person applies force to the solid spring by pushing the spring downward. This energy enters the object and leaves the person. All of that energy is just trapped inside the object as long as the spring is kept in place. The energy changes from potential strain energy to kinetic energy when the spring is released. Some or all of the energy would be lost to deformation if the spring was held down for an extended period of time and its shape changed.

3.4.1. Impact Loading

Consider a rod BC of length L and uniform cross-sectional area A, which is attached at B to a fixed support, and subjected at C to a slowly increasing axial load P. By by plotting magnitude P of the load against the deformation x of the rod, we obtain a certain load-deformation diagram which is characteristics of the rod BC.

Let us now consider the work dU done by the load P as the rod elongates by a small amount dx. This elementary work is equal to the product of the magnitude P of the load and of the small elongation dx. write

$dU = Pdx$

That is equal to the element of area of width d_x located under the load-deformation diagram. The work U_i done by the load as the rod undergoes a deformation X_1 is

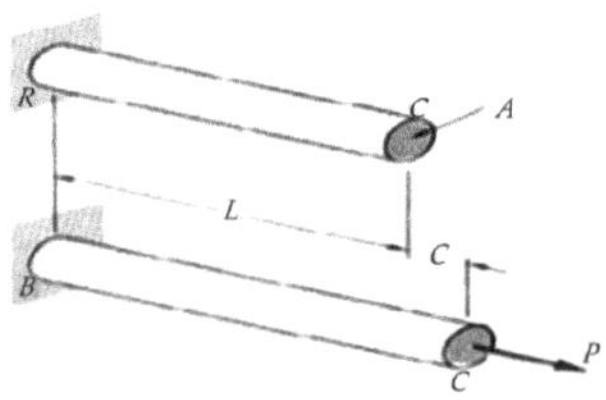

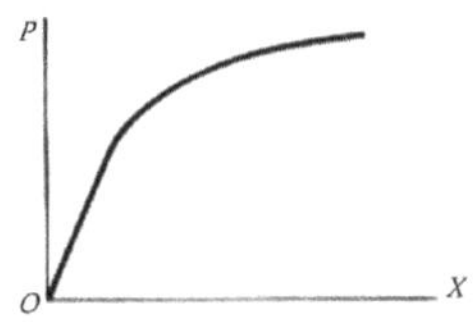

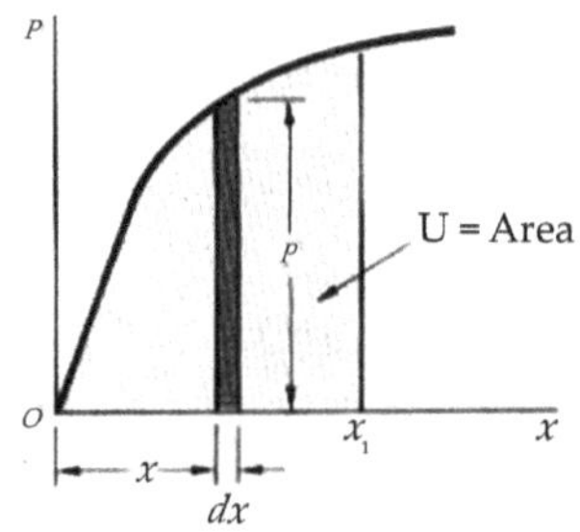

$$U = \int_0^{x_1} Pdx$$

U is equal to the area under the load-deformation diagram between and x =0 and x= x_1.

The work done by the load P as it is slowly applied to the rod must result in the increase of some energy associated with the deformation of the rod. This energy is referred to as the strain energy of the rod. We have, by definition,

$$\text{Strain energy} = U_i = \int_0^{x_1} Pdx$$

Work and energy should be expressed in units obtained by multiplying units of length by units of force. Thus, if SI metric units are used, work and energy are expressed in N.m; this unit is called, a joule (J.). If U.S. customary units are used, work and energy are expressed in ft.lb or in.lb.

In the case of a linear and elastic deformation, the portion of the load-deformation diagram involved may be represented by a straight line, of equation P = kx. Substituting for P in the above equation, we get

$$U_i = \int_0^{x_1} kxdx = \frac{1}{2}kx_1^2 = \frac{1}{2}P_1x_1$$

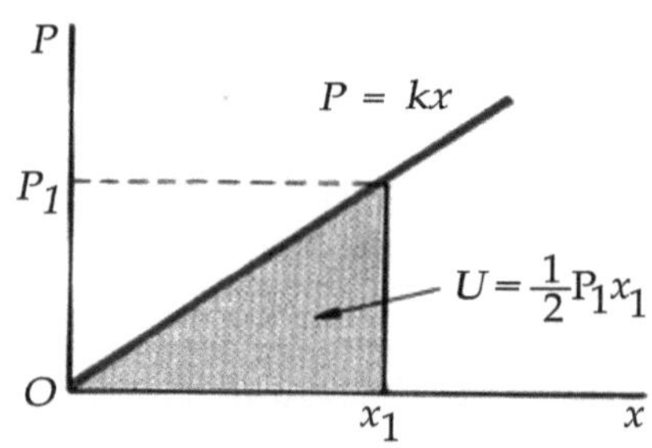

where P_1 is the value of the load corresponding to the deformation x_1. The concept of strain energy is particularly useful in the determination of the effects of impact loadings on structures or machine components.

3.4.2. Strain Energy Density

The load-deformation diagram for a rod BC depends upon the length L and the cross-sectional area A of the rod. The strain energy U_i will also depend upon the dimensions of the rod. In order to eliminate the effect of size, direct our attention to the properties of the material. Consider the strain energy per unit volume, dividing the strain energy U_i by the volume V = AL of the rod.

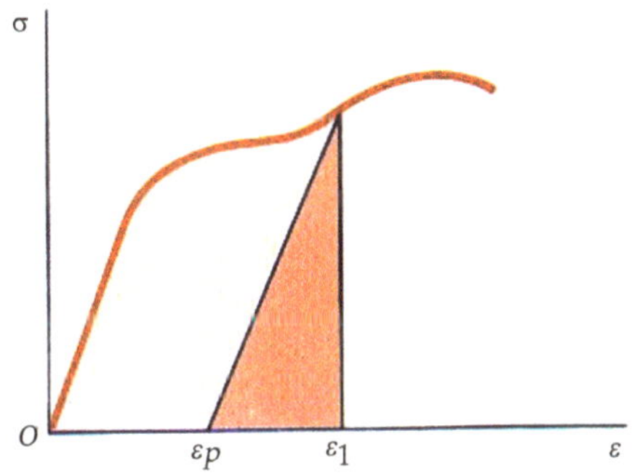

$$\frac{U_i}{V} = \int_0^{x_1} \frac{P}{A}\frac{dx}{L}$$

P/A represents the normal stress σ_x in the rod, and x/L the normal stain ε_x, we write

$$\frac{U_i}{V} = \int_0^{\varepsilon_1} \sigma_x d\varepsilon_x$$

where ε_1 denotes the value of the strain corresponding to the elongation The strain energy per unit volume, U_i/V, is referred to as the strain energy density and will be denoted by the letter u_i. We have, therefore,

$$\text{Strain energy density} = u_i = \int_0^{\varepsilon_1} \sigma_x d\varepsilon_x$$

strain-energy density u_i is expressed in units obtained by dividing its of energy by units of volume. Thus, if SI metric units are used, strain-energy density is expressed in J/m^3 or its multiples kJ/m^3 and MJ/m^3; if U. S. customary units are used, it is expressed in lb/in^3.

The strain-energy density u_i is to the area under the stress-strain curve, measured from $\varepsilon_x = 0$ to ε_1. If the material is unloaded, the stress returns to zero, but there permanent deformation represented by the strain ε_p, and only the .portion of the strain energy per unit volume corresponding to the triangular area may be recovered. The remainder of the energy spent in deforming the material is dissipated in the form of heat.

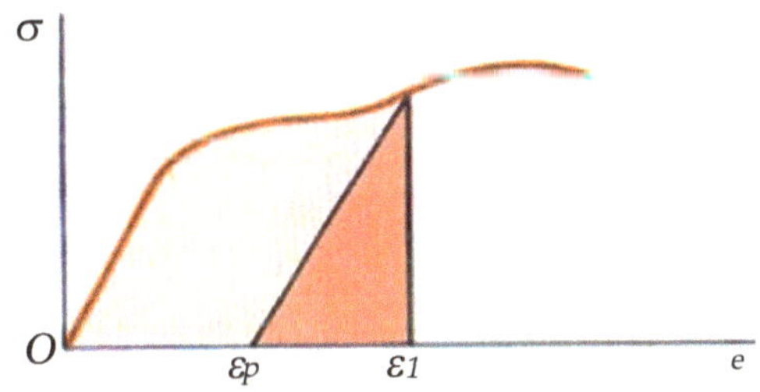

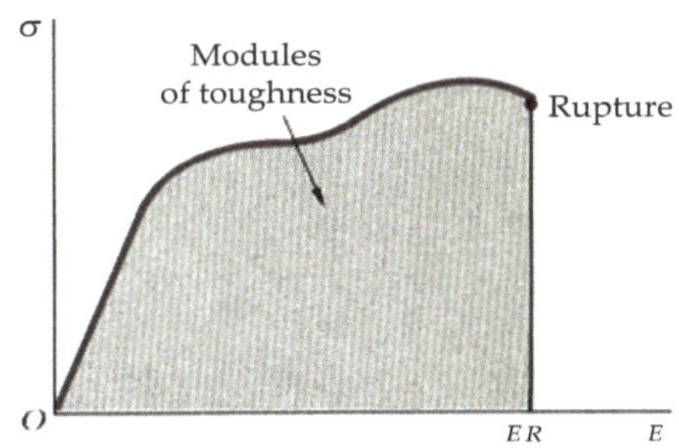

The modulus of toughness of the material is the value of the strain-energy density obtained by setting $\varepsilon_1 = \varepsilon_R$ in, where ε_R is the strain at rupture. It is the area under the entire stress-strain diagram and indicates the amount of energy per unit volume needed to rupture the material. It is evident that the ultimate strength and ductility of a material are correlated with its toughness, and that a structure's ability to bear an impact load is dependent on the material's toughness. Hooke's law applies and writing is permitted if the stress σ_x stays within the material's proportional limit.

$$\sigma_x = E\varepsilon_x$$

Substituting for σ_x into strain energy density equation, we have

$$u_i = \int_0^{\varepsilon_1} E\varepsilon_x d\varepsilon_x = \frac{E\varepsilon_1^2}{2} = \frac{\sigma_1^2}{2E}$$

The value u_y of the strain-energy density obtained by setting $\sigma_1 = S_y$, where S_y is the yield strength, is called the modulus of resilience of the material. Thus we have

$$u_y = \frac{S_y^2}{2E}$$

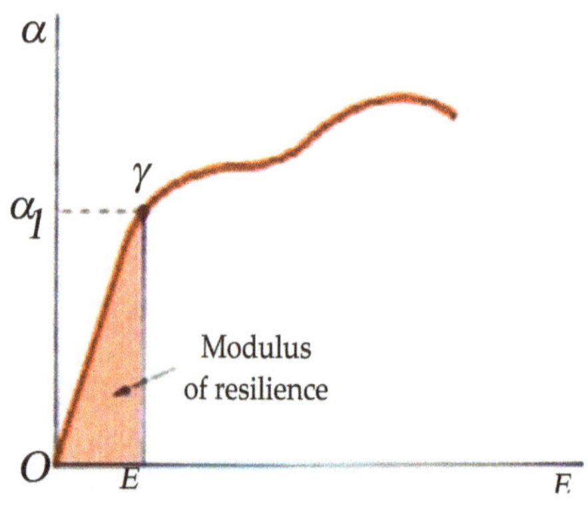

The energy per unit volume that the material may absorb without yielding is represented by the modulus of resilience, which is equal to the area under the straight-line portion OY of the stress strain diagram. It is obvious that the resilience of the material used determines a structure's ability to withstand an impact load without permanently deforming. Since the toughness and resilience moduli are representative values of the strain-energy density of the material under consideration, they are both expressed in units of J/m^3 or its multiples when using SI units, and in lb/in^3 when using U units. Usage of SDOT customary units occurs.

3.4.3. Elastic Strain Energy for Normal Stress

The strain energy density, which can be defined as the ratio U_i/V of the energy U_i and the volume V of the rod, was constant throughout the rod because the previously discussed rod was subject to uniformly distributed stresses. In a structural element or machine part with a non-uniform stress distribution, the strain-energy density u_i may be defined by considering the strain energy density of a small element of material of volume V and thus u_i may be defined by considering the strain energy of a small element of material of volume ΔV and writing.

$$u_i = \lim_{\Delta V \to 0} \frac{\Delta U_i}{\Delta V} \quad \text{or} \quad \frac{dU_i}{dV}$$

The expressions obtained for u_i before is valid and the strain-energy density u_i will generally vary from point to point.

For values of σ_x within the proportional limit $\sigma_x = E\varepsilon_x$ we write

$$u_i = \int_0^{\varepsilon_x} E\varepsilon_x d\varepsilon_x = \frac{E\varepsilon_x^2}{2} = \frac{\sigma_x^2}{2E}$$

The value of the strain energy U_i of a body subjected to uni-axial normal stresses may be obtained as

$$U_i = \int \frac{\sigma_x^2}{2E} dV$$

The expression obtained is valid only for elastic deformation and is referred as the elastic strain energy of the body.

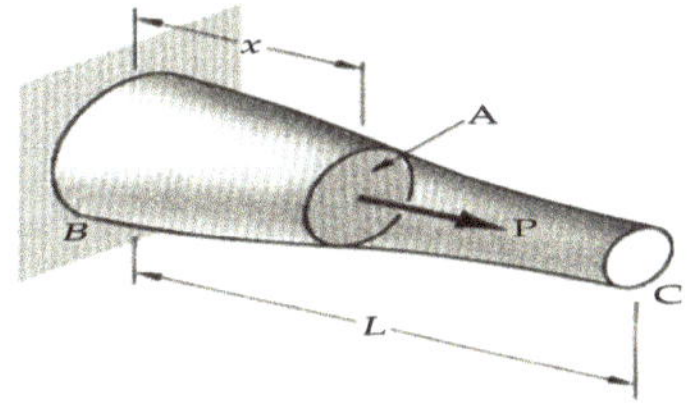

3.4.4. Strain Energy under Uniaxial Loading

When a rod is subjected to centric axial loading, the normal stress σ_x may be assumed uniformly distributed in any given transverse section. Denoting A the area of the section located at a distance x from the end B of the rod and P the internal force in that section, we write $\sigma_x = P/A$.

Thus,

$$U_i = \int \frac{P^2}{2EA^2} dV \qquad dV = A\,dx$$

$$U_i = \int \frac{P^2}{2EA} dV$$

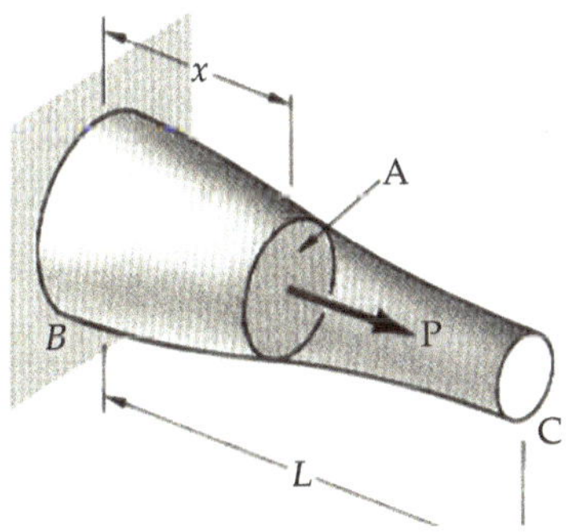

In the case of a rod of uniform cross section subjected at its ends to equal and opposite forces of magnitude P yields

$$U_i = \frac{P^2 L}{2EA}$$

3.4.5. Strain Energy in Bending

Consider a beam AB subjected to a given loading and M be the bending moment at a distance x from end A. Neglecting for the time being, the effect of shear, and taking into account only the normal stresses $\sigma_x = My/I$, we can write

$$U_i = \int \frac{\sigma_x^2}{2E} dV = \int \frac{M^2 y^2}{2EI^2} dV$$

Setting dV = dA dx, where dA represents an element of the cross sectional area and noting that $M^2/2EI^2$ is a function of x alone, we have

$$U_i = \int \frac{M^2}{2EI^2} \left(\int y^2 dA\right) dx$$

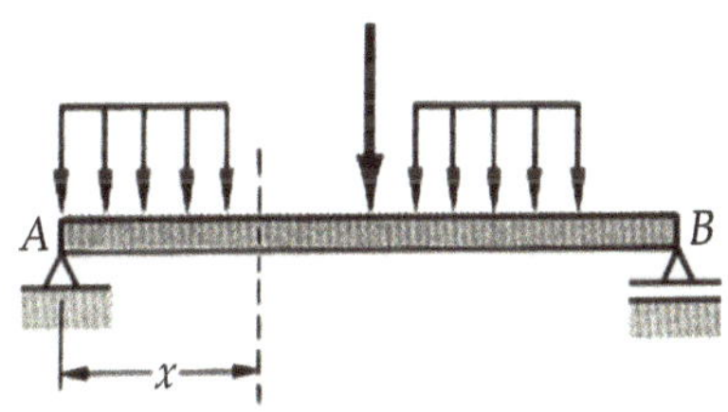

Note that the integral within the parenthesis represents the moment of inertia of the cross section I about its neutral axis. Thus we write

$$U_i = \int_0^L \frac{M^2}{2EI} dx$$

CLASS ACTIVITY:

MEASURING STRAIN IN ROCKS USING DEFORMED GRIDS

- **Objective:** To understand and measure strain in rocks through a hands-on activity that simulates rock deformation using a rubber sheet with a printed grid.
- **Materials Needed:** Rubber sheet, permanent marker, rulers and protractors, data recording sheets, weights or force application tools, and calculator.
- **Assessment**: Facilitate a class discussion on the results. Ask students to compare the strain observed in different parts of the sheet and discuss the uniformity of deformation. Students will gain a hands-on understanding of how rocks can deform under stress.

SUMMARY

- Strain in rocks refers to the deformation or change in shape that occurs due to stress applied over time. Rocks can deform in response to various forces acting upon them, such as compression, tension, and shear.
- Shear strain occurs when rocks are subject to parallel forces acting in opposite directions, causing them to deform by sliding past each other. This type of strain is common along transform plate boundaries where tectonic plates slide horizontally past each other, leading to the formation of faults and earthquakes.
- Rock, in geology, naturally occurring and coherent aggregate of one or more minerals. Such aggregates constitute the basic unit of which the solid Earth is composed and typically form recognizable and mappable volumes.
- Igneous rocks are those that solidify from magma, a molten mixture of rock-forming minerals and usually volatiles such as gases and steam. Since their constituent minerals are crystallized from molten material, igneous rocks are formed at high temperatures.
- Geologic materials—mineral crystals and their host rock types—are cycled through various forms. The process depends on temperature, pressure, time, and changes in environmental conditions in the Earth's crust and at its surface.
- The texture of a rock is the size, shape, and arrangement of the grains (for sedimentary rocks) or crystals (for igneous and metamorphic rocks). Also of importance are the rock's extent of homogeneity (i.e., uniformity of composition throughout) and the degree of isotropy.
- The study of deformation resulting from the strain of rocks in response to stresses is called rock mechanics. When the scale of the deformation is extended to large geologic structures in the crust of the Earth, the field of study is known as geotectonics.
- The deformation of materials is characterized by stress-strain relations. For elastic-behavior materials, the strain is proportional to the load (i.e., the applied stress). The strain is immediate with stress and is reversible (recoverable) up to the yield point stress, beyond which permanent strain results.
- Physical properties of rocks are of interest and utility in many fields of work, including geology, petrophysics, geophysics, materials science, geochemistry, and geotechnical engineering. The scale of investigation ranges from the molecular and crystalline up to terrestrial studies of the Earth and other planetary bodies.
- Density varies significantly among different rock types because of differences in mineralogy and porosity. Knowledge of the distribution of underground rock densities can assist in interpreting subsurface geologic structure and rock type.
- The electrical nature of a material is characterized by its conductivity (or, inversely, its resistivity) and its dielectric constant, and coefficients that indicate the rates of change of these with temperature, frequency at which measurement is made, and so on.

REVIEW QUESTIONS

- What are the different types of rocks?
- How are rocks classified by grain or crystal size?
- What are the stress-strain relationships in rocks?
- What are the applications of the study of rock magnetization?
- How is strain analyzed in materials work?
- What are the mechanics of materials with respect to strain?

REFERENCES

1. Cowan, D. S. (1999). Do faults preserve a record of seismic slip? A field geologist's opinion. Journal of Structural Geology, 21(8–9), 995–1001.
2. Fossen, H., Schultz, R. A., Shipton, Z. K., & Mair, K. (2007). Deformation bands in sandstone: A review. Journal of the Geological Society, 164(4), 755–769.
3. Gueydan, F., Leroy, Y. M., & Jolivet, L. (2004). Mechanisms of lithospheric extension at continental margins: Comparison between the Gulf of Corinth and the Tyrrhenian Sea. Earth and Planetary Science Letters, 225(3–4), 451–465.
4. Lister, G. S., & Hobbs, B. E. (1980). The simulation of fabric development during plastic deformation and its application to quartzite: The influence of deformation history. Journal of Structural Geology, 2(3), 355–370.
5. Marrett, R., & Allmendinger, R. W. (1991). Estimation of strain from the stylolites. Journal of Structural Geology, 13(1), 1–10.
6. Means, W. D. (1976). Stress and Strain: Basic Concepts of Continuum Mechanics for Geologists. Springer-Verlag.
7. Tikoff, B., & Fossen, H. (1999). Three-dimensional reference deformations and strain facies. Journal of Structural Geology, 21(11), 1497–1512.
8. Tikoff, B., & Wojtal, S. (1999). Displacement control of geologic structures. Journal of Structural Geology, 21(8–9), 1107–1123.

CHAPTER

4 Stress

LEARNING OBJECTIVES

After studying this chapter, you will be able to:

- Give an overview of stress and strain
- Focus on stress on a surface
- Discuss about stress components

KEY TERMS FROM THIS CHAPTER

Brittle deformation

Elastic materials

Geology

Plastic deformation

Compressional stress

Elastic strain

Magnitude

Tensional stress

INTRODUCTION

In geology, stress is the force per unit area applied to a rock or sediment that shapes its structural features and deformation. It originates from multiple sources, which have a major impact on geological processes and formations. These sources include tectonic movements, gravitational forces, and human activities. Three main types of stress are commonly identified: shear, tensional, and compressional stress. These three types of stress have distinct effects on rocks. When rocks are compressed together, compressional stress is created, which frequently results in the development of thrust faults and folds. Where tectonic plates collide, at convergent plate boundaries, this kind of stress is more common. As the Indian Plate converges with the Eurasian Plate, for example, compressional forces cause extensive folding and uplift, which leads to the formation of the Himalayas.

Conversely, tensional stress causes rocks to separate and is frequently linked to divergent plate borders. As the East African Rift system demonstrates, this stress results in the formation of typical faults and rift valleys. The African Plate is dividing into two smaller plates here, resulting in a region marked by significant volcanic activity and crustal extension. Rocks slide past one another due to shear stress, which is caused by forces acting parallel to one another but in opposite directions. The San Andreas Fault in California serves as an example of this stress, which is common at transform plate boundaries. Significant seismic activity is produced by the lateral movement along this fault line, and earthquakes frequently occur as a result of the release of accumulated stress.

4.1. OVERVIEW OF STRESS AND STRAIN

In structural geology, stress and strain are basic ideas that explain how rocks react to tectonic forces and other types of deformation. The force per unit area acting on a rock is referred to as stress, and the resulting deformation or change in shape of the rock is referred to as strain.

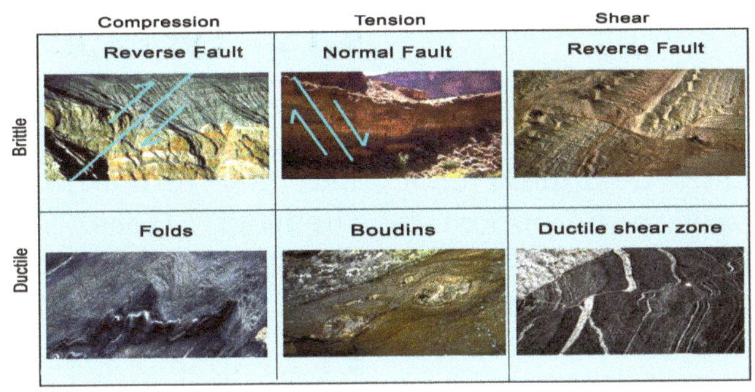

Figure 4.1. Stress and strain are important concepts in structural geology.

Source: https://i0.wp.com/geologyscience.com/wp-content/uploads/2023/03/stress_strain_table.v7.gif?resize=960%2C720&ssl=1.

Three categories of stress exist: shear stress, tensional stress, and compressional stress. Rocks that are compressed or pushed together experience compressional stress, whereas those that are pulled apart or stretched experience tensional stress. When rocks are forced to slide past one another in opposing directions, shear stress is created.

Elastic strain and plastic strain are the two categories into which strain can be divided. When a rock experiences stress and then returns to its original shape after the stress is released, this phenomenon is known as elastic strain. When a rock undergoes stress and does not regain its original shape after the stress is released, this is known as plastic strain. Rather, the rock's deformation persists indefinitely.

In structural geology, stress and strain are crucial ideas because they offer a framework for comprehending the behavior of rocks under various tectonic and geological processes. Geoscientists can learn more about a region's geological past and the likelihood of geological hazards like earthquakes and landslides by examining stress and strain. Furthermore, the development of new technologies and materials, as well as the exploration and extraction of resources, depend on an understanding of stress and strain. All things considered, stress and strain are basic ideas in structural geology and are critical to comprehending the mechanisms forming the Earth's crust.

4.1.1. Definitions, Magnitudes, and Units

The force applied to a material's unit area is called stress. Strain is the term for the physical result of stress. The body can morph due to stress. Stress units can be used to quantify the amount of force that a material experiences. Depending on which direction the deforming forces acting on the body are directed, stress can be divided into three categories.

4.1.1.1. What Is Stress?

An object deforms when the deforming force is applied to it. There will be an opposing force generated inside the object to return it to its original size and shape. The applied deforming force and this restoring force will have the same magnitude and direction of action. Stress is the measurement of this restoring force produced per unit area of the material. Thus, Stress is defined as "The restoring force per unit area of the material." It is a tensor quantity. Denoted by Greek letter σ. Measured using Pascal or N/m2. Mathematically expressed as:

$$\sigma = \frac{F}{A}$$

where,

- F is the restoring force measured in Newton or N.
- A is the area of cross-section measured in m^2.
- σ is the stress measured using N/m^2 or Pa.

4.1.1.2. Stress Units

Stress can be expressed using multiple units. Refer to the table given below for Stress units.

System of units	Stress units
Fundamental units	$Kg.m^{-1}.s^{-2}$
SI (derived units)	N.m2
SI (derived units)	Pa or pascal
SI (mm)(derived units)	M.Pa or $N/(mm)^2$
US unit (ft)	lbf/ft^2
US unit (inch)	Psi ($lbf/inch^2$)

4.1.2. Types of Stress

There are several types of stress in physics but it is mainly categorized into two forms that are Normal Stress and Tangential or Shearing Stress. Some stress types are discussed in the points below.

4.1.2.1. Compressional Stress

One kind of stress that develops when rocks are pushed or squeezed together is called compressional stress. Tectonic processes like plate convergence, in which two plates collide and press against one another, are usually linked to this kind of stress.

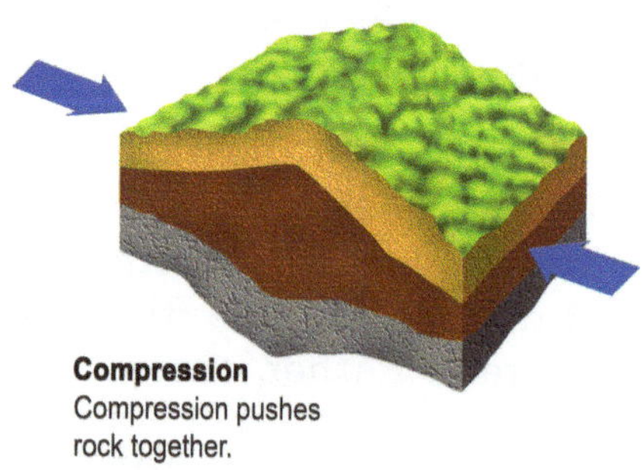

Figure 4.2. Compressional stress.

Source: https://i0.wp.com/mhshonorsenvr.weebly.com/uploads/1/3/4/1/13416460/5643474_orig.png?ssl=1.

Depending on their strength and the amount of stress applied, rocks can deform in a variety of ways when under compressional stress. Folding or faulting, in which the rock layers are compressed and deformed, can be caused by compressional stress in weaker rocks, such as sedimentary rocks. Compressive stress in stronger rocks, like metamorphic or igneous rocks, can cause crushing or fracture. The creation of geological structures like mountain ranges can be significantly impacted by compressional stress. Rocks between two plates are subjected to compressional stress when they converge, which can lead to deformation and uplift. Mountains may form as a result of this process over time. With substantial effects on the formation and deformation of rocks and geological structures, compressional stress is a major form of stress in structural geology. Scientists studying compressional stress and its consequences can learn more about the tectonic processes that sculpt the crust of the Earth.

4.1.2.2. Tensional Stress

One kind of stress that develops when rocks are stretched or torn apart is called tensional stress. Tectonic processes like divergent plate boundaries, where two plates are moving apart, are often linked to this kind of stress.

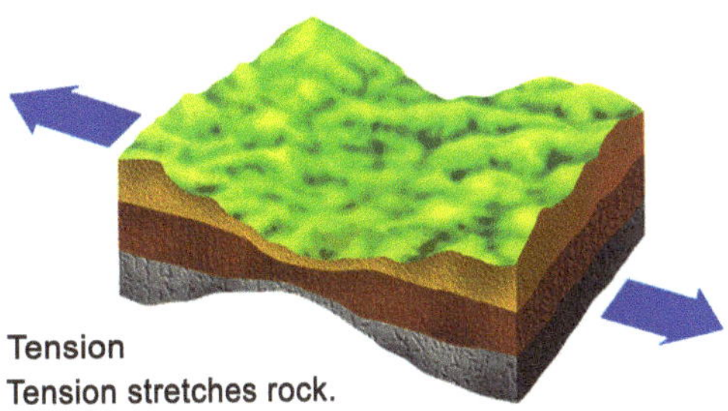

Figure 4.3. Tensional stress.

Source: https://mhshonorsenvr.weebly.com/uploads/1/3/4/1/13416460/4670157_orig.png.

Depending on their strength and the amount of stress applied, rocks can deform in a variety of ways when under tensional stress. Tensional stress in weaker rocks—like sedimentary rocks—can cause the separation of rock layers, leading to the formation of joints or fractures. Tensional stress can cause the rock to thin or stretch in stronger rocks, like igneous or metamorphic rocks. Important effects of tensional stress can also be seen in the development of rift valleys and other geological structures. The tensional stress that results from two plates diverging can cause the rocks in between to stretch and thin. This process has the potential to create rift valleys over time.

With substantial effects on the formation and deformation of rocks and geological structures, tensional stress is a major form of stress in structural geology. Geoscientists can learn more about the tectonic processes that shape the Earth's crust by researching tensional stress and its effects.

4.1.2.3. Shear Stress

Shear stress is a type of stress that occurs when rocks are subjected to forces that cause them to slide past each other in opposite directions. This type of stress is typically associated with tectonic processes such as transform plate boundaries, where two plates are sliding past each other (Figure 4.4).

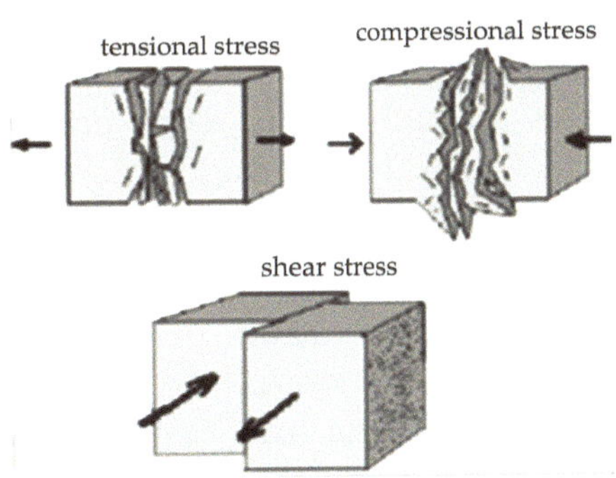

Figure 4.4. Shear stress.

Source: https://i0.wp.com/unacademy.com/content/wp-content/uploads/sites/2/2022/04/shearing-300x230.png?resize=300%2C230&ssl=1.

Depending on their strength and the amount of stress applied, rocks can deform in a variety of ways when subjected to shear stress. Shear stress can cause the formation of faults in weaker rocks, like sedimentary rocks, where the rocks slide past one another along a plane of weakness. Shear stress can cause ductile deformation, in which the rock layers bend or fold, in stronger rocks, such as igneous or metamorphic rocks.

Important effects of shear stress can also be seen in the development of fault zones and other geological structures. Shear stress can cause rocks to develop weak spots that increase the likelihood that they will eventually deform. These areas may eventually turn into fault zones, which could have significant effects on resource development and geological risks like earthquakes. In structural geology, shear stress is a crucial kind of stress that has a big impact on how rocks and geological structures deform and form. Scientists studying shear stress and its consequences can learn more about the tectonic processes that sculpt the Earth's crust.

4.1.2.4. Examples of Each Type of Stress

Here are some examples of each type of stress:

1. Compressional stress
   - Collision of two continental plates, leading to the formation of mountain ranges such as the Himalayas.
   - Compaction of sedimentary rocks, leading to the formation of folds and thrust faults.
   - Impact events, such as meteorite impacts, can cause compressional stress and lead to the formation of deformation structures.
2. Tensional stress
   - Divergence of two tectonic plates, leading to the formation of rift valleys such as the East African Rift Valley.
   - Stretching and thinning of the Earth's crust, leading to the formation of normal faults and grabens.
   - Cooling and solidification of magma, leading to the formation of columnar jointing.
3. Shear stress
   - Transform plate boundaries, such as the San Andreas Fault in California, where two tectonic plates slide past each other.
   - Ductile deformation of rocks due to shear stress, leading to the formation of folds and cleavage.
   - Movement of glaciers, causing shear stress and leading to the formation of glacial striations and other landforms.

These are just a few examples, and there are many other geological processes and structures that can result from different types of stress.

4.1.3. Types of Strain

Strain can be of four different types, and the direction of application of forces generally defines them.

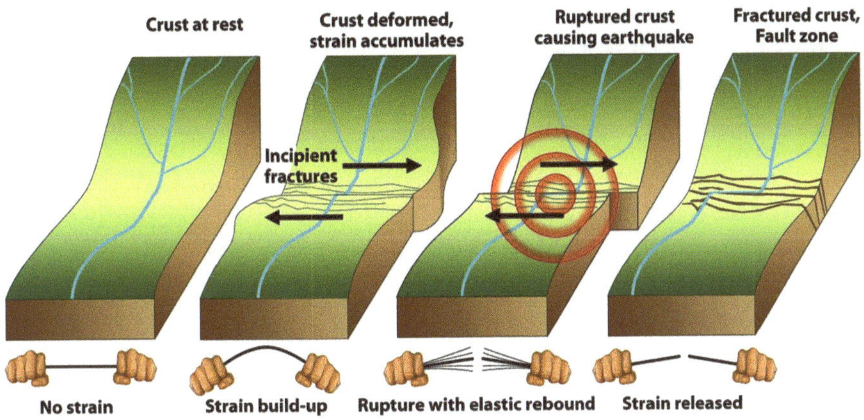

Figure 4.5. Types of strain.

Source: https://i0.wp.com/geophile.net/Lessons/Earthquakes/images/releaseofstress.fw.png?ssl=1

4.1.3.1. Elastic Strain

A material that experiences stress and then regains its original size and shape after the stress is released is said to exhibit elastic strain. This is a result of the material responding to the applied stress in an elastic manner, much like a spring. The bonds that hold together atoms in a material stretch or compress when it is under stress.

These bonds can compress or stretch momentarily in an elastic material, but they will always revert to their initial length when the stress is released. This indicates that there is no long-term deformation or damage to the material.

A material's elasticity or stiffness determines how much elastic strain it can withstand. Before reaching their yield point, or limit of elasticity, more stiff or elastic materials, like some metals, can withstand greater elastic strain. When the material reaches its yield point, it may experience plastic deformation, which is a permanent deformation that prevents the material from returning to its original shape even after the stress is released.

In structural geology, elastic strain is a crucial idea because it clarifies how rocks behave under stress and how they change over time. Geoscientists can gain a better understanding of how rocks respond to various forms of stress and how they play a role in the formation of geological structures like faults, folds, and other deformation features by examining the elastic properties of rocks.

4.1.3.2. Plastic Strain

When a material experiences stress above its elastic limit, it can deform in a way known as plastic strain. Plastic strain, in contrast to elastic strain, is permanent and irreversible, meaning that once the stress is released, the material does not revert to its initial size and shape. The bonds separating atoms in a material start to break and reorganize when the material is stressed beyond its elastic limit. Because the bonds cannot revert to their initial state after the stress is released, this causes the material to permanently deform.

A material's capacity for plastic strain is determined by its structure, composition, and the kind and degree of stress it is subjected to. While some materials—like metals and certain kinds of rocks—can withstand considerable plastic strain without breaking or fracturing, others might fracture more readily. Because it causes many geological structures, including folds, faults, and shear zones, to permanently deform, plastic strain is a key idea in structural geology. Geoscientists can gain a better understanding of how rocks deform under various types and amounts of stress as well as how geological structures change over time by examining the plastic properties of rocks.

4.1.4. Relationship Between Stress and Strain

In structural geology, stress and strain are closely related concepts because stress is the force exerted on a material and strain is the material's subsequent deformation as a result of that force. Elasticity can be used to explain the relationship between stress and strain.

The property of a material known as elasticity allows it to distort under stress and then return to its original size and shape when the stress is released. The amount of deformation is directly proportional to the applied stress in an elastic material because the relationship between stress and strain in this material is linear.

This relationship can be described by a mathematical equation known as Hooke's Law:

$$\sigma = E\varepsilon,$$

where σ is the stress, E is the elastic modulus (a measure of the material's stiffness), and ε is the strain. Hooke's Law states that the stress in a material is proportional to the strain, with the constant of proportionality being the elastic modulus.

The yield point is the upper limit of the linear relationship between stress and strain, though. The material starts to experience plastic deformation beyond the yield point, at which point the relationship between stress and strain becomes non-linear. The kind and amount of stress applied, together with the composition and structure of the material, all affect how much plastic deformation takes place.

In elastic materials, the amount of deformation is directly correlated with the applied stress and has a linear relationship with strain. The material experiences plastic deformation beyond the yield point, at which point the relationship becomes non-linear. Knowing this relationship is crucial to comprehending the deformation of rocks and the formation of geological structures like faults and folds.

4.1.5. Deformation Mechanisms

The mechanisms that cause a material to deform under stress are known as deformation mechanisms. Knowing these processes is crucial to understanding how rocks deform in structural geology and how geological structures like folds, faults, and shear zones form.

There are several deformation mechanisms that can occur in different materials and under different types and amounts of stress.

Some of the most common mechanisms include:

1. Dislocation: This is the movement of atoms within a crystal lattice in response to stress. Dislocations can occur along a plane within the lattice, causing the material to deform.
2. Twinning: This is a deformation mechanism that occurs in certain types of crystals, where a portion of the crystal lattice mirrors another portion, resulting in a change in shape.
3. Grain boundary sliding: This occurs in polycrystalline materials, where grains slide past each other along their boundaries in response to stress.
4. Fracture: This is the breaking of a material due to stress, which can occur in brittle materials such as rocks.
5. Ductile flow: This is a deformation mechanism that occurs in materials that can undergo plastic deformation, such as metals or some types of rocks. Ductile flow involves the permanent deformation of the material under stress, without fracturing.

A material's particular deformation mechanism is determined by a number of variables, such as the kind and magnitude of stress applied, the material's composition and structure, and the ambient temperature and pressure. Geoscientists can gain a better understanding of how rocks deform under various stresses and how geological structures evolve over time by comprehending these mechanisms.

4.1.5.1. Brittle Deformation

When high stresses are applied to rocks and other materials over a brief period of time, brittle deformation is one kind of deformation that results. The development of faults or fractures, which happen when the material fractures in reaction to applied stress, is what distinguishes this kind of deformation (Figure 4.6).

Figure 4.6. Brittle deformation is a type of deformation

Source: https://i0.wp.com/opengeology.org/textbook/wp-content/uploads/2017/01/Faults_in_Moenkopi_Formation_Moab_Canyon_Utah_USA_01-e1494992495928.jpg?ssl=1.

Rocks near the Earth's surface experiencing low temperatures and pressures are prone to brittle deformation. This can also happen in rocks facing sudden stress changes, like during earthquakes. Under high stress, rocks can break along weak planes, forming fractures or faults. Fractures involve breaks without much movement, while faults show significant displacement on both sides of the break.

Brittle deformation can also result from other kinds of stress, like those brought on by quarrying or mining operations, tunnel or other subterranean structure excavation, in addition to earthquakes. Comprehending brittle deformation is crucial for anticipating and minimizing the possible effects of these operations on the nearby geology and ecosystem.

4.1.5.2. Ductile Deformation

Rocks and other materials that are exposed to high stresses over an extended length of time may experience ductile deformation. The material will always be bending, flowing, or stretching without breaking under this kind of deformation (Figure 4.7).

Figure 4.7. Ductile deformation is a type of deformation that occurs in rocks and other materials

Source: https://upload.wikimedia.org/wikipedia/commons/e/e3/Spaghetti_Rock.jpg.

High temperatures and pressures, like those found deep within the Earth's crust, are known to cause ductile deformation in rocks. It can also happen in rocks that have been exposed to long-term, gradual changes in stress. Rocks that experience ductile deformation can acquire characteristics like cleavage planes, folds, and lineations. These characteristics are the outcome of the rock's ongoing deformation under stress.

Ductile deformation, as opposed to brittle deformation, is the result of the atoms or molecules inside the material permanently rearranging themselves as opposed to their bonds breaking. As previously mentioned, processes like dislocation, twinning, or grain boundary sliding can cause this rearrangement.

Comprehending ductile deformation is essential for deciphering the geological chronicles of an area and forecasting the potential behavior of rocks under diverse stress conditions. Since it sheds light on how materials behave under extreme stress and over extended periods of time, it is also crucial for a wide range of applications in engineering and materials science.

4.1.5.3. Factors that Influence Deformation Mechanisms

Deformation mechanisms are influenced by a variety of factors, including:

1. Temperature: Temperature has a significant impact on deformation mechanisms. At low temperatures, deformation is typically brittle, while at high temperatures, deformation is typically ductile.
2. Pressure: Pressure also plays a role in deformation mechanisms. High pressure tends to favor ductile deformation, while low pressure favors brittle deformation.
3. Strain rate: The rate at which a material is deformed can also affect the deformation mechanism. Fast deformation rates tend to favor brittle deformation, while slow deformation rates tend to favor ductile deformation.
4. Composition: The composition of the material being deformed can also influence the deformation mechanism. Materials with high amounts of brittle minerals, such as quartz, tend to exhibit brittle deformation, while materials with high amounts of ductile minerals, such as mica or feldspar, tend to exhibit ductile deformation.
5. Grain size: The grain size of a material can also affect the deformation mechanism. Smaller grain sizes tend to favor ductile deformation, while larger grain sizes tend to favor brittle deformation.
6. Fluids: The presence of fluids, such as water, can also influence deformation mechanisms. Fluids can lubricate grain boundaries, making it easier for them to move and deform, and can also facilitate chemical reactions that can alter the properties of the material being deformed.
7. Time: The duration of the stress also plays a role in deformation mechanisms. Slow, sustained stress tends to favor ductile deformation, while rapid, short-term stress tends to favor brittle deformation.

It can be challenging to anticipate which deformation mechanism will take place in a particular scenario because all of these variables can interact with one another in intricate ways. Nonetheless, geologists and engineers can more accurately forecast how rocks and other materials will respond to various forms of stress by comprehending the variables that affect deformation mechanisms.

4.1.6. Stress and Strain in Rocks

The concepts of stress and strain are crucial in comprehending how rocks behave when they are deformed. Rocks experience stress from different sources like tectonic forces, gravity, and variations in temperature and pressure. When rocks are under stress, they can undergo deformation, causing a transformation in their shape or size. The correlation between stress and the resulting strain plays a significant role in understanding how rocks behave.

Three different types of stress exist in rocks: shear, tensional, and compressional stress. When two tectonic plates collide or other forces squeeze rocks together, compressional stress is created. When two tectonic plates move apart, for example, rocks are stretched apart and cause tensional stress. When rocks are pushed in opposing directions and slide past one another, shear stress is created. Rocks can fracture, deform plastically, or undergo elastic deformation when under stress. Elastic deformation is the deformation of a rock under stress that eventually returns to its original shape upon release of the stress. Plastic deformation is the result of a rock permanently changing under stress without breaking. When a rock is under more stress than it can withstand, it fractures and breaks into pieces.

A stress-strain curve is commonly used to illustrate the relationship between stress and strain in rocks. The response of the rock to increasing stress is depicted by this curve, which can be used to estimate when the rock will fracture or experience plastic deformation. Three regions are usually seen on the stress-strain curve for rocks: elastic deformation, plastic deformation, and fracture. Geology, engineering, and materials science are just a few of the disciplines that benefit from an understanding of stress and strain in rocks. Scientists and engineers can better predict how structures and materials will perform in a range of conditions by understanding how rocks behave under various types and levels of stress. They can also develop strategies for mitigating damage and preventing failure.

4.2. STRESS ON A SURFACE

Josiah Willard Gibbs (1839–1903) initially defined surface stress as the quantity of reversible work per unit area required to elastically stretch a pre-existing surface. This is an example of atomistic versus continuum definitions; depending on the convention applied, the area is either the original, unstretched one that represents a constant number of atoms, or occasionally it is the final area. It is important to take caution to make sure that the definition is in line with the elastic strain energy, as there have been misunderstandings and conflicts in the literature (Vermaak et al., 1968).

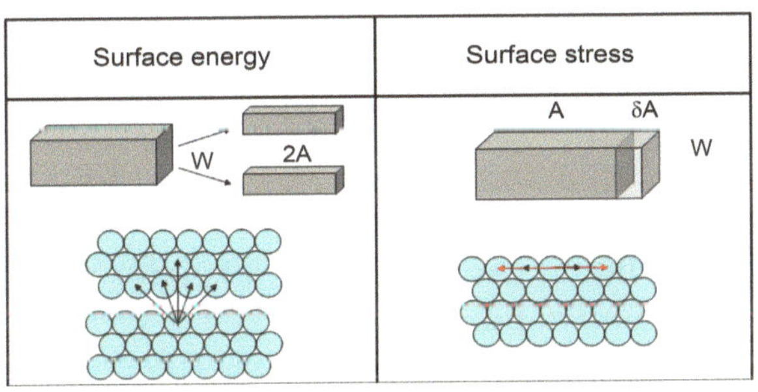

Figure 4.8. Comparison of surface energy, creating new surface on the left, and surface stress due to elastic deformation

Source: https://upload.wikimedia.org/wikipedia/commons/thumb/3/32/Surface_Stress_versus_Surface_Energy.tif/lossless-page1-800px-Surface_Stress_versus_Surface_Energy.tif.png.

4.2.1. Thermodynamics of Surface Stress

The continuum definition of surface free energy is the amount of reversible work dw performed to create new area dA of surface, expressed as:

$$dw = \gamma dA$$

In this definition the number of atoms at the surface is proportional to the area. Gibbs was the first to define another surface quantity, different from the surface free energy γ, that is associated with the reversible work per unit area needed to elastically stretch a pre-existing surface. In a continuum approach one can define a surface stress tensor f_{ij} that relates the work associated with the variation in γA the total excess free energy of the surface due to a strain tensor e_{ij}.

$$Af_{ij} = d(\gamma A)/de_{ij} = Ad\gamma/de_{ij} + \gamma dA/de_{ij}$$

In general there is no change in area for shear, which means that for the second term on the right i = j and $dA/de_{ij} = A\delta_{ij}$ using the Kronecker delta. Cancelling the area then gives $f_{ij} = d\gamma/de_{ij} + \delta_{ij}\gamma$ called the Shuttleworth equation.

An alternative approach is an atomistic one, which defines all quantities in terms of the number of atoms, not continuum measures such as areas. This is related to the ideal of using Gibb's equimolar quantities rather than continuum numbers such as area, that is keeping the number of surface atoms constant. The surface stress is defined as the derivative of the surface energy with strain, that is (deliberately using a different symbol)

$$g_{ij} = d\gamma/de_{ij}$$

This second definition is more convenient in many cases. A conventional liquid cannot sustain strains, so in the continuum definition the surface stress and surface energies are the same, whereas in the atomistic approach the surface stress is zero for a liquid. So long as care is taken the choice of the two does not matter, although this has been a little contentious in the literature.

4.2.2. Physical Origins of Surface Stress

It will be advantageous if they desire to get closer because an expansive or tensile strain will raise the surface energy. The derivative is positive for many metals and negative for others, such as solid argon and certain semiconductors. The molecules adsorbed on the surface can also have a significant influence on the sign. There will be a drawback if these desire to be farther apart. If they want to be closer, then $d\gamma/de_{ij}$ will be positive—a tensile or expansive strain will increase the surface energy.

For many metals the derivative is positive, but in other cases it is negative, for instance solid argon and some semiconductors. The sign can also strongly depend upon molecules adsorbed on the surface. If these want to be further apart that will introduce a negative component

4.2.3. Surface Stress Values

4.2.3.1. Theoretical Calculations

The most popular technique for determining surface stresses is to compute the derivative of the surface free energy with respect to elastic strain. Several techniques have been employed, the most popular being density functional theory, including first principles, atomistic potential calculations, and molecular dynamics simulations. Typical values of the surface energies are 1-2 Joule per meter squared (Jm^{-2}), with the trace of the surface stress tensor g_{ij} in the range of -1 to 1 Jm^{-2}. Some metals such as aluminum are calculated to have fairly high, positive values (e.g., 0.82) indicating a strong propensity to contract, whereas others such as calcium are quite negative at -1.25, and others are close to zero such as cesium (-0.02).

4.2.3.2. Surface Stress Effects

Surface stresses can be significant whenever there is a balance between a surface energy term and a bulk elastic energy contribution. Surface stress effects are frequently significant at the nanoscale because surface contributions are more significant at smaller sizes.

4.2.3.3. Surface Structural Reconstruction

As discussed earlier, atoms on a surface often have a preference to be either closer or further apart. However, the atoms underneath (substrate) have a fixed spacing in the plane that the surface must align with. To minimize energy, additional atoms may be added or removed from the surface. For example, the gold (111) surface has a higher surface density by about 5% when it undergoes reconstruction. This causes a misalignment with the underlying bulk, which is managed by partial dislocations between the first two layers. The silicon (111) surface also exhibits a similar pattern with a 7x7 reconstruction that includes extra atoms in the plane and adatoms on top. In contrast, anatase (001) surfaces have a different behavior.

4.2.3.3.1. Adsorbate-Induced Changes in the Surface Stress

Two processes can cause the surface stress to alter when atoms or molecules are adsorbed on it. The first is a shift in the atoms' electron density at the surface, which modifies the in-plane bonding and, consequently, the surface stress. The second is caused by interactions between the adsorbed molecules or atoms, which may desire to move in opposite directions from the surface's atomic spacings. Note that the surface stress tensor will exhibit a similar dependence as adsorption frequently depends heavily on the environment, such as temperature and gas pressure.

4.2.3.3.2. Lattice Parameter Changes in Nanoparticles

The volume of a spherical particle scales as a cube, but the surface area scales as the square of the size. As a result, surface contributions to energy can become significant in nanoparticles at small sizes. The entire particle can be shrunk if the energy of the surface atoms decreases with increasing distance. An energy cost for the shrinking (deformation) that scales with volume will offset the energy gain from the surface stress, which will scale with area. When combined, these result in a size-inverse shift in the lattice parameter. This has been measured using x-ray or electron diffraction for a variety of materials. This phenomenon has occasionally been described as having a surface tension and being comparable to the Laplace pressure, also known as the capillary pressure. This is incorrect because liquids are covered by these terms. One issue is that when surface segregation is possible or when nanoparticles with more complex shapes have more involved forms due to changes in the lattice parameter.

4.2.3.3.3. Stabilization of Decahedral and Icosahedral Nanoparticles

Regarding nanoparticles as well, surface stress can be important for stabilizing icosahedral twins and decahedral nanoparticles. Lower energy surface energy facets are the result of an internal twin boundary arrangement in both situations. There are nominal angular gaps, or disclinations, to balance this, and an elastic deformation eliminates them. Although the strain energy and the external surface energy make up the majority of the energy contributions, the surface stress links the two and can be crucial to the stability of the system as a whole.

4.2.3.3.4. Deformation and Instabilities at Surfaces

Surface stress acts as a coupling term to combine the surface energy and internal strain that can be in balance during thin film growth. A morphological instability can happen and the film can begin to become highly uneven instead of growing continuously. This is often caused by a breakdown of the equilibrium between surface and elastic energies. Similar wrinkling in nanowires and morphological instability in thin films can result from surface stress.

4.2.4. Stress at a Point

Because the description of the stress at a point in a material involves two directions (and thus two free indices), it requires nine components in total. The orientation of the surface under stress is indicated by one of the directions, while the direction of the force acting on that surface is indicated by the other.

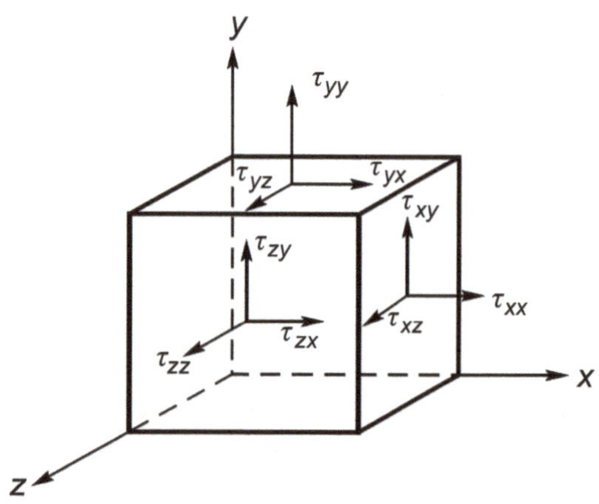

Figure 4.9. Stress components on an infinitesimal rectangular parallelepiped.

Source: https://www.iist.ac.in/sites/default/files/people/stress.pdf.

Consider an infinitesimal rectangular parallelepiped with faces perpendicular to the coordinate axes as shown in Figure 4.9. On each face there is a normal stress and a tangential stress. The tangential stress can be further resolved into two components in the directions of the axes. The complete stress matrix is given below:

$$\bar{\bar{\tau}} = \begin{pmatrix} \tau_{xx} & \tau_{xy} & \tau_{xz} \\ \tau_{yx} & \tau_{yy} & \tau_{yz} \\ \tau_{zx} & \tau_{zy} & \tau_{zz} \end{pmatrix}$$

The sign convention for the stress is that, on a surface whose outward normal points in the positive direction of a coordinate axis, the normal and shear stresses are positive, if they point in the positive direction of the axes. The $(ij)^{th}$ component of the stress tensor $\bar{\bar{\tau}}$ is given by τ_{ij}. The first index of τ_{ij} indicate the direction of the normal to the surface on which the stress is considered, and the second index indicates the direction in which the stress acts. The diagonal elements τ_{xx}, τ_{yy}, τ_{zz} are the normal stresses and the off-diagonal elements are the tangential or shear stresses. Although a parallelepiped is shown, its size actually is zero and the figure really shows the stresses on three orthogonal planes passing through a point.

4.3. STRESS COMPONENTS

The stress components are essential parts of the stress tensor, representing the internal forces per unit area within a material. They play a crucial role in understanding how materials behave under different loading scenarios. In a three-dimensional space, stress at a specific point is shown by a second-order tensor with nine components, which can be divided into normal and shear stresses.

a. Normal Stress Components: These are the components of stress acting perpendicular to the material's surface. They describe the tensile or compressive forces per unit area.
   - σ_{xx}: Normal stress in the x-direction.
   - σ_{yy}: Normal stress in the y-direction.
   - σ_{zz}: Normal stress in the z-direction.

b. Shear Stress Components: These are the components of stress acting parallel to the material's surface. They describe the shearing forces per unit area, which can cause deformation or change in shape.
   - τ_{xy} (or σ_{xy}): The shear stress acting on the x-plane in the y-direction. It represents the component of stress parallel to the surface.
   - τ_{yx} (or σ_{yx}): The shear stress acting on the y-plane in the x-direction. Due to the symmetry of the stress tensor, $\tau_{xy} = \tau_{yx}$.
   - τ_{xz} (or σ_{xz}): The shear stress acting on the x-plane in the z-direction. It describes the tangential force per unit area acting along the z-axis.
   - τ_{zx} (or σ_{zx}): The shear stress acting on the z-plane in the x-direction. Similarly, $\tau_{xz} = \tau_{zx}$.
   - τ_{yz} (or σ_{yz}): The shear stress acting on the y-plane in the z-direction. It indicates the tangential force per unit area acting along the z-axis.
   - τ_{zy} (or σ_{zy}): The shear stress acting on the z-plane in the y-direction. Correspondingly, $\tau_{yz} = \tau_{zy}$.

The stress components can be organized into a 3x3 matrix known as the stress tensor:

$$\sigma = \begin{bmatrix} \sigma_{xx} & \tau_{xy} & \tau_{xz} \\ \tau_{yx} & \sigma_{yy} & \tau_{yz} \\ \tau_{zx} & \tau_{zy} & \sigma_{zz} \end{bmatrix}$$

The diagonal components (σ_xx, σ_yy, σ_zz) represent normal stresses, while the off-diagonal components (τ_xy, τ_xz, τ_yz, and their symmetric counterparts) represent shear stresses. This tensor fully characterizes the internal forces within a material under external loads, essential for analyzing stress and strain relationships, predicting material failure, and designing resilient structures.

4.3.1. The Stress Tensor (Matrix)

Stress is defined as force per unit area. If we take a cube of material and subject it to an arbitrary load we can measure the stress on it in various directions (Figure 4.10). These measurements will form a second rank tensor; the stress tensor.

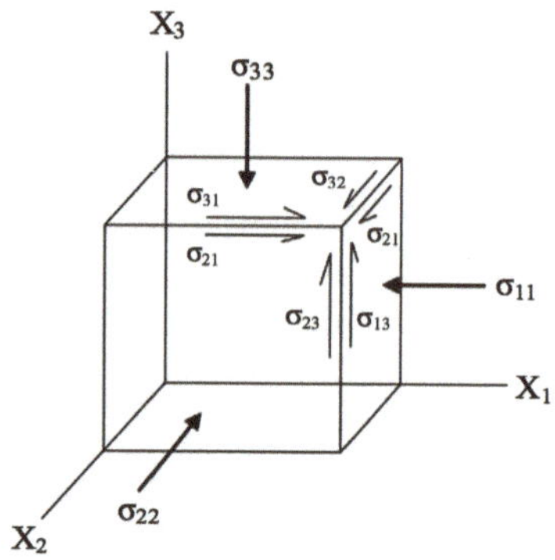

$$[\sigma_{ij}] = \begin{pmatrix} \sigma_{11} & \sigma_{12} & \sigma_{13} \\ \sigma_{21} & \sigma_{22} & \sigma_{23} \\ \sigma_{31} & \sigma_{32} & \sigma_{33} \end{pmatrix}$$

Figure 4.10. The Stress Tensor.

Source: https://cdn.serc.carleton.edu/images/NAGTWorkshops/mineralogy/mineral_physics/fig_4.v3_700.webp

The Eigen values of σ_{ij}; represented as σ_1, σ_2, σ_3 are referred to as the principle stresses.

$$\begin{pmatrix} \sigma_1 & 0 & 0 \\ 0 & \sigma_2 & 0 \\ 0 & 0 & \sigma_3 \end{pmatrix}$$

The Eigen vectors are the principle stress directions known as the maximum, intermediate and minimum principle stresses respectively; in geology compression is considered positive and the maximum compressive stress is referred to as σ_1. However, in engineering and physics, tension is considered positive so the maximum compressive stress is referred to as $\sigma_{3.}$ Therefore, it is important to be aware of which sign convention is being used. A cube with its edges parallel to the principle stress directions experiences no sheer stresses across its faces.

An important property of the stress tensor is that it is symmetric:

$\sigma_{ij} = \sigma_{ji}$

Intuitively, this can be seen if one images shrinking the cube in Figure 4.10 to a point. If the cube is infinitesimally small, the forces across each face will be uniform. If the cube is to remain stationary the normal forces on opposite faces must be equal in magnitude and opposite in direction and the shear tractions which would tend to rotate it must balance each other. So for example, if σ_{13} is not equal in magnitude to $\sigma_{31,}$ the cube will spin around the X_2 direction. Therefore, it is only necessary to find 6 of the components of the tensor.

Important concepts are often used are deviatoric stress and hydrostatic pressure. Any stress tensor may be broken into two parts

$$\begin{pmatrix} \sigma_{11} & \sigma_{12} & \sigma_{13} \\ \sigma_{21} & \sigma_{22} & \sigma_{23} \\ \sigma_{31} & \sigma_{32} & \sigma_{33} \end{pmatrix} = \begin{pmatrix} (\sigma_{11}-p) & \sigma_{12} & \sigma_{13} \\ \sigma_{21} & (\sigma_{22}-p) & \sigma_{23} \\ \sigma_{31} & \sigma_{32} & (\sigma_{33}-p) \end{pmatrix} + \begin{pmatrix} p & 0 & 0 \\ 0 & p & 0 \\ 0 & 0 & p \end{pmatrix}$$

where

$p = (\sigma_{11} + \sigma_{22} + \sigma_{33})/3)$

Using the Kronecker delta notation this may be written as

$\sigma_{ij} = \sigma_{ij(dev)} + p\delta_{ij}.$

4.3.2. Deviatoric Stress and Mean Stress

Deviatoric stress and mean stress are essential concepts in the study of material mechanics because they help us comprehend how materials behave under load. The part of the total stress that results in distortion or a change in shape without changing the material's volume is referred to as deviatoric stress. It is calculated by deducting the mean stress from the total stress, making sure that only the elements causing shear and not volumetric changes are taken into account. Since it directly affects the beginning of yielding and flow in ductile materials, this kind of stress is essential for assessing failure mechanisms and plastic deformation in materials.

Conversely, mean stress—which is the average of the three principal stresses acting on a material—is also referred to as hydrostatic stress or volumetric stress. It is calculated by dividing the total principal stresses by three. The material's volume is impacted by mean stress, which causes expansion or contraction but not distortion. Understanding phenomena like cavitation in fluids and volumetric strain in solids depends on this stress component. The overall structural integrity and stability of materials are greatly impacted by mean stress, even though it does not cause shear deformation. This is especially true for brittle materials where high mean stresses can cause fracture.

Deviatoric and mean stresses combined offer a thorough knowledge of the stress state in materials. Deviatoric stress provides information about the material's resistance to shape changes and possible yielding, whereas mean stress emphasizes the effect on volume and fracture susceptibility. By ensuring that materials can withstand both shear and volumetric stresses under operating loads, engineers can design more resilient structures and increase longevity by preventing failure and improving material performance. In domains like materials science and civil engineering, where knowledge of stress distributions guarantees dependability and safety, this dual analysis is essential.

4.3.3. Mohr Circle and Diagram

A two-dimensional graphic representation of the transformation law for the Cauchy stress tensor is called a Mohr's circle. Mohr's circle is frequently used in calculations related to the strength of materials in mechanical engineering, soil strength in geotechnical engineering, and built structure strength in structural engineering. By breaking down stresses into their vertical and horizontal components, it can also be used to calculate stresses in a variety of planes. These are known as principal planes, and it is from these that principal stresses are computed. One of the simplest methods for locating the principal planes and principal stresses in a graphical representation is to utilize Mohr's circle.

The elements of the Cauchy stress tensor at a specific material point are known with respect to a coordinate system after completing a stress analysis on a material body assumed as a continuum. Following that, the stress components acting on a rotated coordinate system, i, are graphically determined using the Mohr circle i.e., acting through that point on a plane that is oriented differently.

The abscissa and ordinate (σ_n, τ_n) of each point on the circle are the magnitudes of the normal stress and shear stress components, respectively, acting on the rotated coordinate system. In other words, the circle is the locus of points that represent the state of stress on individual planes at all their orientations, where the axes represent the principal axes of the stress element.

The first person to develop a graphical representation for stresses while taking longitudinal and vertical stresses in horizontal beams during bending was the German engineer Karl Culmann in the 19th century. Christian Otto Mohr, a fellow German engineer and the man who gave the stress circle its name, was motivated by his work to expand its application to two- and three-dimensional stresses and create a failure criterion based on the stress circle. Other graphical techniques for representing the stress state at a point are the Cauchy's stress quadric and Lamé's stress ellipsoid. Any symmetric 2x2 tensor matrix, including the strain and moment of inertia tensors, can be fitted with the Mohr circle.

CASE STUDY

TENSIONAL STRESS: THE BASIN AND RANGE PROVINCE

The Basin and Range Province in the western United States is a prime example of a geological region shaped predominantly by tensional stress. This area stretches across Nevada, Utah, and parts of California and Arizona, characterized by its distinct topography of alternating mountain ranges and valleys.

Geological Background

The Basin and Range Province is the result of extensional tectonics that began around 17 million years ago and continues today. The region lies within the North American plate, which experiences significant stretching due to the movement of the Pacific plate. The extensional forces are believed to be driven by a combination of plate boundary forces and gravitational collapse of an over-thickened crust.

Mechanism of Tensional Stress

Tensional stress in this region is caused by the pulling apart of the Earth's crust. This type of stress is oriented horizontally, causing the crust to thin and stretch. As the crust extends, it fractures and creates normal faults, leading to the characteristic basin (valley) and range (mountain) topography.

Formation and Features

1. **Normal Faulting:**
   - Tensional stress causes the formation of normal faults, where the hanging wall block moves downward relative to the footwall block.
   - Examples include the Wasatch Fault in Utah and the Sierra Nevada Fault in California.
2. **Horst and Graben Structures:**
   - The alternating mountain ranges (horsts) and valleys (grabens) are direct results of the crustal extension and normal faulting.
   - Horsts are uplifted blocks, while grabens are down-dropped blocks.
3. **Volcanism and Crustal Thinning:**
   - As the crust thins, it facilitates volcanic activity, evident in features like the Great Basin and the Snake River Plain.
   - This thinning also reduces pressure on the underlying mantle, leading to partial melting and magma formation.

Impacts and Implications

- **Seismic Activity:**
  - The Basin and Range Province is seismically active due to the ongoing extensional tectonics, with frequent earthquakes along the normal faults.
- **Resource Distribution:**
  - The region's geology influences the distribution of natural resources, including mineral deposits and geothermal energy sources.
- **Land Use and Hazards:**
  - Understanding tensional stress and the resulting faulting patterns is crucial for land use planning, infrastructure development, and hazard mitigation.

Conclusion

The Basin and Range Province provides a clear example of how tensional stress shapes the Earth's crust. Studying this region helps geologists understand the processes and consequences of crustal extension, including fault formation, seismic activity, and resource distribution. This case study underscores the importance of recognizing and analyzing tensional stress in geological contexts to predict and manage its impacts effectively.

CLASS ACTIVITY

For a class activity on stress in structural geology, divide students into small groups and provide them with different rock samples and modeling clay. Each group should apply various stresses (compressional, tensional, shear) to their samples and document the resulting deformations. Then, use Mohr's Circle diagrams to analyze the stress states. Students should present their findings, discussing how each type of stress affected the material. This hands-on activity reinforces theoretical concepts, demonstrates real-world applications, and encourages collaborative learning and critical thinking about geological stress and deformation processes.

SUMMARY

- Stress is a common phenomenon experienced by individuals in various stages of life, characterized by physical, mental, or emotional strain or tension.
- Stress and strain are closely related concepts in structural geology, as stress is the force applied to a material, while strain is the resulting deformation of the material under that force.
- Deformation mechanisms are the processes that lead to the deformation of a material under stress. In structural geology, understanding these mechanisms is important for understanding how rocks deform and how geological structures such as folds, faults, and shear zones form.
- An alternative approach is an atomistic one, which defines all quantities in terms of the number of atoms, not continuum measures such as areas.
- The stress at a point in a material needs nine components for its complete description because two directions (and therefore, two free indices) are involved in its description.
- Stress components are the elements of the stress tensor that describe the internal forces per unit area within a material.
- Mohr's circle is often used in calculations relating to mechanical engineering for materials' strength, geotechnical engineering for strength of soils, and structural engineering for strength of built structures.

REVIEW QUESTIONS

1. What are the principal stresses?
2. How is deviatoric stress different from mean stress?
3. Explain the concept of stress tensor in structural geology.
4. What is Mohr's Circle, and how is it used in stress analysis?
5. Differentiate between compressional, tensional, and shear stress.
6. Describe the difference between lithostatic stress and tectonic stress.

REFERENCES

1. Askeland, D., & Wright, W. J. (2016). The Science and Engineering of Materials (7th ed.). Boston, MA: Cengage Learning.
2. Bottomley, D. J., Makkonen, L., & Kolari, K. (2009). Incompatibility of the Shuttleworth equation with Hermann's mathematical structure of thermodynamics. Surface Science, 603(1), 97–101.
3. Lee, J. Y., Punkkinen, M. P. J., Schönecker, S., Nabi, Z., Kádas, K., Zólyomi, V., Koo, Y. M., Hu, Q. M., Ahuja, R., Johansson, B., Kollár, J., Vitos, L., & Kwon, S. K. (2018). The surface energy and stress of metals. Surface Science, 674, 51–68.

4. Muller, P. (2004). Elastic effects on surface physics. Surface Science Reports, 54(5–8), 157–258.
5. Müller, P., Saùl, A., & Leroy, F. (2013). Simple views on surface stress and surface energy concepts. Advances in Natural Sciences: Nanoscience and Nanotechnology, 5(1), 013002.
6. Oehl, N., Knipper, M., Parisi, J., Plaggenborg, T., & Kolny-Olesiak, J. (2015). Size-dependent lattice distortion in ε-Ag3Sn alloy nanoparticles. The Journal of Physical Chemistry C, 119(25), 14450–14454.
7. Tirella, A., & Ahluwalia, A. (2014). Strain rate viscoelastic analysis of soft and highly hydrated biomaterials. Journal of Biomedical Materials Research, 102(10), 3352–3360.
8. Vermaak, J. S., Mays, C. W., & Kuhlmann-Wilsdorf, D. (1968). On surface stress and surface tension. Surface Science, 12(2), 128–133. https://doi.org/10.1016/0039-6028(68)90118-0.
9. Yuan, W., Wang, Y., Li, H., Wu, H., Zhang, Z., Selloni, A., & Sun, C. (2016). Real-time observation of reconstruction dynamics on TiO_2 (001) surface under oxygen via an environmental transmission electron microscope. Nano Letters, 16(1), 132–137.

CHAPTER

5 Extensional Regimes

LEARNING OBJECTIVES

After studying this chapter, you will be able to:

- Define extensional faults
- Identify and describe different types of fault systems
- Define low-angle faults and core complexes
- Describe ramp-flat-ramp geometries
- Describe footwall versus hanging-wall collapse
- Explain the rifting
- Understand the half-grabens and accommodation zones
- Define pure and simple shear models
- Explain the stretching estimates, fractals, and power-law relations
- Define passive margins and oceanic rifts
- Understand orogenic extension and orogenic collapse

KEY TERMS FROM THIS CHAPTER

Biaxial extension
Deformation
Extensional viscosity
Planar extension
Rheology
Capillary breakup
Elongational flow
Filament stretching
Polymer solutions
Strain hardening

INTRODUCTION

Historically, the attention paid to contractional structures has been greater than that of extensional structures. But in the 1980s, the tide began to turn when it was discovered that many of the faults and shear zones that were previously believed to be thrusts actually contained evidence of being low-angle extensional structures. It has now been established that extensional faults and shear zones are common in the majority of orogenic belts, having first been identified in the western USA's Basin and Range province. Most people would concur that our understanding of orogens and orogenic cycles has changed significantly as a result of the study of extensional structures. The fact that rift settings are home to a large portion of the world's offshore hydrocarbon resources and that normal faults control a large portion of hydrocarbon traps is another factor contributing to the current interest in extensional faults. Furthermore, the majority of hydrocarbon reservoirs require a solid understanding of extensional faults, including their characteristics and intricacies, in order to be developed.

5.1. EXTENSIONAL FAULTS

In deformed rocks, extensional faults result in the extension of the crust or certain reference layers. Figure 5.1b shows an example of an extensional fault that affects horizontal layers. It is easy to determine the sense and amount of displacement because that fault exhibits a displacement that is comparable to the layering thickness. Not quite as much as the many hundreds of kilometers estimated on some thrust faults and strikeslip faults, but still quite significant, other extensional faults have accumulated up to a hundred kilometers of offset.

The crust itself is our logical choice of reference at this scale. Extension occurs in the direction of deformation if there is an increase in the distance between two points on Earth's surface, one on each side of the fault. However, depending on the relative positions of the two points, this can also happen across a strike-slip fault. Therefore, in order to determine whether a fault is a true extension fault, we must assess extension perpendicular to the fault strike. Similar to how it is the principal shortening direction for contractional dip-slip faults, this is the principal extension direction across extensional dip-slip faults (Figure 5.1b). There is no variation in length perpendicular to a fault in a perfect strike-slip fault (Fletcher & Bartley, 1995).

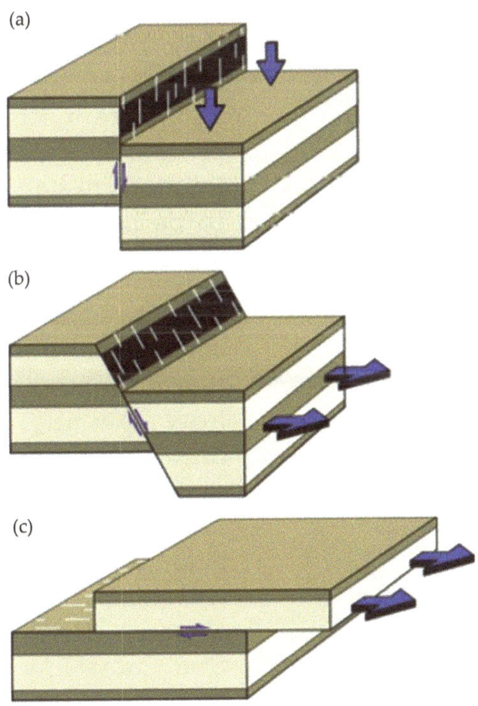

Figure 5.1. Using the surface of the Earth as a reference, extensional faults (b) are a spectrum of normal faults between vertical (a) and horizontal (c). Vertical and horizontal faults are neither extensional nor contractional.

Source: Haakon Fossen, structural geology, ISBN 978-1-10705764-7.

Regardless of the layering orientation, faults that extend a specific reference layer can be referred to as extensional faults for smaller faults. This means that if the faults extend the reference layer, reverse faults may also be extensional faults. Figure 5.2 illustrates an extensional reverse fault and provides an explanation of how it developed through rotation. As a result, it is helpful to designate a reference surface, for instance by referring to the layer parallel extension and crustal extension.

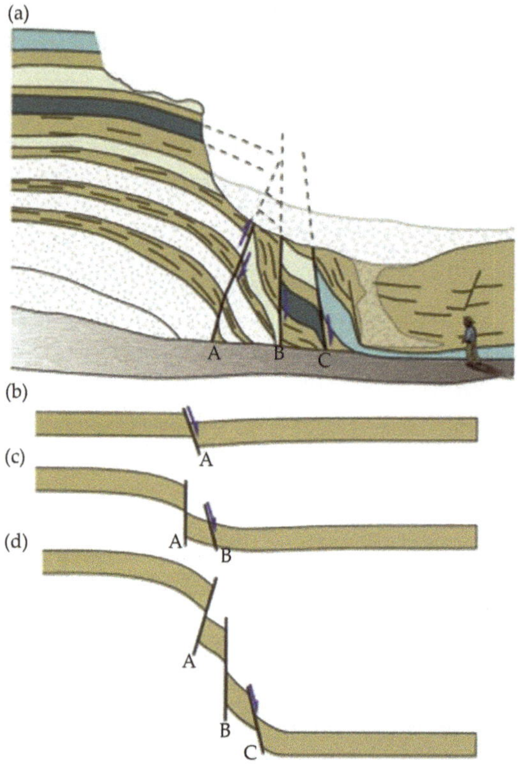

Figure 5.2. (a) Reverse faults and normal faults coexisting in a folded layer along a fault. Both the reverse and normal faults are extensional faults because they extend the layers. The reverse faults probably formed as normal faults (b) before being rotated during the folding (c, d).

Source: Haakon Fossen, structural geology, ISBN 978-1-10705764-7.

Fault dip is likewise significant. Extensional faults span a range of fault dips, from vertical (Figure 5.1a) to horizontal (Figure 5.1c), as seen in Figure 5.1. Vertical faults suggest only vertical motions in the crust, not its extension or shortening (Figure 5.1a). Vertical block faulting can be compared on a larger scale to playing the piano, where the keyboard moves vertically but stays the same length. During the Cretaceous Laramide phase, such vertical tectonics dominated much of the Colorado Plateau in the western United States. Faults oriented perpendicular to a layer typically do not cause the layer to shorten or stretch. The other end-member is represented by horizontal (or layer-parallel) faults, which neither shorten nor lengthen horizontal (or fault-parallel) layers (Figure 5.1c). Layer-parallel faults can be found as the curved extensional or contractional faults' flat sections. The most straightforward explanation is that almost all rocks possess an anisotropy that was carried over from previous stages of deformation. Therefore, strike-slip faults or reactivated joints are a straightforward explanation for extremely steep faults. Keep in mind that joints form perpendicular to σ_3, which tends to be horizontal in the upper crust, making them nearly vertical.

In a similar vein, it has been suggested that low-angle extension faults may have been reactivated thrust faults, which is how low-angle normal faults originate. Simultaneously, field observations and experiments suggest that certain low- and high-angle extensional faults developed during a single phase of extension without the assistance of weak structures that were already there. Specifically, it is believed that some low-angle normal faults formed directly with little rotation, while other low-angle faults must have rotated from original high-angle faults to low-angle structures. We will begin examining these and related observations using the domino model, a basic fault rotation model.

5.2. FAULT SYSTEMS

5.2.1. The Domino Model

A sequence of rotated fault blocks arranged akin to overturned books in a partially filled bookshelf or domino bricks are commonly observed in sections through rifted portions of the upper crust (Figure 5.3a). The term bookshelf tectonics, also known as the (rigid) domino model, originated from this analogy.

The rigid domino model describes a series of rigid fault blocks that rotate simultaneously in a uniform sense.

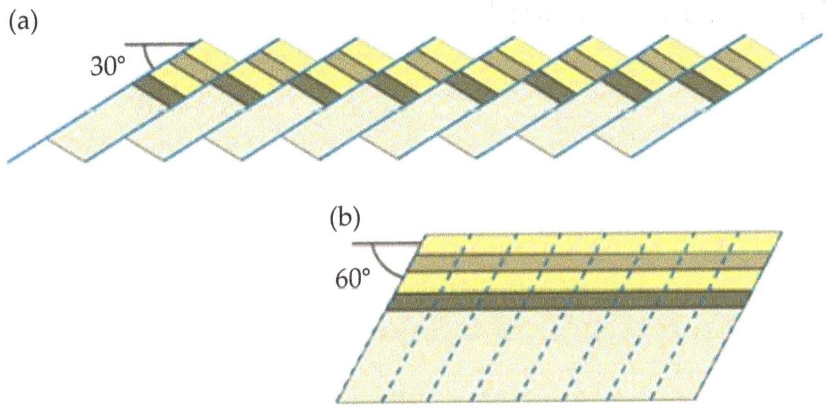

Figure 5.3. (a) Schematic illustration of rigid domino-style fault blocks. (b) Such fault blocks can be restored by rigid rotation until layering is horizontal. In this case we have applied 30° rotation and displacement removal.

Source: https://www.geokniga.org/bookfiles/geokniga-structural-geology_1.pdf.

As always, boundary conditions and environmental compatibility must be taken into account when applying models. Open spaces and overlaps in particular are not acceptable. What is happening at each end of the domino system presents the first difficulty. As illustrated in Figure 5.4, this is quite elegantly resolved by introducing an appropriate listric fault. On the opposing side, a set of domino fault blocks that dip in the opposite direction may be positioned alongside a listric fault. In order to join the two sets, a graben will then be required in the center.

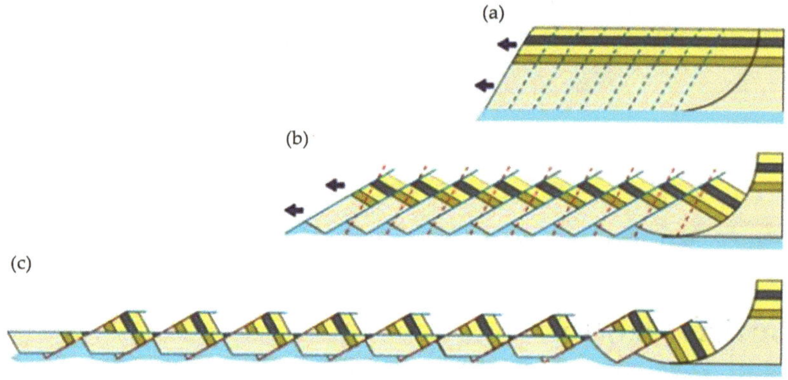

Figure 5.4. Schematic illustration of the development of a domino system. (a) The transition to the undeformed footwall is accommodated by a listric fault. (b) A new set of faults develops at high extension. (c) The resulting fault pattern can become quite complex.

Source: Haakon Fossen, structural geology, ISBN 978-1-10705764-7.

5.2.2. The Soft Domino Model

It is rare for fault blocks to exhibit rigidity, especially in rift systems that contain relatively weak rocks or unconsolidated sediments. Furthermore, it has been observed that faults arise in populations in which the displacement of faults varies along each fault. All faults must have the same length and displacement according to the rigid domino model; no displacement gradient may be seen.

These inherent divergences from the rigid domino model lead to the definition of a soft domino model, which allows internal strain to build up inside blocks. This permits folds in layers, variances in fault sizes, and variations in fault displacement. Blocks that rotate rigidly backward or employ simple line balancing are no longer effective methods for determining the amount of extension. Rather, the internal fault-block deformation representative model needs to be selected.

5.2.3. Why Do Domino Systems Form?

A more or less symmetric horstand-graben system or one of the domino systems covered above can arise from crustal extension. The fault arrangement is dependent on how the rocks react to the deformation, even though the total extension and crustal thinning may be the same i.e., how the crust accommodates strain (see Figures 5.3 and 5.5). The existence of a weak lowangle layer or structure is undoubtedly one of the most significant elements that encourages the development of asymmetric domino-style fault systems. This could be an overly stressed formation, a salt layer, a mobile clay or stone, or an old fault that is prone to reactivation. The lack of a weak layer or décollement that dips gently, on the other hand, promotes a more symmetric horst-and-graben system. Numerous physical experiments have shown the significance of a dipping (as opposed to horizontal) basal soft layer or décollement, as illustrated in Figure 5.6.

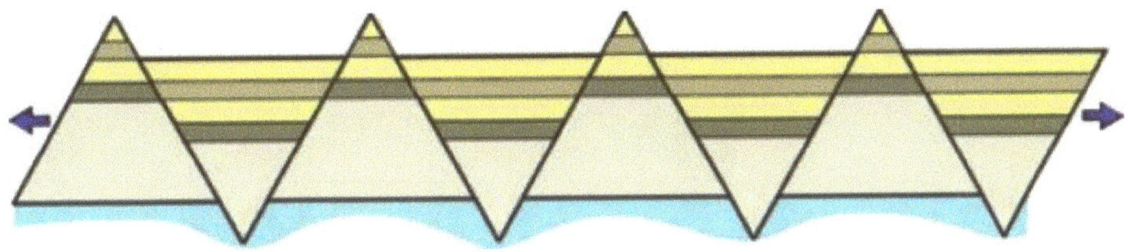

Figure 5.5. The alternative to domino-style stretching is the development of horst-graben systems. This deformation style is, ideally, symmetric and an overall pure shear strain.

Source: https://www.geokniga.org/bookfiles/geokniga-structural-geology_1.pdf.

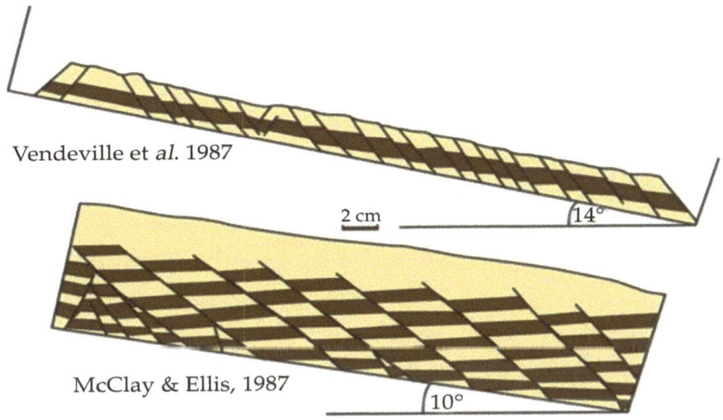

Figure 5.6. Sandbox experiments where the base of the model is tilted prior to extension. The tilting may be the cause for the uniform dip and dip direction of the faults.

Source: Haakon Fossen, structural geology, ISBN 978-1-10705764-7.

5.2.4. Multiple Fault Sets In Domino Systems

A new set of faults forms (Figure 5.4b, red faults) when a domino system is exposed to high extension, as shown in Figure 5.4. The faults will rotate so far from their original and favorable orientation. This occurs when the critical shear stress of the deforming rock, which is determined by the mechanical characteristics (friction) of the faults and the strength of the host rock, is reached below the shear stress along the initial faults. It can be demonstrated that new faults are expected after 20–45° rotation for realistic frictional values. This indicates that domino faults can rotate to a dip of 40–15° while they are still active, given an initial fault dip of 60°. After that, the initial domino faults stop rotating and are replaced by new faults that sever the previous ones and continue to rotate the entire system. There is field evidence supporting this model from regions with high crustal extension, like the western USA's Basin and Range province.

5.3. LOW-ANGLE FAULTS AND CORE COMPLEXES

Low-angle faults with large displacements were almost exclusively mapped in fold-and-thrust belts for a very long time, and all of them were thought to be contractional faults there. It was widely accepted that normal faults are high-angle structures that usually have dips of about 60°. Rock mechanics considerations support this somewhat simplistic distinction, which was applied less rigidly after field mapping in the Basin and Range province in the 1970s and later in many other parts of the world revealed that, despite the majority of normal faults being steeper, extensional low-angle faults are fairly common in several extensional regions. The interpretation of contemporary seismic imagery, in conjunction with numerical and physical modeling, has also demonstrated that low-angle structures may be the occurrence of extensional faults. Whether they were active as low-angle structures or if they were just steeper normal faults that rotated is sometimes unclear. (Kusznir, N. J. and Ziegler, P. A., 1992)

5.3.1. The Problem with Low-Angle Normal Faults

Since a considerable number of low-angle normal faults were mapped and interpreted in the 1970s and 1980s, structural geologists have debated these faults. They make kinematic sense because, in contrast to steep normal faults, they are very effective at accommodating extensional strain. Although many low-angle normal faults have been reported from different regions of the world, their formation mechanism and motivation remain unknown.

Low-angle normal faults have both a mechanical and a seismic issue. Mechanically speaking, when the maximum stress is vertical and the rock is mechanically isotropic, normal faults are expected to form with initial dips of about 60°. Data on the focal mechanism for earthquakes is related to the seismic problem. It is possible to interpret the lack of recorded moderate-to-large earthquake ruptures (magnitude > 5.5) on low-angle normal faults as evidence against low-angle normal faults that are seismically active. This is shown by global seismic data. This could indicate that these faults are either high-angle normal faults that rotated during deformation or that they just don't cause a lot of big earthquakes.

5.3.2. Rotated Normal Faults

A model for the formation of low-angle normal faults is already available to us thanks to the fault rotation model displayed in Figure 5.4. A later (red) set of faults in this figure rotated the earlier (blue) sets of faults to a horizontal orientation. Nevertheless, a lot of large-scale low-angle extensional faults don't split up into younger faults like the red faults in Figure 5.4c, which are persistent. As such structures defy explanation, we need to explore for alternative models. The model that gained traction at the end of the 1980s is one that extends to the crust as a whole. The mobile base of the model is necessary for the isostatic effects of extension, but physical experiments with a fixed base, like the ones in Figure 5.6, do not meet this requirement.

Figure 5.7 provides an illustration of the model's basic idea. In the upper crust, a listric normal fault develops and flattens along a weak detachment zone close to the brittle–plastic transition. There are instances where the footwall is referred to as the lower plate and the hanging wall to large detachments as the upper plate.

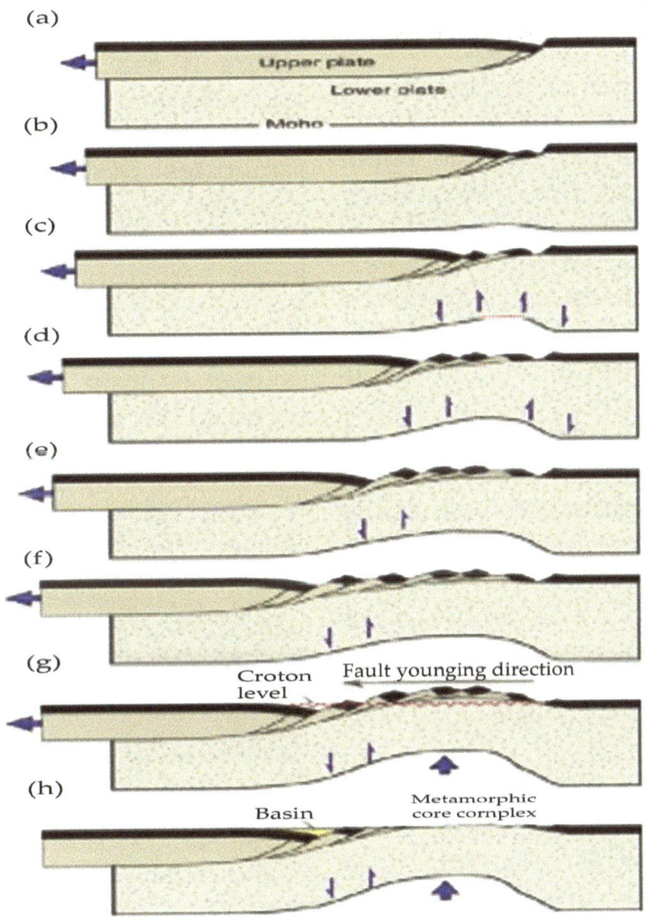

Figure 5.7. Development of a metamorphic core complex during crustal-scale extension and isostatic compensation. Note how new wedge-shaped fault blocks are successively torn off the hanging wall. Also note how isostatic compensation is accommodated by means of vertical shear.

Source: https://www.geokniga.org/bookfiles/geokniga-structural-geology_1.pdf.

A new fault in the hanging wall develops after a certain amount of extension, and the initial fault is rendered inactive. Isostatic uplift of the thinned crustal portion is one of the causes of inactivation. The original fault is rotated by this uplift to the point where it is mechanically advantageous to form a new fault in the hanging wall. Until a sequence of rotated domino-style blocks and associated half-grabens are established, this process is repeated. The youngest and active fault in this model will have the steepest fault (compare these faults' geometry with that of horses in a thrust duplex).

5.3.3. Rolling Hinges and Metamorphic Core Complexes

Simple shear deformation of the lower and middle crust is depicted in the model in Figure 5.7. Moho is raised and the upper crust is thinned during this deformation. This effect is exacerbated by erosion of the upper plate's uplifted fault complex, which ultimately exposes rocks from the lower plate. There is brittle deformation in the upper plate, but initially there is plastic shearing along the horizontal detachment. But when the detachment is raised, this deformation becomes more brittle. The detachment will eventually show up as a window through upper plate rocks with brittle structures overlying a core of metamorphic and mylonitic rocks (Figure 5.7h). A metamorphic core complex, which includes the lower plate rocks in this window, was initially identified and documented in the Arizona and Nevada Basin and Range region. Since then, similar complexes have been discovered in many other regions of the world, including those connected to mid-oceanic rifts, as well as the whole Cordillera of western North America.

5.3.4. Active Low-Angle Normal Faults

Classical models that build on Anderson's idealized assumption of vertical and horizontal principal stresses and isotropic rocks, the Mohr–Coulomb fracture criterion and Byerlee's law, predict that normal faults lock up and become inactive when they obtain a low dip. In simple terms, this happens because the shear stress on a fault is reduced as its angle to σ_1 increases. The width of the sector depends on the strength (internal friction) of the preexisting structure and the state of stress. For common coefficients of friction the critical dip values are around 30–40°.

Nonetheless, there are several instances where it can be demonstrated that low-angle normal faults, whether or not they were rotated, were active low-angle structures. Microseismic observations are included in this, i.e., data gathered from tiny earthquakes that happen near low-angle structures. Cataclastic fault rocks characterize late faults along a low-angle detachment zone, where no significant syn- or post-faulting rotation appears to have occurred. Figure 5.8 illustrates one example of a low-angle normal fault.

Figure 5.8. The low-angle normal fault underneath the Hornelen Devonian basin in the Scandinavian Caledonides separates Devonian sandstones and conglomerates from mylonitic rocks of the Nordfjord-Sogn Detachment. This fault is brittle, located between mylonitic fabrics of the extensional Nordfjord-Sogn Detachment and the overlying supradetachment basin.

Source: Haakon Fossen, structural geology, ISBN 978-1-10705764-7.

We depart from the idealized Andersonian scenario due to heterogeneities represented by preexisting low-angle normal faults. We can apply the overpressure defense we used for thrusts in this situation. Put another way, a fault zone or low-angle stratigraphic layer with high fluid pressure can be incredibly weak and have very little friction. Overpressured shale layers are a good example of sedimentary sequences. Another is thin salt layers, both of which support low-angle extensional detachments on continental slopes, such as those offshore Brazil and Nigeria/Angola. Weak structures in igneous and metamorphic rocks are characterized by preexisting faults and shear zones that are subject to excessive pressure, especially during seismic activity.

Through a process called microseismicity, certain low-angle normal faults that are currently active have been seismically identified. Microseismicity is the term used to describe small but frequent earthquakes that are usually generated along preexisting faults and fractures and can be recorded in large numbers (hundreds) over short periods of time (months). Microseismic activity and seismic reflections interpreted as low-angle detachments are well correlated, as demonstrated by microseismicity data across the Gulf of Corinth and the Altotiberina fault in the Italian Appennines. Therefore, there is proof of low-angle normal fault activity without significant earthquakes. Immediately, two explanations come to mind: either many of the large earthquakes are so weak that creep and small earthquakes completely dominate over stick-slip behavior that creates major seismic events, or the large earthquakes have such long recurrence intervals that they have not yet been detected. Minerals that are mechanically weak, such as talc or platy minerals (clay minerals, micas, and chlorite), should be present in a weak fault core. We are aware that the San Andreas Fault exhibits aseismic (creep) behavior over a sizable portion, with only minor (microseismic) earthquakes. It should be easier for a low-angle fault zone to collect fluids and produce overpressure, which strengthens the fault rock. However, there is a problem with low-angle faults' seismic silence worldwide.

It's intriguing that experiments can create low-angle normal faults by either décollémenting or severing previous high-angle normal faults. Low-angle faults can still occur even in the absence of a predefined weak layer or plastic sheet, which makes décollement faults easy to create in the lower or basal region of the model. Figure 5.9 provides an example, where the red Fault 5 very effectively accommodates the subsequent extension as a lowangle fault after cutting the steeper Fault 1 at a certain point. It should be noted that the red fault (5) forms by joining two steeper segments, and that it is low-angle only for a limited portion of its entire length. Such a linking fault may potentially be guided by a weak low-angle layer or fault in the natural world. A fault that starts with a ca is illustrated by fault (6) in Figure 5.10. Its intermediate and lower portions have a 25° angle (the upper part is steeper). In this instance, a rigid basement might be in charge of the fault geometry.

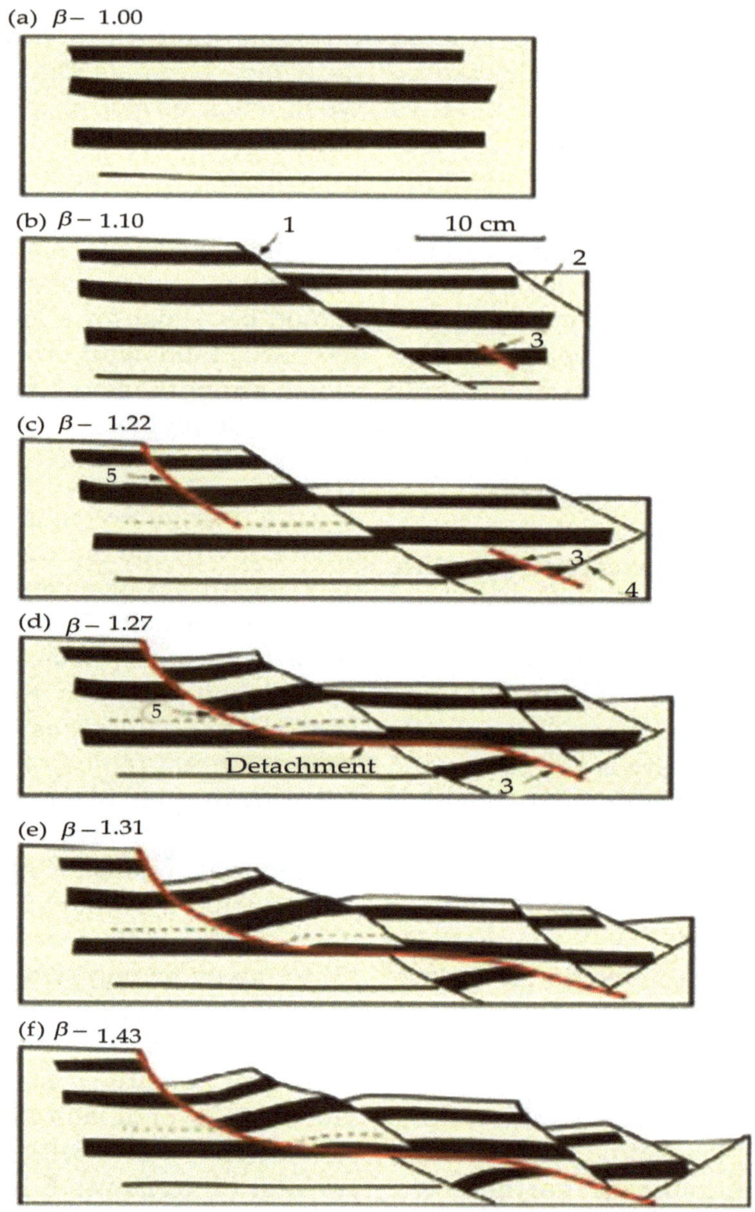

Figure 5.9. Plaster experiment showing how a low-angle fault can form at a relatively late stage during an extensional deformation history. Note that the late fault (red) cuts preexisting high-angle faults and that a horst is about to establish above the flat fault segment. Black pseudolayering was painted on the side walls prior to extension.

Source: Haakon Fossen, structural geology, ISBN 978-1-10705764-7.

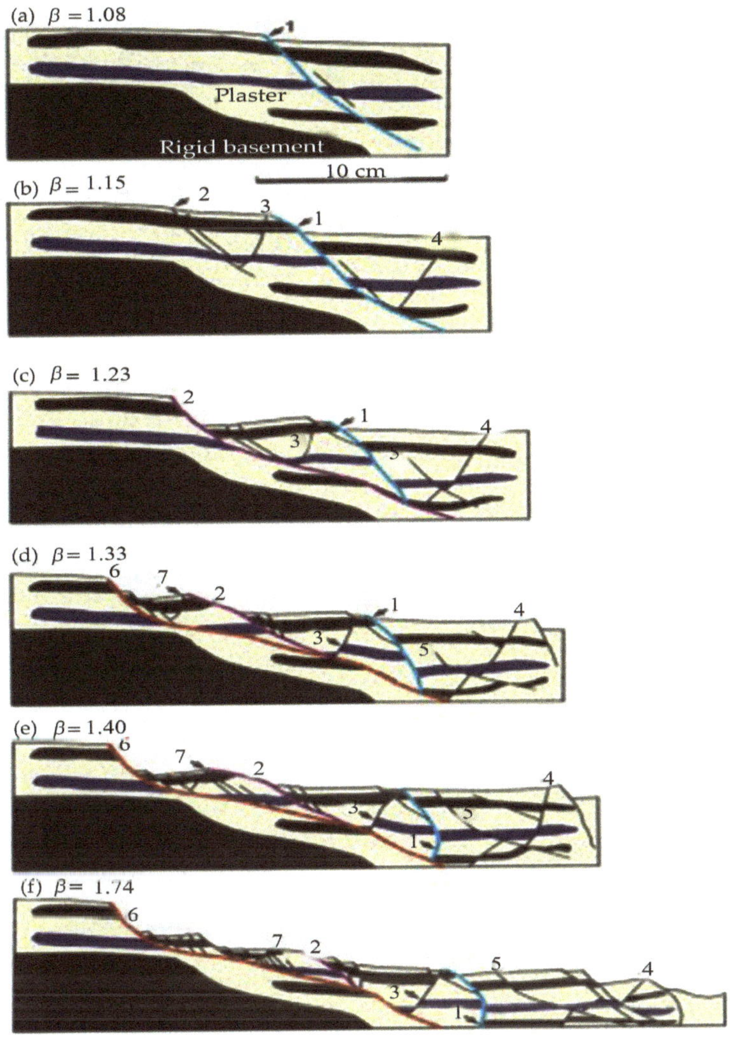

Figure 5.10. Footwall collapse as seen in a plaster experiment. Note how new faults sequentially form in the footwall at the same time as the dip of the controlling fault is reduced.

Source: https://www.geokniga.org/bookfiles/geokniga-structural-geology_1.pdf.

Apart from the potential existence of weak structure within the crust, it is also possible that σ1 is non-vertical in other places as well, such as near a basement structure similar to that shown in Figure 5.10 or near a weak low-angle fault, like an existing thrust. This is another significant departure from the Andersonian conditions that may cause low-angle normal faults to form or become active, and more research is required to fully comprehend the subsurface variations in principal stresses.

5.4. RAMP-FLAT-RAMP GEOMETRIES

Ramp-flat-ramp geometries create significant hanging-wall strain as the hanging wall needs to adapt to the fault's shape as it moves. An illustration of a ramp-flat-ramp geometry can be seen in Figure 5.9 (red fault), where the flat portion is formed by connecting two steeper fault segments.

Figure 5.11 shows that above the ramp-flat-ramp fault, a sequence of wedge-shaped fault blocks may form, where the faults either die out upward or reach the surface (also see the hanging wall to Fault 6 in Figure 5.10f). Similar to the contractional regime, an extensional imbrication zone is a grouping of these faults or fault blocks. A group of lenses that come together to form an extensional duplex is a related kind of extensional structure. Similar to contractional duplexes, extensional duplexes have faults in the roof and floors (Rosendahl, 1987).

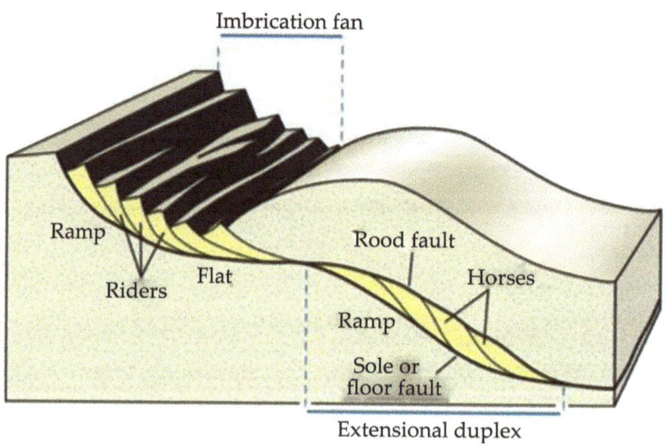

Figure 5.11. Schematic illustration of extensional imbrication and a duplex structure. Horses and riders are shown in yellow.

Source: Haakon Fossen, structural geology, ISBN 978-1-10705764-7.

In rift systems like the North Sea and in experiments, a horst complex is occasionally observed to form above the flat (Figure 5.10). As the footwall collapses, domino-style fault arrays typically form behind such horsts. A horst complex repeats its development in Figure 5.10. Faults 1 and 3 define the first horst, and Faults 2 and 7 define the second.

5.5. FOOTWALL VERSUS HANGING-WALL COLLAPSE

Extensional imbrication zones are commonly thought to form as a result of faults and slices in the hanging wall forming one after the other. This view is supported by the general observation that the hanging wall is typically more strained than the footwall. This phenomenon, known as hanging-wall collapse, is undeniably real. Figure 5.10, where an increasing number of faults are added to the hanging wall of Fault 6 during the final (d–f) portion of the experiment, provides an illustration of hanging-wall collapse. Figure 5.7 provides an additional example. Experiments like the one demonstrated in Figure 5.10 indicate that footwall collapse, which involves the creation of new faults in the footwall, can happen. This process is observed in the initial phase of the experiment, where a synthetic fault (2) emerges in the footwall of Fault 1 (Figure 5.10a–c). Subsequently, Fault 6 is formed further into the footwall, leading to the collapse of its hanging wall into an imbrication zone.

Footwall collapse occurs frequently in rift systems where large fault blocks are formed and rotated, causing the elevated crests of these blocks to collapse due to the effects of gravity and tectonic forces. Sometimes, gravity alone is enough to cause footwall collapse. This leads to gravitational collapse and slumping on curved fault surfaces, resulting in complex stratigraphic relationships that pose challenges in petroleum exploitation. The presence of weak layers like mud, over pressured sediments, or salt often controls slumping, leading to the development of contractional folds and faults in the toe zone, as well as extensional faults in the central and rear parts (see Figure 5.12).

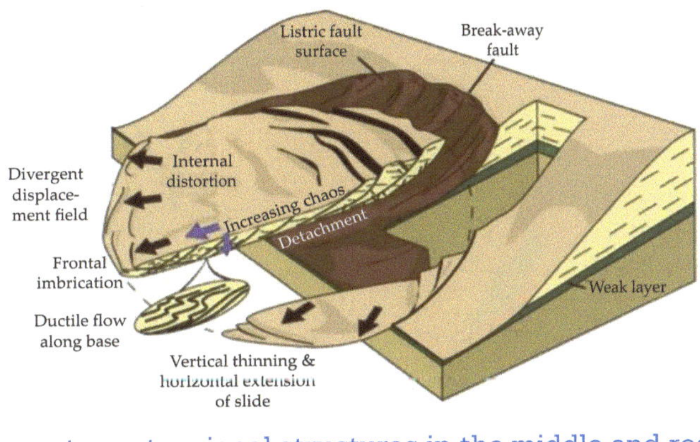

Figure 5.12. Slumping generates extensional structures in the middle and rear part, and contractional structures in the frontal part.

Source: https://www.geokniga.org/bookfiles/geokniga-structural-geology_1.pdf.

5.6. RIFTING

Where tectonic forces pull apart the crust, a rift forms. A rift can form for a variety of reasons, and there are two end-member models that describe this process: active and passive rifting. According to the active rifting model, doming and additional tensile stresses are added to the domed area as a result of rising hot mantle material or plumes in the asthenospheric mantle that create the rift. As a result, there is a rift that is more magmatism-dominated than possibly much extension. According to the passive rifting model, far-field stresses connected to plate tectonics cause the rift to form. Reactivated contractional structures along former orogenic zones are examples of zones of inherited weakness in the lithosphere where passive rifts typically form. Both of these models are often present in many natural rifts. In a simplified scenario, large-scale crustal doming could be the cause of initial rifting (Figure 5.13a). At this point, steep fracture systems start to form and have the potential to go deep enough to encourage the formation of magma and mantle intrusions. Stretching is the next and primary stage, during which the crust thins vertically and extends laterally (Figure 5.13b). This stage is when major faults and fault blocks form. The last stage of subsidence is reached once stretching stops (Figure 5.13c). Postrift sediments are deposited, the basement is deepened, and the crust cools. The majority of faults are restricted to those created by differential compaction.

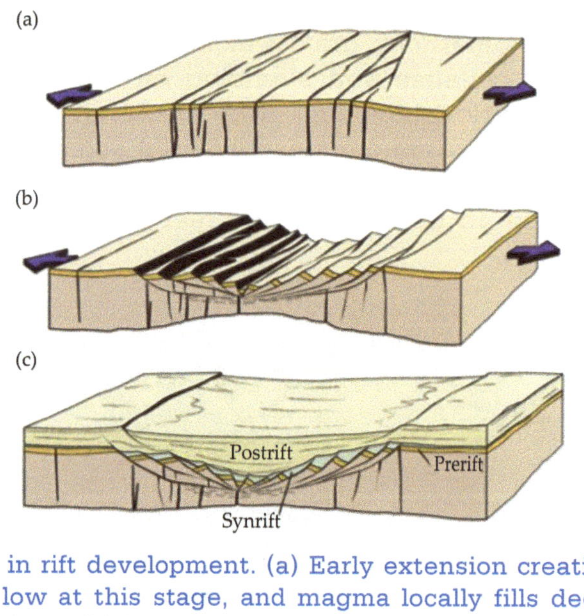

Figure 5.13. Three stages in rift development. (a) Early extension creating or rejuvenating deep-going fractures. Strain is low at this stage, and magma locally fills deep fractures as dikes. (b) The stretching phase, during which major fault complexes and arrays form. Synrift sediments not shown. (c) Postrift subsidence and sedimentation. Compactional faults in postrift sequence due to differential compaction.

Source: Haakon Fossen, structural geology, ISBN 978-1-10705764-7.

The sedimentary record of a rift reflects its extensional development. The sequence of layers deposited before extension is known as the prerift sequence, while the layers deposited during rifting make up the synrift sequence. Within the synrift sediments, there are variations in thickness and facies across growth faults, with characteristics such as hanging-wall thickening and footwall thinning or non-deposition. The postrift sequence is influenced by the geometry of fault blocks and thermal subsidence following the end of extension.

One has to do with the domino model, in which the ultimate rift width is predetermined at the outset in order to allow the faults that form the dominoes to accumulate displacement and rotate with the domino blocks as extension builds up. In the alternative end-member model, there is a significant footwall collapse. According to this model, faults get younger as one moves away from the central graben as the rift widens from the rift axis. Naturally occurring rifts are intricate, incorporating aspects from multiple models. In general, rift development is influenced by a number of variables, such as the crust's mechanical structure (the distribution and orientation of preexisting weak structures) and the mantle's processes and thermal structure.

5.7. HALF-GRABENS AND ACCOMMODATION ZONES

Most rifts have a master fault on one of the flanks, and symmetric rifts are uncommon. The master fault's location may appear to flip as we proceed along a rift, shifting from one side to the other. This is a typical rift system feature that has been thoroughly studied by Bruce R. Rosendahl and associates in the Lake Tanganyika region of the East African rift system. Here, the rift develops into a sequence of half-grabens that dip in opposite directions.

Each half-graben is shaped like a half-moon and has a curved geometry. Half-grabens usually take over where one ends and another dips in the opposite direction. Basinal highs (horsts) or lows (grabens) may form depending on how the grabens and secondary faults are arranged in their hanging walls (Figure 5.14). This kind of half-graben overlap structure is occasionally referred to as an accommodation zone. It is important to note that, as Figure 5.14 illustrates, only the zone between two oppositely dipping half-grabens can yield (nearly) symmetric sections.

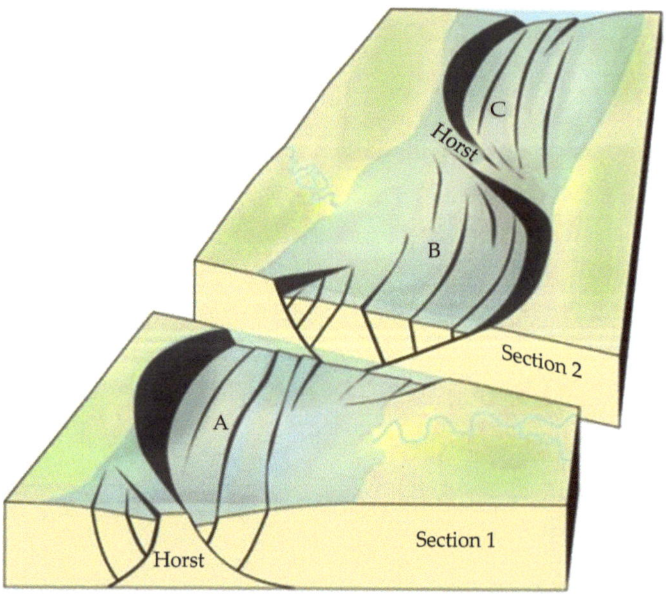

Figure 5.14. Rift system composed of interfering and overlapping halfgrabens. The overlaps were called accommodation zones by Rosendahl and co-workers in the 1980s, based on observations from the East African Rift. Different types of half-graben arrangement occur. Accommodation zones may contain horsts (section 1) or grabens (section 2)

Source: Haakon Fossen, structural geology, ISBN 978-1-10705764-7.

5.8. PURE AND SIMPLE SHEAR MODELS

The pure shear and simple shear models are sometimes used to discuss crude stretching in rift zones (Figures 5.15 and 5.16). The simple shear model is sometimes referred to as the Wernicke model, after the authors who published the corresponding models in the 1970s and 1980s, while the pure shear model is also known as the McKenzie model. (McKenzie, D., 1978)

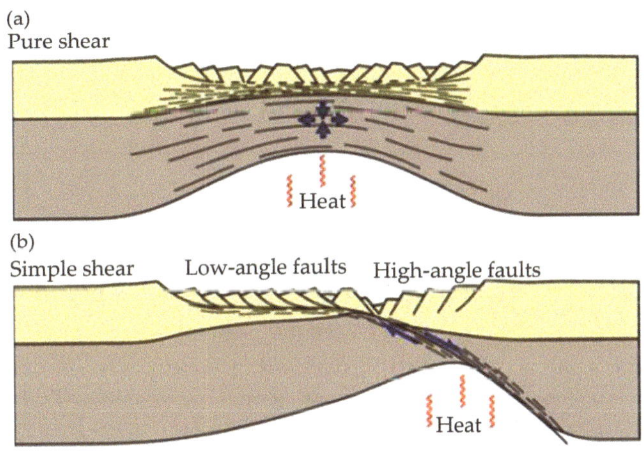

Figure 5.15. Two idealized models for crustal extension and rifting. The pure shear model is symmetric, with maximum heat underneath the middle of the rift. The simple shear model is dominated by a generally low-angle shear zone that produces an asymmetry on the rift.

Source: https://www.geokniga.org/bookfiles/geokniga-structural-geology_1.pdf.

In the pure shear model, which is the older of the two models, the cumulative effect of individual faults in the rift results in a symmetrical thinning of the crust. The dominant strain is pure shear, with horizontal extension matched by vertical thinning. Plastic deformation mechanisms cause the lower crust to thin, while brittle faulting causes deformation in the upper crust.

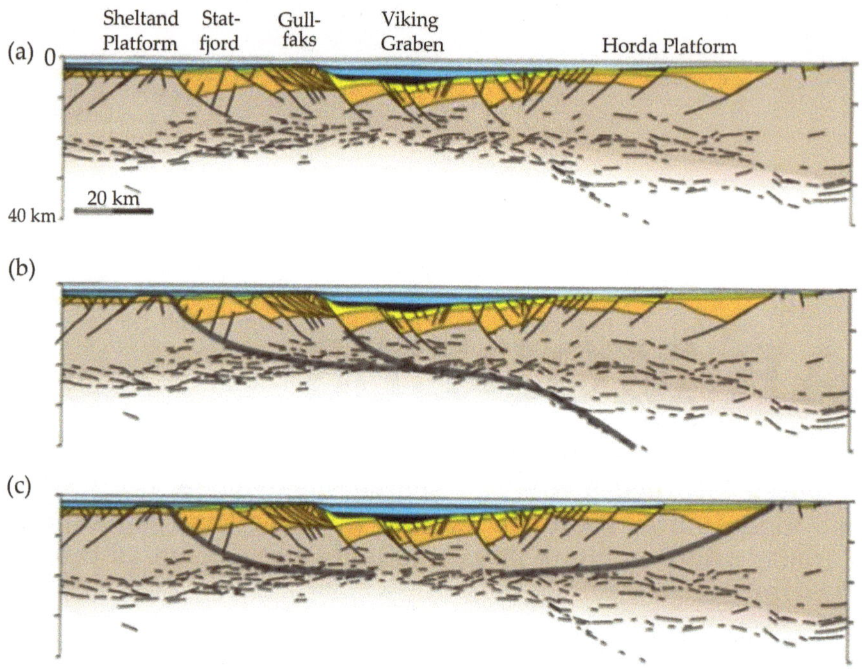

Figure 5.16. Section based on a deep seismic line across the northern North Sea. (a) This section has been interpreted in terms of simple shear (b) as well as pure shear (c) and may be considered to contain elements of both models.

Source: Haakon Fossen, structural geology, ISBN 978-1-10705764-7.

This specific model is governed by a dipping detachment fault or shear zone that crosses the crust and potentially the entire lithosphere, hence the term "simple shear." The term "simple shear model" is appropriate because the detachment involves a localized simple shear that is significant enough. The thermal structure and the two sides of a rift that is under the control of a dipping detachment differ geometrically. The greatest temperature gradient is located beneath the basin's center in the pure shear model, but it is offset in the simple shear model (Figure 5.15). This affects basin development through patterns of uplift and subsidence, and various iterations of the basic shear model produce varying outcomes.

5.9. STRETCHING ESTIMATES, FRACTALS, AND POWER-LAW RELATIONS

The amount of extension or stretching across a rift can be found in several different ways. Geologic sections based on deep seismic profiles that portray the crustal thinning beneath the rift can be used. If area conservation is assumed, i.e., no material transport in or out of the section and no assimilation of crustal material by the mantle, then extension can be estimated by restoring the section to the point where the crustal thickness is constant and equal to that of the present rift margins.

Summing fault heaves across a reference horizon provides a purely structural estimate of the extent of extension. The total extension along the section should be represented by the sum if block rotation is minimal. Nevertheless, a discrepancy is usually discovered between the strain estimate computed from area balancing of the crust and that derived from summing fault heaves. The summing of heaves typically yields the lower stretching estimate. The two estimates don't always match up by a factor of two. Why is there this disparity?

One potential reason for this discrepancy is that the lower section of the crust being pulled apart may be absorbed into the mantle. This means that the assumption of keeping the same total area does not apply, resulting in an overestimation of the extension. However, the absorption of crustal material does not appear to be significant enough to account for the difference, except in cases of active rifts with a large amount of magma present.

When the fault displacement model is examined closely, it is clear that the subseismic faults' contribution, i.e., Fault heaves are summed from a geologic section that excludes structures that are too small to show up there. We might assume that faults that are too tiny to show up on a geologic section or seismic line would not really matter. But if there are a lot of these little errors, they can add up to a substantial amount of extension. However, if the faults are below the resolution of observation, how can we adjust for fault extension?

During the 1980s and 1990s, researchers extensively studied a problem that led them to realize that fault offsets in many fault populations follow a power-law relation. To study this phenomenon, data on fault displacement was collected from seismic lines, geologic maps, and outcrops. The data was then plotted on cumulative plots with logarithmic axes, with fault offset measured for each fault along a chosen horizon. The offset values were sorted on a spreadsheet and plotted on the horizontal axis, while the cumulative number was plotted on the vertical axis. The largest offset was plotted first with a cumulative number of 1, followed by the second largest offset with a cumulative number of 2, and so on.

In cumulative loglog diagrams, real data rarely define a perfectly straight line, and Figure 5.17's example is no different. Nonetheless, a central straight segment that is terminated by curved segments at both ends may be identified (between 10 and 100 m in Figure 5.17). The underrepresentation of very small displacement faults (the resolution problem) and very large displacement faults (the profiles do not always intersect the largest faults in an area) is the cause of the truncation. This phenomenon is occasionally denoted as the censoring effect. The straight segment can be extended upward into the domain of small displacements in Figure 5.17, which would compensate for the lower truncation effect. However, it cannot last forever, and this relationship will eventually come to an end. This may happen in porous rocks and sediments when the grain size of the deformed rock approaches, if not before.

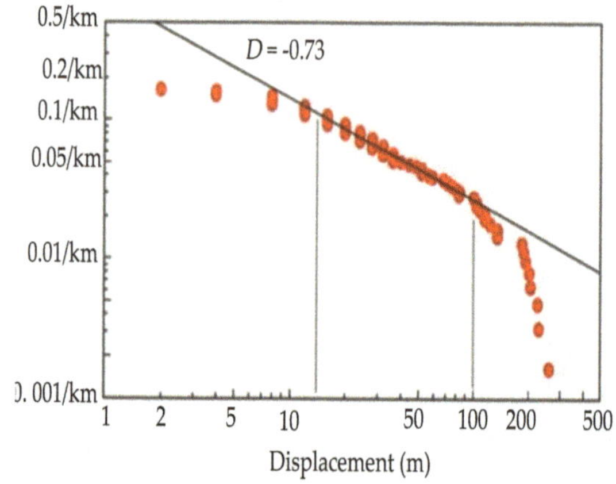

Figure 5.17. Log-log plot of fault displacement against cumulative number (normalized against length). The data define a fairly straight line in the central part. Undersampling of small and large faults explains the deviation from the straight line at each end. Data from a subpopulation of faults in the North Sea Gullfaks Field.

Source: Haakon Fossen, structural geology, ISBN 978-1-10705764-7.

For instance, by balancing the crustal area (crustal thinning), the Mesozoic extension across the northern North Sea rift is estimated to be nearly 100 km. A 50 km extension is indicated by summing the fault heaves from regional seismic line interpretations. If the power-law relation can be extended to small subseismic faults, it has been demonstrated that subseismic faults can explain most or all of this difference.

5.10. PASSIVE MARGINS AND OCEANIC RIFTS

The entire crust will fracture and be replaced by oceanic crust if a continental rift is allowed to spread sufficiently. The rift, which is now in oceanic crust, is then marked by the establishment of a passive margin on both sides. Stretching associated with the late Jurassic–early Cretaceous rift stopped at about 150% ($\beta = 1.5$) in the Viking Graben in the northern North Sea. When stretching exceeds 1.5, magmatism and volcanism typically occur first, followed by increasingly more magmatism until oceanic crust begins to form at a stretching factor of about 3. Passive margin cross-sections usually display a rotated fault block bounded by normal faults that dip oceanward and flatten or terminate at the base of the upper crust, which is brittle. This rotation of fault blocks is a consequence of the previous history of rifting.

Limited seismic activity occurs in passive margins following periods of rifting. In these settings, fault movement is primarily influenced by gravity, leading to slumping and the development of large extensional fault systems located on weak salt or clay layers, indicated by a black layer in Figure 5.18.

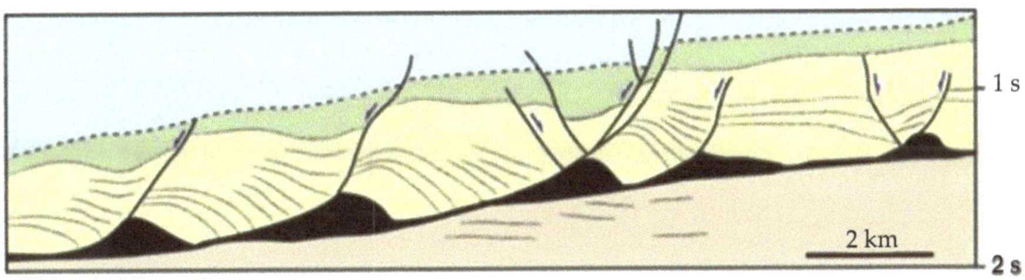

Figure 5.18. Gravity-driven synsedimentary extension above a lowangle detachment in the Kwanza Basin on the passive west African continental margin (Angola). Fault blocks are sliding on a thin and weak salt layer and are therefore detached from the substrata. The salt is flowing plastically and accommodates area problems caused by the fault block rotation.

Source: https://www.geokniga.org/bookfiles/geokniga-structural-geology_1.pdf.

Passive continental margins slowly lower and accumulate clastic sediments, while oceanic rifts experience significant tectonic activity. One key distinction between continental and oceanic rifts is the greater amount of magma and heat present in the latter. Moreover, erosion of fault crests in oceanic rifts is minimal, except for instances of gravity sliding.

The combination of hot magma and thin lithosphere in the oceanic rift area creates an elevated structure with a narrow graben along the central axis. This high elevation releases potential energy through listric normal faults. Mid-oceanic ridges display low-angle detachments with normal movement, resembling metamorphic core complexes found in areas of continental extension. Oceanic metamorphic core complexes expose mantle rocks through submarine windows, resulting from tectonic unroofing by normal faulting. This scenario is similar to Figure 5.8, but without erosion.

The inaccessibility of oceanic ridges hinders our understanding of their structural geology, but recent, high-quality seismic and deep-ocean topographic data are shedding light on the structural processes that occur there.

5.11. OROGENIC EXTENSION AND OROGENIC COLLAPSE

Extension is not restricted to passive margins and rift zones. In orogenic zones where plate convergence has stopped and in active mountain belts, there are some of the most striking examples of extensional faults and shear zones. The Wilson cycle has multiple stages, one of which is orogeny. Put another way, orogenic belts usually rift again at a later stage, having been built on older divergent plate boundaries or rifts. Back-arc rifting is the mechanism by which extension happens in the early phases of a typical orogenic cycle, when there is still an ocean between two converging continents (Figure 5.19a). Stretching also happens in the upper oceanic crust as it passes through the island arc's subduction zone.

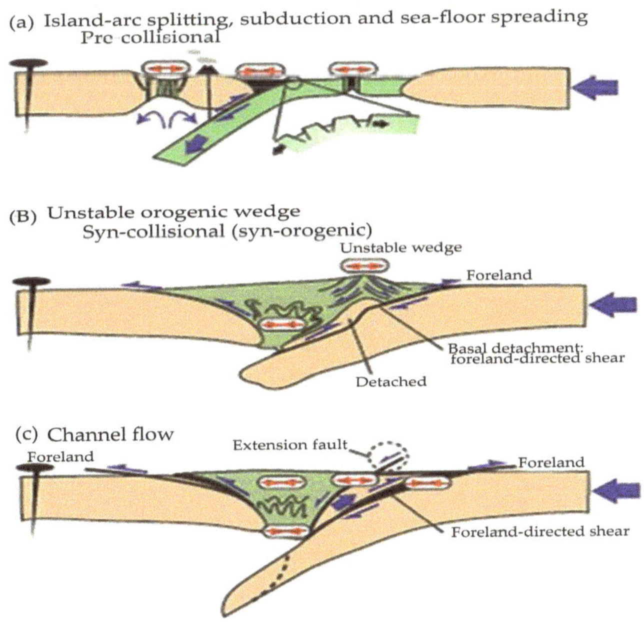

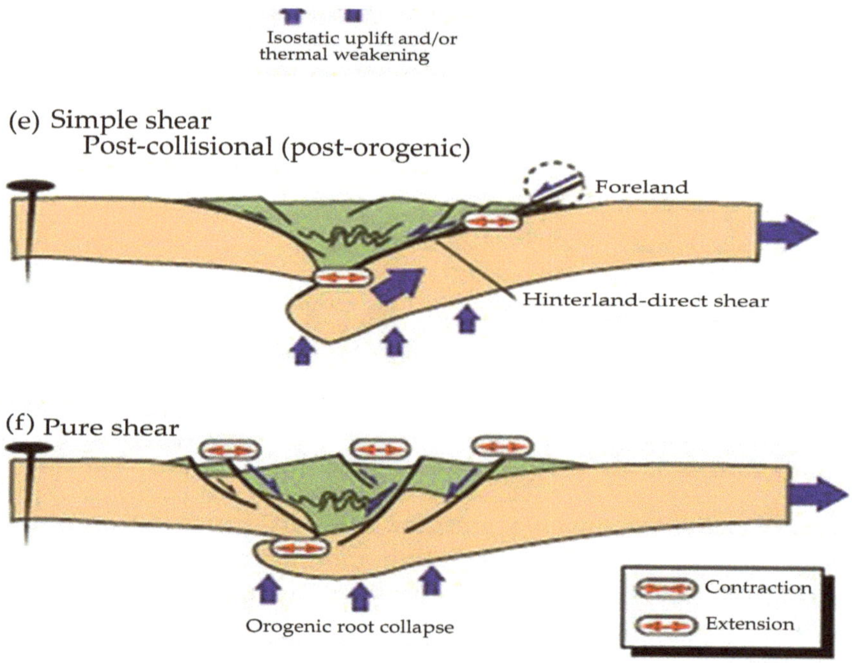

Figure 5.19. Different types of extension connected with an orogenic cycle.

Source: Haakon Fossen, structural geology, ISBN 978-1-10705764-7.

Extensional faults and shear zones may form in the orogenic wedge where and when the wedge becomes unstable during the later stages of continent-continent collision. The wedge thickens excessively and produces shear zones or normal faults in response to the instability caused by the incorporation of a large basement slice (Figure 5.19b). A detached and heated basement slice has been proposed to have low enough density to ascend buoyantly with the formation of a thrust fault on its lower side and a normal fault on its upper side, with reference to the active Himalayan orogen (Figure 5.19c). The basement slice's geometry, which thins the hinterland, would encourage the extrusion of these slabs toward the foreland.

The operation of this model is illustrated in Figure 5.20. The slice could potentially be either a rigid block or have a soft, internally flowing composition. The process of light and hot basement material being extruded in the hinterland and flowing towards the foreland through low-viscosity channels is known as channel flow. The top of the channel creates a shear zone, developing within a regime of overall contraction.

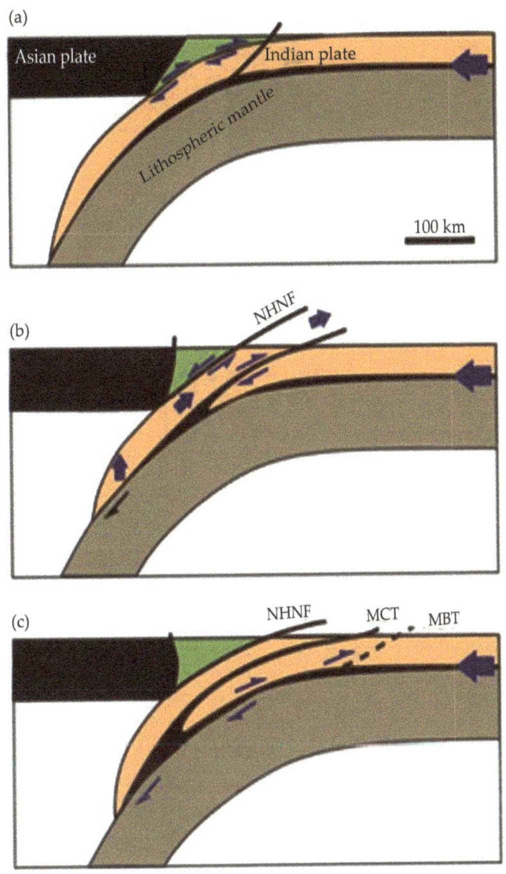

Figure 5.20. Model of the formation of a major normal fault (NHNF) in the Himalayan orogen. A slice of the continental crust is detached and uplifted by a combination of compression and buoyancy. The slice is underlain by a thrust (MCT) and overlain by a normal fault (NHNF).

Source: https://www.geokniga.org/bookfiles/geokniga-structural-geology_1.pdf.

Alterations to the thermal structure of the lower crust and the lithospheric mantle are the subject of a third model for synconvergent extension. Warm, subducted crustal material is produced when two continents collide. Along extensional faults and shear zones, heating weakens the crust, possibly to the point where it collapses under its own weight (Figure 5.19d). This model can be referred to as gravitational orogenic collapse.

Although the mountainous upper portion of an orogenic structure is typically thought of as collapsing due to gravitational forces, it may also be crucial to take into account the processes occurring at the crust's base. The force of continental subduction can push the crust down to a depth of approximately one hundred kilometers. While the lower portion of the plate, the lithospheric mantle, is (for a while) colder than its surroundings and possibly denser, a large portion of the subducted crust is lighter than the surrounding lithosphere and therefore buoyant.

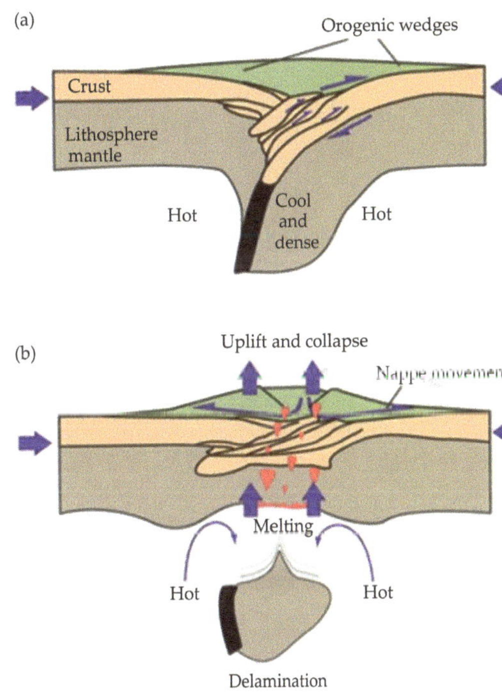

Figure 5.21. The delamination model causing orogenic collapse and orogenic root collapse. (a) Cold and dense lower root pulls down continental crust. (b) Delamination and descent of dense root by delamination, causing upward collapse of deep continental crust, uplift and collapse in the upper part of the collision zone (in the orogenic wedge) with high-level transport of rocks from the hinterland toward the foreland.

Source: Haakon Fossen, Structural Geology, ISBN 978-1-10705764-7.

Any oceanic crust that is joined to the continent's leading edge shares the same characteristics. Furthermore, the density of rocks in the root zone may rise due to phase transitions, such as eclogitization. An orogenic root's densest portion can probably be extracted by sinking into the mantle below. Delamination model (Figure 5.21) this model causes partial melting, increased magmatic activity, and faster heating of the (remaining) orogenic root.

Orogenic uplift results from the delamination of a dense root, which releases the buoyant portion of the root (Figure 5.19d). This model, which collapses the root upward, is known as an orogenic root collapse (Figure 5.19f). This may occur as a result of the root spreading laterally along the crust's base due to a coaxial strain mechanism. It most likely could also influence channel flow. The upper orogenic edifice collapses gravitationally as a natural result of a root collapse and regional uplift. Root delamination, root collapse, and related models have been used to explain what is happening in the Tibet Plateau at the moment.

5.12. POSTOROGENIC EXTENSION

The sense of movement on the basal thrust is always toward the foreland during the divergent history of an orogenic belt, as exemplified by the Caledonian example depicted in Figure 5.22a. The orogen kinematically enters the divergent or postorogenic stage when the sense of shear is reversed and the orogenic wedge advances toward the center of the collision zone (Figure 5.22b). At all crustal levels, extensional deformation is dominant at this point. Reversal of the basal thrust and higher-level thrusts within the orogenic wedge are two examples of postorogenic extensional deformation (Figures 5.19e and 5.22b).

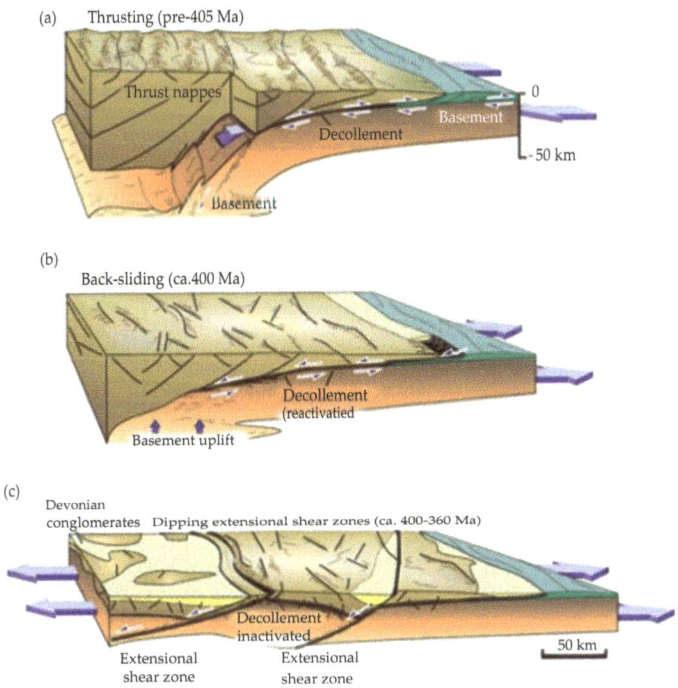

Figure 5.22. The development of low-angle extensional faults and shear zones in the southern Scandinavian Caledonides. (a) Emplacement of nappes during plate convergence. (b) Back-sliding of the orogenic wedge, causing reversal of the sense of shear along the basal detachment. (c) Formation of hinterland-dipping shear zones and faults that cut both the thrusts and the basement.

Source: Haakon Fossen, structural geology, ISBN 978-1-10705764-7.

Another method of extension occurs when hinterland-dipping shear zones are created by cutting through the crust. These zones can form after the hinterland has been uplifted and orogenic thrusts have been rotated to orientations that are not conducive for extensional reactivation. These hinterland-dipping shear zones usually impact the entire crust, transect it, and further rotate the reversed basal thrust. Well-defined instances of these structures have been observed and studied in the Scandinavian Caledonides.

CASE STUDY

THE EAST AFRICAN RIFT SYSTEM (EARS)

Overview

The East African Rift System (EARS) is one of the most prominent extensional regimes on Earth. It extends from the Afar Triple Junction in the northeast to Mozambique in the south, covering a distance of approximately 3,000 kilometers. The rift system is a prime example of continental rifting, where a continent is being pulled apart.

Geological Characteristics

1. **Rift Valleys:**
   - The EARS is characterized by a series of rift valleys, which are elongated depressions formed by the stretching and thinning of the crust.
   - Major rift valleys in the EARS include the Eastern Rift (Kenya Rift) and the Western Rift (Albertine Rift).
2. **Normal Faulting:**
   - Normal faults are prevalent in the EARS, resulting from the extensional forces. These faults create the characteristic rift valley topography, with uplifted flanks and subsiding central blocks.
3. **Volcanism:**
   - Volcanic activity is common along the rift, with numerous volcanoes such as Mount Kilimanjaro, Mount Kenya, and the Virunga Mountains.
   - The rift is associated with significant geothermal activity due to the upwelling of magma from the mantle.
4. **Seismicity:**
   - The EARS experiences frequent seismic activity, including earthquakes of varying magnitudes. This seismicity is a direct result of the ongoing tectonic extension.

Geological Processes

1. **Crustal Extension:**
   - The primary process in the EARS is crustal extension, where tectonic forces pull the lithosphere apart. This extension is driven by the divergent movement of the African Plate and the Somali Plate.

2. **Faulting and Subsidence:**
   - As the crust extends, it fractures, creating normal faults. The central blocks of these faults subside, forming rift valleys, while the flanks are uplifted.
3. **Magma Upwelling and Volcanism:**
   - The thinning crust allows magma from the mantle to rise, leading to volcanic eruptions. This magma upwelling also contributes to the formation of geothermal fields.
4. **Sedimentation:**
   - Rift valleys often become basins for sediment accumulation. Lakes such as Lake Tanganyika and Lake Malawi have formed in the rift valleys, collecting sediments over time

Implications

1. **Geothermal Energy:**
   - The geothermal activity in the EARS provides significant potential for geothermal energy production, which is being harnessed in countries like Kenya.
2. **Mineral Resources:**
   - The rift system is associated with various mineral resources, including geothermal minerals and possibly hydrocarbons in sedimentary basins.
3. **Seismic Hazards:**
   - The frequent seismic activity poses risks to communities living in and around the rift. Understanding the tectonic processes can help in hazard mitigation.
4. **Biodiversity:**
   - The unique geological features of the EARS create diverse habitats, contributing to the region's rich biodiversity. The rift valleys and associated lakes support unique ecosystems.
5. **Human Settlement:**
   - The fertile soils and availability of water in rift valleys have historically attracted human settlements. However, the geological instability poses challenges.

Conclusion

The East African Rift System exemplifies the dynamic processes and features of an extensional regime. It highlights the interplay between tectonic forces, geological formations, and human activities. Understanding such extensional regimes is crucial for resource management, hazard mitigation, and appreciating the geological complexity of our planet.

CLASS ACTIVITY

EXPLORING EXTENSIONAL REGIMES

Objective:

Understand the characteristics, causes, and consequences of extensional regimes in the Earth's crust. Analyze real-world examples to identify common features and geological structures associated with these regimes.

Materials Needed:

- World map or geological map
- Access to online geological databases or textbooks
- Paper and pens for notes and drawings
- Computer with internet access

Step

Explain the key characteristics of extensional regimes, including crustal thinning, faulting, and the formation of rift valleys and basins.

Discuss the geological processes involved, such as tectonic plate movements and mantle dynamics.

SUMMARY

- Extensional deformation structures in the form of normal faults and shear zones or detachments are extremely common in many tectonic settings and also in non-tectonic settings (e.g., gravitational sliding). They control rift structures, post-contractional stages of mountain belt cycles and are important also during orogeny.
- The easiest explanation is that most or all rocks have an anisotropy inherited from earlier phases of deformation. Thus, a simple explanation for very steep faults is that they represent reactivated joints or strike-slip faults. Recall that joints are close to vertical because they form perpendicular to σ_3, which tends to be horizontal in the upper crust.
- Fault blocks seldom or never behave as rigid objects, and certainly not in rift systems with relatively weak rocks or unconsolidated sediments. Besides, we have already seen that faults occur in populations where fault displacement varies along each fault. The rigid domino model requires all faults to be of equal length and displacement, not showing any displacement gradient.
- Low-angle normal faults have been a matter of discussion among structural geologists since they were mapped and interpreted in significant numbers in the 1970s and 1980s. They are very effective in accommodating extensional strain, as opposed to steep normal faults, and hence make kinematic sense. A number of low-angle normal faults have been described from various parts of the world, but it is still not clear how and why they form.
- Classical models that build on Anderson's idealized assumption of vertical and horizontal principal stresses and isotropic rocks, the Mohr–Coulomb fracture criterion and Byerlee's law, predict that normal faults lock up and become inactive when they obtain a low dip. In simple terms, this happens because the shear stress on a fault is reduced as its angle to σ_1 increases.
- Footwall collapse is common where huge fault blocks form and rotate in rift systems, and the elevated crests of these blocks collapse under the influence of gravity and tectonic forces. In some cases gravity alone causes collapse of the footwall.
- A rift forms where the crust is pulled apart by tectonic forces. There can be several factors leading to the formation of a rift, and two end-member models are known as active and passive rifting. In the active rifting model, the rift is generated by rising hot mantle material or plumes in the asthenospheric mantle, causing doming and adding tensile stresses to the domed area.

REVIEW QUESTIONS

1. What are the primary characteristics of extensional faults, and how do they differ from compressional faults?

2. Explain how fault systems interact within a tectonic setting and describe an example of a well-known fault system.
3. Describe the formation process of a metamorphic core complex and the role of low-angle faults in its development.
4. What is a ramp-flat-ramp geometry in fault systems, and how does it influence the overall structure and deformation patterns of an area?
5. Differentiate between footwall collapse and hanging-wall collapse. What geological features or evidence would you look for to identify each type of collapse?
6. What are the main stages of rifting, and how does the process lead to the formation of new ocean basins?
7. How do half-grabens form, and what is the significance of accommodation zones in the development of extensional basins?
8. Explain the concept of stretching estimates in extensional tectonics. How do fractals and power-law relations help in understanding the distribution and scaling of faults?

REFERENCES

1. Brun, J. P., & Choukroune, P. (1983). Normal faulting, block tilting, and décollement in a stretched crust. Tectonics, 2(4), 345–356.
2. Ebinger, C. J. (2005). Continental break-up: The East African perspective. Astronomy & Geophysics, 46(2), 2.16–2.21.
3. England, P. C., & Houseman, G. A. (1988). The mechanics of the Tibetan plateau. Philosophical Transactions of the Royal Society of London Series A, 326(1589), 301–320.
4. Fletcher, J. M., Bartley, J. M., Martin, M. W., Glazner, A. F., & Walker, J. D. (1995). Large-magnitude continental extension: An example from the central Mojave metamorphic core complex. Geological Society of America Bulletin, 107(11), 1468–1483.
5. Jackson, J., & McKenzie, D. (1983). The geometrical evolution of normal fault systems. Journal of Structural Geology, 5(5), 472–483.
6. Kusznir, N. J., & Ziegler, P. A. (1992). The mechanics of continental extension and sedimentary basin formation: A simple-shear/pure-shear flexural cantilever model. Tectonophysics, 215(1–2), 117–131.
7. McKenzie, D. (1978). Some remarks on the development of sedimentary basins. Earth and Planetary Science Letters, 40(1), 25–32.
8. Nur, A., Ron, H., & Scotti, O. (1986). Fault mechanics and the kinematics of block rotations. Geology, 14(8), 746–749.
9. Ring, U. (2014). The East African Rift System. In R. D. G. (Ed.), Regional Geology and Tectonics: Principles of Geologic Analysis (pp. 141–161). Elsevier.

10. Ring, U., & Schlische, R. W. (Eds.). (2003). Extensional Tectonics and Structural Styles (Vol. 208). Geological Society, London, Special Publications.
11. Scott, R. J., & Lister, G. S. (1992). Detachment faults: Evidence for a low-angle origin. Geology, 20(9), 833–836.
12. Wernicke, B. (1985). Uniform-sense normal simple shear of the continental lithosphere. Canadian Journal of Earth Sciences, 22(1), 108–125.
13. Wernicke, B., & Axen, G. J. (1988). On the role of isostasy in the evolution of normal fault systems. Geology, 16(9), 848–851.
14. Wernicke, B., & Burchfiel, B. C. (1982). Modes of extensional tectonics. Journal of Structural Geology, 4(2), 105–115.

CHAPTER

6

Fracture and Brittle Deformation

LEARNING OBJECTIVES

After studying this chapter, you will be able to:

- Describe brittle deformation mechanisms
- Explain about brittle deformation
- Discuss about tensile cracking
- Explain the processes of brittle faulting
- Illustrate the formation of shear fractures
- Understand predicting initiation of brittle deformation

KEY TERMS FROM THIS CHAPTER

Axial stress
Cataclasis
crack surface
extension fractures
Fracture surface
Host rock
Shear fracture
Brittle fault
Cataclastic deformation
Ductile strain
Failure envelope
Granular flow
Particulate flow
Shear stress

INTRODUCTION

Fracture and brittle deformation are critical concepts in the fields of geology, materials science, and engineering, describing how materials break under stress. These processes play a vital role in understanding the behavior of natural and synthetic materials under various conditions, impacting everything from structural integrity to earthquake dynamics.

Fracture refers to the breaking or cracking of a material when subjected to stress, leading to the formation of new surfaces. It is typically categorized into two types: brittle and ductile. In brittle fracture, materials break suddenly without significant prior deformation, releasing stored energy rapidly. This type of fracture is characterized by a lack of plastic deformation, resulting in a sudden and often catastrophic failure. In contrast, ductile fracture involves significant plastic deformation before breaking, allowing for energy dissipation over a longer period.

Brittle deformation, on the other hand, occurs when a material deforms elastically up to a point and then fractures without significant plastic deformation. This process is prominent in materials that are strong but lack ductility, such as certain metals at low temperatures, ceramics, and rocks. In geological contexts, brittle deformation is significant in the formation of faults and fractures in the Earth's crust, which can lead to earthquakes and influence fluid flow and mineral deposition in rocks.

The mechanics of brittle deformation are governed by the principles of stress and strain. When a material is subjected to stress beyond its elastic limit, it may experience microcracking and eventual macroscopic failure. The conditions under which brittle deformation occurs are influenced by factors such as temperature, pressure, strain rate, and the presence of flaws or defects. For example, low temperatures and high strain rates tend to promote brittle behavior in materials.

6.1. BRITTLE DEFORMATION MECHANISMS

An unfractured rock may develop permanent strain via plastic flow if the differential stress in the rock surpasses a particular threshold. When the rock reaches its rupture strength, it will fracture and deform in the frictional or brittle regime. Grain reorganization and crushing occur during brittle fracturing, resulting in more localized strain (displacement).

The physical circumstances that favor brittle deformation mechanisms like frictional sliding along grain contacts, grain rotation, and grain fracture are known as the brittle regime. It is sometimes necessary to describe the degree of fracture present in a deformed rock, and a distinction is made between brittle deformation that involves fracture and that which does not. In porous rocks and sediments that are relatively poorly consolidated, frictional deformation without the formation of fractures usually takes place (soils). As seen in Figure 6.1a, frictional slip happens along preexisting grain boundaries in these rocks and sediments, and pore space permits grains to move in relation to one another. As a result, in order to account for frictional grain boundary slip, the grains translate and rotate; this process is known as particulate flow or granular flow. Grain boundary sliding is affected by friction, as it always is in the brittle regime; for this reason, the process is known as frictional sliding. This indicates that for frictional sliding to happen, a specific friction-controlled resistance against sliding needs to be overcome. This is not to be confused with the non-frictional sliding of grain boundaries that occurs in the plastic regime. The friction between individual sand grains in loose sand controls the angle of repose. The angle of repose increases with increasing intergranular friction. In this instance, grain contact areas are subject to a vertical force from gravity, and the orientation of the surfaces will determine the shear stress.

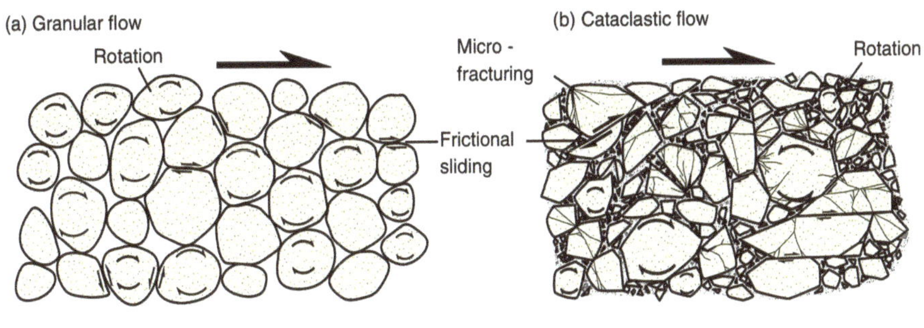

Figure 6.1. Brittle deformation mechanisms. Granular flow is common during shallow deformation of porous rocks and sediments, while cataclastic flow occurs during deformation of well-consolidated sedimentary rocks and non-porous rocks.

Source: Haakon Fossen, 2010, Structural Geology, ISBN-13 978-0-511-77282-5

Grain friction sliding can occur in large, diffuse areas across a rock volume as well as in narrow, millimeter- to decimeter-wide zones or bands. A ductile shear zone created by granular flow allows lamination to be continuously traced from one side of the zone to the other. Brittle deformation mechanisms control this kind of ductile shear zone.

In other situations, during deformation, new fractures develop. This always occurs in non-porous rocks during brittle permanent deformation, but it can also happen in porous rocks if the stress on grain contact areas rises to a certain level. Intragranular fractures, or fractures confined to a single grain, are frequently found in porous rocks. Intergranular fractures, which are found in brittlely deformed low-porosity or non-porous rocks, are fractures that span several grains. Cataclasis is the term used to describe the breaking and crushing of grains along with the frictional sliding along grain contacts and rotation of grains. In thin zones along slip or fault surfaces where there is a significant reduction in grain size, intense cataclasis takes place. In slightly wider brittle or cataclastic shear zones, more moderate cataclastic deformation may take place. In this instance, the broken pieces from the crushing of the grain flow during shearing. We call this process "cataclastic flow."

While cataclastic flow also includes grain fracturing or cataclasis, particulate flow also involves grain rotation and frictional sliding between grains. At the mesoscopic level, both can result in structures that seem ductile. Pulverization is the term for strong grain crushing that has also been observed in the absence of shear offset. Although the pulverization process is poorly understood, it appears to be associated with strain rates exceeding >100 s^{-1} and could be linked to significant earthquake events that result in high rupture rates.

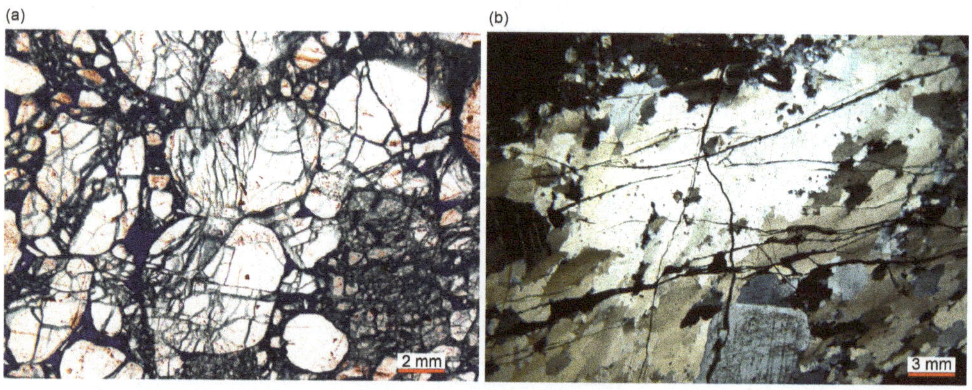

Figure 6.2. (a) Intragranular fractures in cataclastically deformed porous sandstone (Mesa Verde Group, Salina, Utah). Dark blue color is epoxy-filled pore space. (b) Intergranular fractures in metamorphic rock.

Source: Haakon Fossen, 2010, Structural Geology, ISBN-13 978-0-511-77282-5

6.2. TYPES OF FRACTURES

6.2.1. What is a Fracture?

According to strict definitions, a fracture is any planar or subplanar discontinuity that forms as a result of external factors and is extremely narrow in one dimension relative to the other two tectonic) or internal stress, such as residual or thermal. Where rocks or minerals are broken, there are discontinuities in displacement and mechanical properties known as fractures. Most fractures are characterized by a reduction in cohesion. Although they are frequently referred to as surfaces, there is always some thickness present at some scale. Shear fractures, or slip surfaces, and opening or extension fractures, or joints, fissures, and veins, are two different categories of fractures. Furthermore, fractures that close or contract can be identified.

Fractures are extremely narrow zones that are frequently regarded as surfaces and are connected to discontinuities in mechanical properties (strength or stiffness) and displacement.

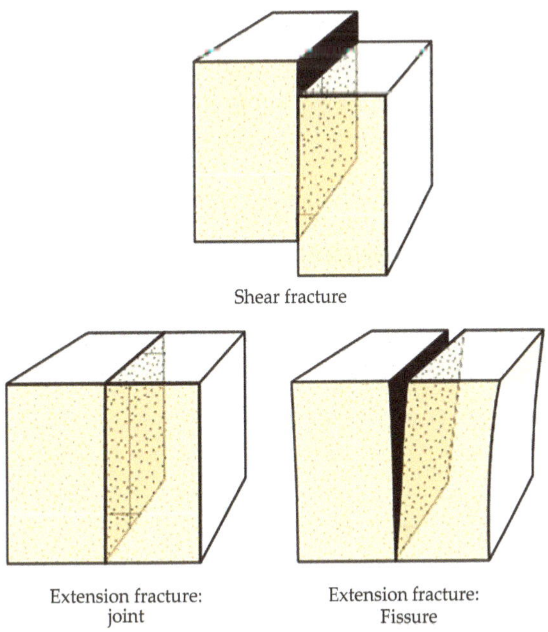

Figure 6.3. Three types of fracture.

Source: Haakon Fossen, 2010, Structural Geology, ISBN-13 978-0-511-77282-5.

A fracture where the relative movement is parallel to the fracture is known as a shear fracture or slip surface. While the term fault is more frequently limited to discontinuities with larger offset, the term shear fracture is used for fractures with small (mm- to dm-scale) displacements. Regardless of the degree of displacement, fractures with fracture-parallel movements are referred to as slip surfaces, which is consistent with the term fault's conventional usage. In literature pertaining to material science and rock mechanics, fractures are frequently referred to as cracks.

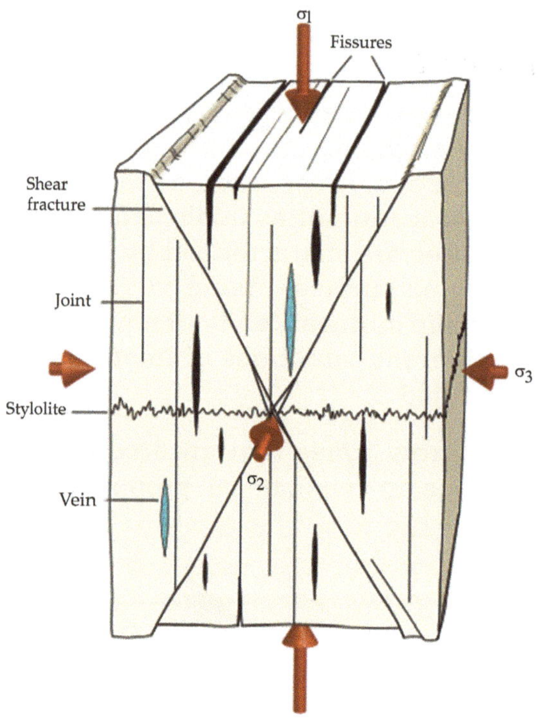

Figure 6.4. The orientation of various fracture types with respect to the principal stresses.

Source: Haakon Fossen, 2010, Structural Geology, ISBN-13 978-0-511-77282-5.

Fractures exhibiting extension perpendicular to the walls are called extension fractures. Most joints have a tiny amount of extensional displacement across the joint surfaces, which is why they are categorized as true extension fractures even though there is little to no macroscopically detectable displacement in the joints. Minerals, gas, or liquids can fill extension fractures. We call it a fissure when it is filled with fluid or air. Veins are extension fractures filled with minerals, whereas dikes are fractures filled with magma. Extension fractures can occur in veins, fissures, or joints.

Anticracks are planar features filled with immobile rock remnants that have contractional displacements across them. Stylolites are compactional structures with highly irregular surfaces as opposed to planar ones. Since stylolites neatly define one of three end-members in a complete kinematic fracture framework along with shear and extension fractures, some geologists now refer to them as contraction fractures or closing fractures. The technical literature refers to these structures as anticracks.

A convenient stage for studying aspects of fracture formation is provided by rock mechanics experiments conducted at different differential stresses and confining pressures; in fact, we will make several references to experimental rock deformation in this chapter.

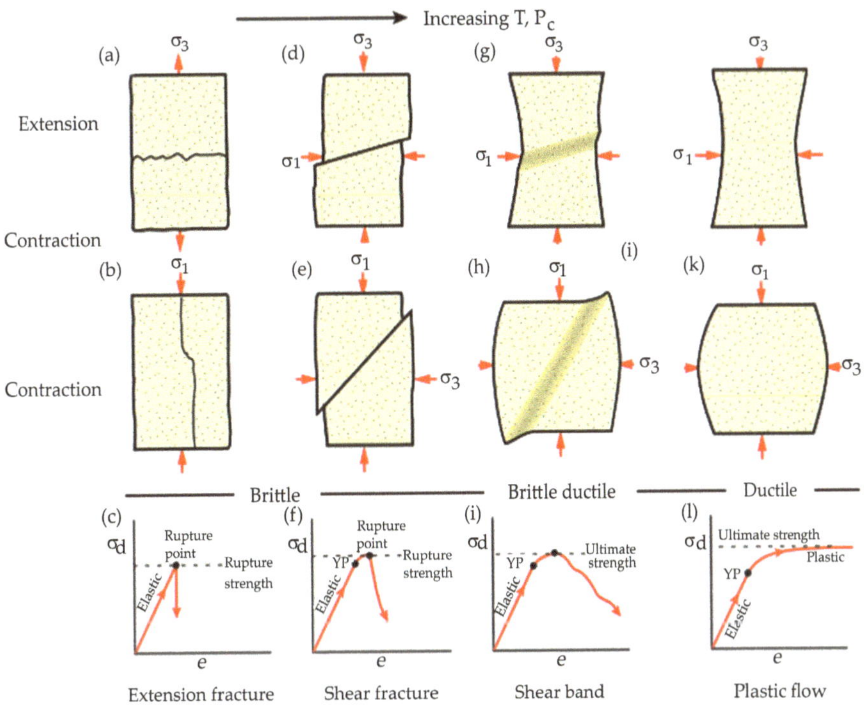

Figure 6.5. Experimental deformation structures that develop under extension and contraction. Initial elastic deformation is seen for all cases, while ductility increases with temperature (T) and confining pressure (P_c). YP, yield point.

Source: Haakon Fossen, 2010, Structural Geology, ISBN-13 978-0-511-77282-5

In a similar vein, numerical modeling has contributed significantly to our knowledge of fracture growth, especially in the area of linear elastic fracture mechanics. Three distinct modes are commonly used to categorize the displacement field of fractures or cracks in the field of fracture mechanics. The opening (extension) mode, or mode I, is characterized by displacement that is perpendicular to the crack walls. Slip (shear) perpendicular to the edge is represented by Mode II (sliding mode), and slip parallel to the crack's edge is represented by Mode III (tearing mode). It may be confusing to refer to Modes II and III cracks as separate fractures because they occur along different sections of the same shear fracture. Hybrid cracks or fractures are combinations of tension (Mode I) and shear (Mode II or III) fractures. Furthermore, contractional features like stylolites are occasionally referred to as Mode IV (closing mode). One crucial factor to consider, for example, when fluid flow through rocks is a concern, is the mode of displacement on fractures.

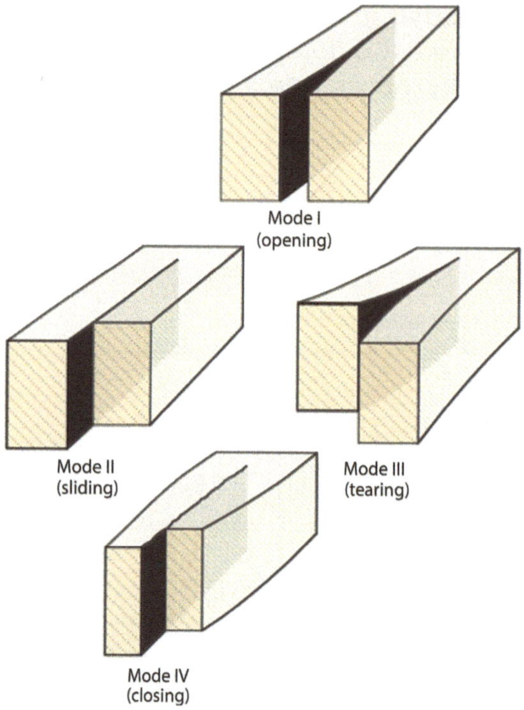

Figure 6.6. Mode I, II, III and IV fractures.

Source: Haakon Fossen, 2010, Structural Geology, ISBN-13 978-0-511-77282-5

6.2.2. Extension Fractures and Tensile Fractures

Extension fractures ideally develop perpendicular to σ_3 and thus contain the intermediate and maximum principal stresses ($2\theta = 0°$). In terms of strain, they develop perpendicular to the stretching direction under tensile conditions, and parallel to the compression axis during compression tests. Because of the small strains associated with most extension fractures, stress and strain axes more or less coincide.

Joints involve very small strains and are the most common type of extension fracture at or near the Earth's surface. Fissures, which are more asymmetrical than joints in their extension, are typical of the uppermost few hundred meters of the solid crust and can extend up to several kilometers.

Extension fractures form at low differential stress and are typical for deformation under low or no confining pressure. Extension fractures are considered to be true tensile fractures if they occur in situations where at least one stress axis is tensile. Such conditions are typically found close to the surface, where the likelihood of negative values of σ_3 is higher. They can also happen further down in the lithosphere, where the effective stress is lessened by high fluid pressure. Numerous additional joints are most likely connected to the cooling and unloading of rocks.

6.2.3. Shear Fractures

Several experiments under confined compression have shown that shear fractures typically develop at 20–30° to σ_1 and exhibit fracture-parallel slip. These experiments also demonstrate that they frequently form conjugate pairs, divided by the symbol σ_1. Shear fractures occur in environments with confining pressures and temperatures similar to those found in the upper crust. They can also form close to the transition between brittle and plastic materials, where they typically expand into broader bands or cataclastic flow zones. These shear fractures cause strain patterns that are characteristic of plastic deformation.

While extension fractures open perpendicular to σ_3, shear fractures are oblique to σ_3 by an angle that depends mostly on rock properties and state of stress.

Stress-strain curves for brittle and plastic deformation differ; the more ductile the deformation, the more plastic deformation there is before fracturing. The relationship between the strain regime (contractional or extensional) and confining pressure (depth) is also noteworthy. According to the experimental results, under extension rather than contraction, the brittle–plastic transition takes place at a higher confining pressure. Other significant variables are temperature and strain rate.

Figure 6.7. Fissures formed in Thingvellir, Iceland, along the rift axis between the Eurasian and Laurentian plates. The fissures are open extension fractures in basalt, but the vertical displacement (right-hand side down) indicates a connection with underlying faults.

Source: Haakon Fossen, 2010, Structural Geology, ISBN-13 978-0-511-77282-5

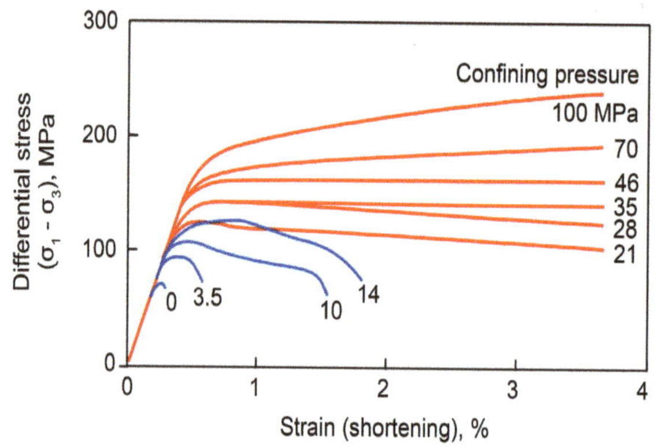

Figure 6.8. Stress–strain curves for triaxial compression of marble for a range of confining pressures. Increasing the confining pressure increases the differential stress that the rock can sustain before failure (blue curves). Above a critical confining pressure the rock retains its strength as it deforms plastically (red curves). From Paterson (1958).

Source: Haakon Fossen, 2010, Structural Geology, ISBN-13 978-0-511-77282-5

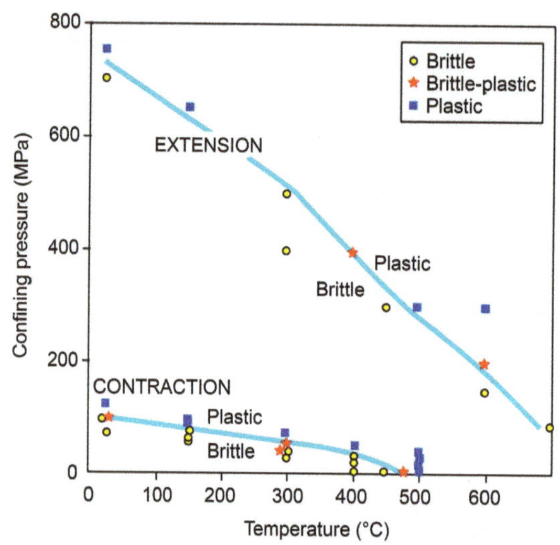

Figure 6.9. Variation of the brittle–plastic transition as a function of confining pressure and temperature for the Solenhofen limestone. From Heard (1960).

Source: Haakon Fossen, 2010, Structural Geology, ISBN-13 978-0-511-77282-5

6.3. BRITTLE DEFORMATION

Brittle deformation is the permanent alteration that happens to a solid material as a result of fractures growing or, if fractures have already formed, sliding on them. Any discontinuous surface—that is, a surface across which the material is no longer bonded—is considered a fracture according to this definition. A fracture is a vein if it fills with minerals that precipitate out of a hydrous solution; a dike forms when it fills with sedimentary or igneous rock that originates elsewhere. A joint is an unmeasurable shear displacement that occurs naturally as a fracture in rock. Joints are also known as cracks or tensile fractures due to the lack of shear involved in joint formation. On the other hand, shear fractures are mesoscopic fractures that have experienced displacement. Geologists will sometimes refer to a fracture as a "shear fracture" rather than a "fault" in order to suggest that the shear displacement on the fracture is negligible and that the shear displacement occurred simultaneously with the formation of the fracture in previously intact rock.

Figure 6.10. A geologist measuring fractures in an outcrop, near Tuross Point on the southeastern coast of Australia. The more intensely fractured rock is a fine-grained mafic intrusive and the less fractured rock is a coarse-grained felsic intrusive.

Source: https://www.geokniga.org/bookfiles/geokniga-es2e-ch06.pdf.

A fault is, broadly speaking, any surface or zone that has experienced quantifiable displacement. Geologists limit the use of the term "fault" to a fracture surface where sliding has occurred. With this limited definition of fault, we use the term "fault zone" to describe either the zone of rock that has fractured during faulting or a band of finite width across which the displacement is divided among numerous smaller faults. A band of finite width where the ductile shear strain is noticeably higher than in the surrounding rock is referred to as a "shear zone." Cataclasis (distributed fracturing, crushing, and frictional sliding of rock grains or rock fragments), diffusion, and crystal-plastic deformation mechanisms (dislocation glide, dislocation climb) can all result in movement in shear zones. A fracture, no matter what kind, does not go on forever in every direction. Whereas some fractures end inside a body of rock, others cross over its surface. The fracture front is the line that divides the portion of the rock that has fractured from the portion that has not, and the fracture trace is the line that shows where a fracture intersects with a rock body's surface. The fracture tip is the location where the fracture trace ends on the rock's surface. Certain fractures have uneven surfaces in three dimensions, while other fractures have geometries that resemble coins or blades in general.

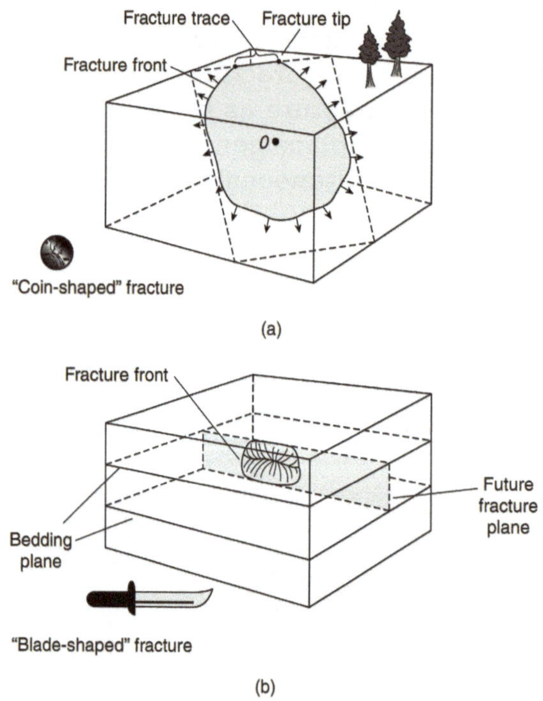

Figure 6.11. (a) A block diagram illustrating that a fracture surface terminates within the limits of a rock body. The top surface of the block is the ground surface; erosion exposes the fracture trace. Note that the trace of the fracture on the ground surface is a line of finite length (with a fracture tip at each end). The arrows indicate that this particular fracture grew radially outward from an origin, labeled O, the dot in the center of the fracture plane. (b) A blade fracture which has propagated in a sedimentary layer and terminates at the bedding plane.

Source: https://www.geokniga.org/bookfiles/geokniga-es2e-ch06.pdf.

Brittle deformation requires an examination of the atomic structure of materials. Atoms or ions make up solids; these particles are joined to one another by chemical bonds, which resemble tiny springs. Every chemical bond has an equilibrium length, and there is an equilibrium value to the angle formed between any two chemical bonds that are attached to the same atom. The bonds that hold the atoms in a solid together stretch, shorten, and/or bend during elastic strain, but they do not break! The elastic strain disappears when the stress is removed because the bonds return to their equilibrium conditions. Elastic strain, then, is fully recoverable. It is impossible for rocks to accumulate significant elastic strains; you can't expect a rock to return to its original shape after being stretched twice its original length.

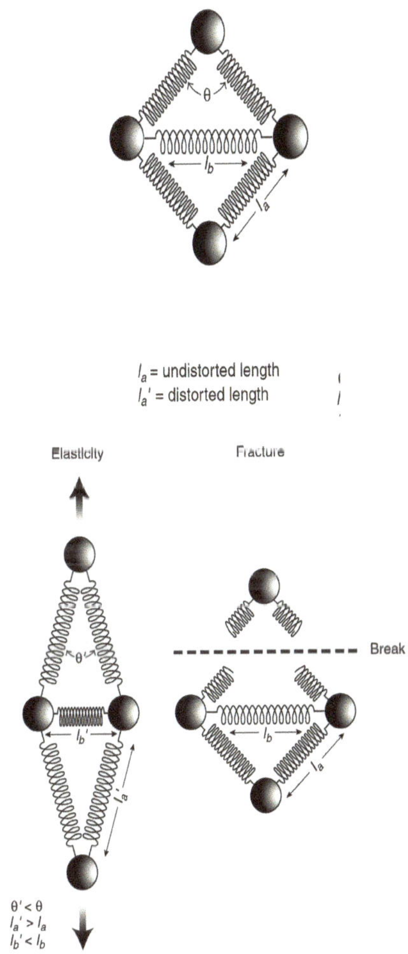

Figure 6.12. A sketch illustrating what is meant by stretching and breaking of atomic bonds. The chemical bonds are represented by springs and the atoms by spheres. (a) Four atoms arranged in a lattice at equilibrium. (b) As a consequence of stretching of the lattice, some bonds stretch and some shorten, and the angle between pairs of bonds changes. (c) If the bonds are stretched too far, they break, and elastic strain is released.

Source: https://www.geokniga.org/bookfiles/geokniga-es2e-ch06.pdf.

Table 6.1. Categories of Brittle Deformation Processes

Cataclastic flow	This type of brittle deformation refers to macroscopic ductile flow as a result of grain-scale fracturing and frictional sliding distributed over a band of finite width.
Frictional sliding	This process refers to the occurrence of sliding on a preexisting fracture surface, without the significant involvement of plastic deformation mechanisms.
Shear rupture	This type of brittle deformation results in the initiation of a macroscopic shear fracture at an acute angle to the maximum principal stress when a rock is subjected to a triaxial compressive stress. Shear rupturing involves growth and linkage of microcracks.
Tensile cracking	This type of brittle deformation involves propagation of cracks into previously unfractured material when a rock is subjected to a tensile stress. If the stress field is homogeneous, tensile cracks propagate in their own plane and are perpendicular to the least principal stress (σ_3).

Source: https://www.geokniga.org/bookfiles/geokniga-es2e-ch06.pdf.

A rock can experience elastic distortion that results in a few percent strain at most. There are two possible outcomes when a rock experiences more stress than it can withstand elastically: either the rock becomes brittle and breaks, or it deforms ductilely, allowing it to stretch without breaking.

In other words, chemical bonds break when the stress is high enough to stretch or bend them to the point where the atoms are no longer attracted to one another. This can cause a fracture to form or a preexisting fracture to slip. Brittle deformation, in contrast to elastic strain, is irreversible, meaning that even after the stress is released, the distortion persists. Once more, earthquakes will be used as an example. In this instance, failure and displacement occur because elastic strain is exceeded at the focus. Because brittle deformation involves multiple processes, the pattern of breakage is contingent upon both the stress conditions and the material properties of the rock. .

6.4. TENSILE CRACKING

6.4.1. Stress Concentration and Griffith Cracks

Figure 6.14 depicts a crack in rock at the atomic level. One possible way for such a crack to form is if all the chemical bonds on the crack surface break simultaneously. The tensile stress required for this to happen equals the strength of each chemical bond multiplied by the total number of bonds that were once present across the crack area. By knowing the strength of a single chemical bond, you can calculate the necessary stress to break all the bonds by simply multiplying the bond strength by the number of bonds. Realistic values for elasticity (Young's modulus, E) and small strain (<10%) can be used in these calculations. Research has shown that rocks in the Earth's crust tend to fracture at crack-normal tensile stresses below 10 MPa when the confining pressure is low, a much lower value compared to the theoretical strength of rock. This raises the question of how natural rocks are able to fracture at stresses significantly lower than their theoretical strength.

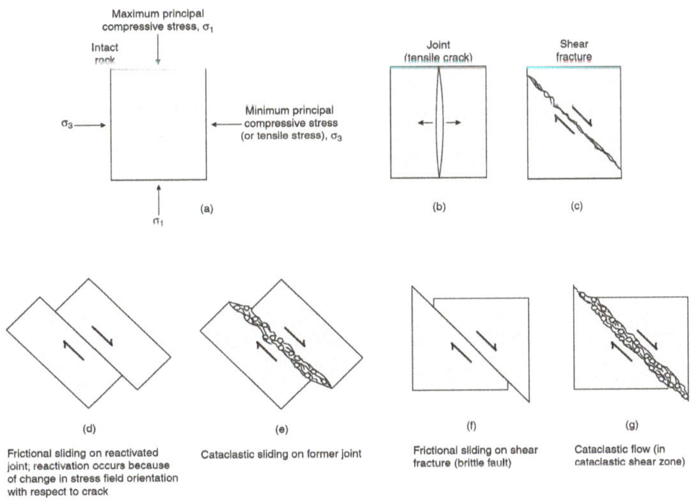

Figure 6.13. Types of brittle deformation. (a) Orientation of the remote principal stress directions with respect to an intact rock body. (b) A tensile crack, forming parallel to σ_1 and perpendicular to σ_3 (which may be tensile). (c) A shear fracture, forming at an angle of about 30° to the σ_1 direction. (d) A tensile crack that has been reoriented with respect to the remote stresses and becomes a fault by undergoing frictional sliding. (e) A tensile crack which has been reactivated as a cataclastic shear zone. (f) A shear fracture that has evolved into a fault. (g) A shear fracture that has evolved into a cataclastic shear zone.

Source: https://www.geokniga.org/bookfiles/geokniga-es2e-ch06.pdf.

When engineers studying the theory of elasticity discovered that the remote stress—that is, stress resulting from a load applied at a distance from a region of interest—gets concentrated at the sides of flaws, they took the first step toward solving the strength paradox , holes) within a substance. For instance, when a circular hole forms at the ends of a vertical elastic sheet that is subjected to tensile stress, the local stress (i.e., stress at the point of interest) is three times the remote stress magnitude (•$_r$)tangent to the hole's sides. At both the top and bottom of the hole, the local tangential stress equals the remote stress in magnitude, but its sign is opposite (i.e., it is constrictive. The amount of stress concentration, C, equals 2b/a + 1 if the hole is elliptical in shape rather than circular, with a and b representing the elliptical's long and short axes, respectively.

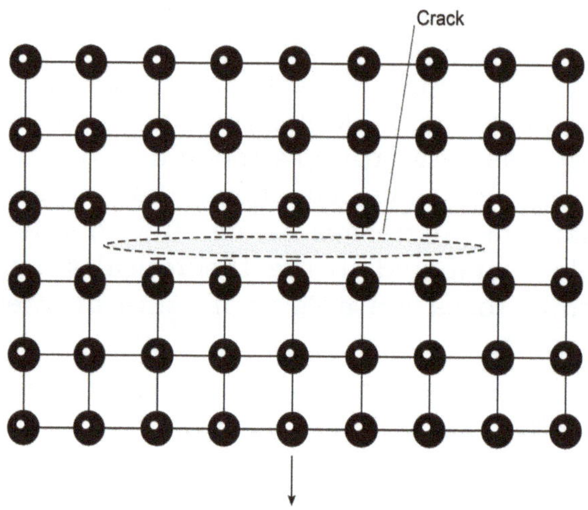

Figure 6.14. A cross-sectional sketch of a crystal lattice (balls are atoms and sticks are bonds) in which there is a crack. The crack is a plane of finite extent across which all atomic bonds are broken.

Source: https://www.geokniga.org/bookfiles/geokniga-es2e-ch06.pdf.

As a result, the elliptical hole's axial ratio determines the values for stress concentration at its ends: the higher the axial ratio, the higher the stress concentration. Keeping this in mind, stress is concentrated by a factor of 17 at the ends of an elliptical hole with an axial ratio of 8:1, and magnified by a factor of approximately 100 in a 1 µm × 0.02 µm crack. W. Griffith applied the idea of stress concentration at the ends of elliptical holes to fracture development in the 1920s, which marked a significant advancement toward the resolution of the strength paradox. According to Griffith, all materials have innate microcracks or flaws where stress concentrations naturally form. These cracks propagate and grow larger despite the host rock being subjected to relatively low remote stresses, because of the stress concentrations that form at their tips.

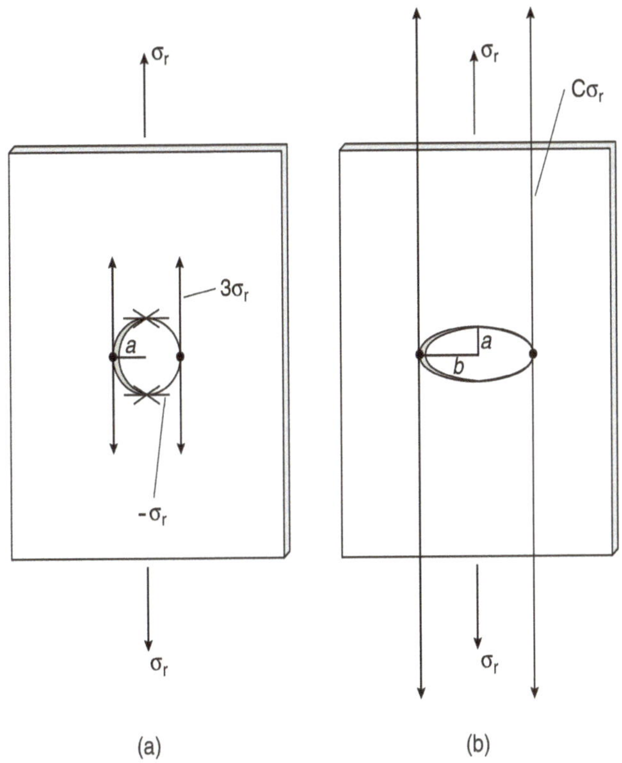

Figure 6.15. Stress concentration adjacent to a hole in an elastic sheet. If the sheet is subjected to a remote tensile stress at its ends (• $_r$), then stress magnitudes at the sides of the holes are equal to C• $_r$, where C, the stress concentration factor, is (2b/a) + 1. (a) For a circular hole, C = 3. (b) For an elliptical hole, C > 3.

Source: https://www.geokniga.org/bookfiles/geokniga-es2e-ch06.pdf.

He found that the crack with the largest axial ratio will probably propagate first in a material with cracks of different axial ratios. Put another way, even though the remote stress is relatively small, the stress at the tips of preexisting cracks can grow to the point where it breaks the chemical bonds that hold the minerals together at the tip and causes the crack to propagate. Named after him, preexisting microcracks and imperfections in a rock, such as grain-scale fractures, pores, and grain boundaries, are now referred to as Griffith cracks. By discovering that rocks in the crust have Griffith fractures, which make them relatively weak, we are able to resolve the strength paradox.

While Griffith's concept shed light on cracking, it failed to address how crack characteristics like shape, length, and orientation impact the cracking process. Engineers later introduced linear elastic fracture mechanics, which assumes cracks have nearly infinite axial ratio, meaning they are very sharp. This theory suggests that, all other factors being equal, a longer crack will propagate further than a shorter crack. The explanation for this will be explored in the upcoming discussion on failure criteria.

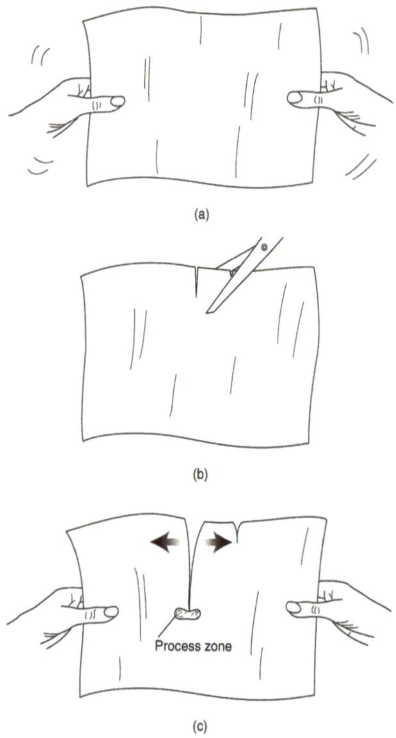

Figure 6.16. Illustration of a home experiment to observe the importance of preexisting cracks in creating stress concentrations. (a) An intact piece of paper is difficult to pull apart. (b) Two cuts, a large one and a small one, are made in the paper. (c) The larger preexisting cut propagates. In the shaded area, a region called the process zone, the plastic strength of the material is exceeded and deforms.

Source: https://www.geokniga.org/bookfiles/geokniga-es2e-ch06.pdf.

With a straightforward experiment, we can investigate the impact of preexisting cracks on the stress threshold required for tensile cracking. Grab a piece of paper and tug on either end. The paper will only tear if you pull very firmly. Next, make two cuts in the sheet's edge near its center, about 0.5 cm and 2 cm long, respectively, and pull again. The strength of the paper is exceeded at the point where the pull is concentrated due to the preexisting cuts. You'll discover that the paper tears apart by growing from the longer preexisting cut and that it requires significantly less force to tear. Because a crack-tip process zone, where the material deforms plastically, blunts the tips of real cracks, sharp cracks do not propagate under extremely small stresses.

Our explanation of crack propagation implies that a crack does not form its entire area instantly; rather, it starts at a small flaw and spreads outward from there. You are well aware of this fact if you have ever stepped out onto thin ice that covers a pond. As you get farther from the coast, you hear a sound that sounds like the echo of a gunshot; this is the sound of an ice fracture developing as a result of the pressure from your weight. If you are brave enough to observe how the crack starts in such dangerous conditions, you will see that it starts beneath your boot and spreads outward into intact ice at a limited speed. This indicates that only the chemical bonds at the crack tip are breaking at any given time. To put it another way, not every bond that is broken by a fracture is broken at once, so the methodology we initially employed to determine theoretical strength is not accurate.

6.4.2. Exploring Tensile Crack Development

Let's examine what transpires in a lab setting when we perform axial stretching—a technique that involves stretching a rock cylinder along its axis at a relatively low confining pressure. Preexisting microcracks in the sample open slightly as soon as the remote tensile stress is applied, and the remote stress is amplified to produce greater local stresses at the crack tips. A crack eventually starts to widen when the stress at its tip surpasses the rock's strength. The crack will continue to grow and may eventually reach the sample's margins if the remote tensile stress remains constant after it starts to propagate. This results in the sample failing, which causes it to split into two disconnected pieces.

Additionally, axial compression applied to a rock cylinder at low confining pressure can cause tensile fracture. Longitudinal splitting is the process by which mesoscopic tensile fractures parallel to the cylinder axis occur under such stress conditions. Similar to tensile cracking, longitudinal splitting occurs when there is uniaxial compression; however, cracks parallel to the compression direction can open up, while those that are not parallel to the σ_1 direction remain closed. Imagine an envelope balanced on its edge to get a sense of this.

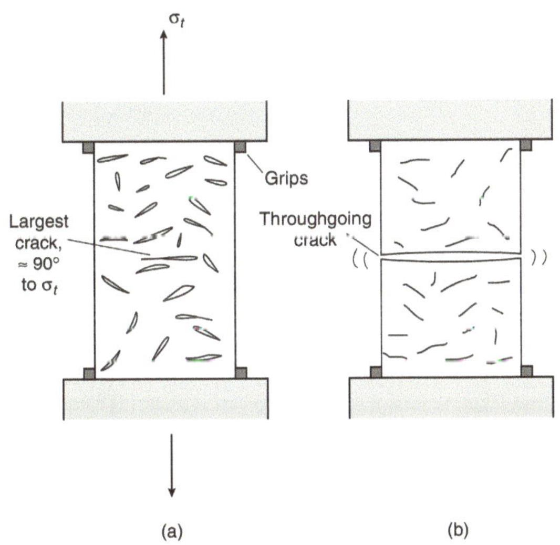

Figure 6.17. Development of a throughgoing crack in a block under tension. (a) When tensile stress (σ_t) is applied, Griffith cracks open up. (b) The largest, properly oriented cracks propagate to form a throughgoing crack.

Source: https://www.geokniga.org/bookfiles/geokniga-es2e-ch06.pdf.

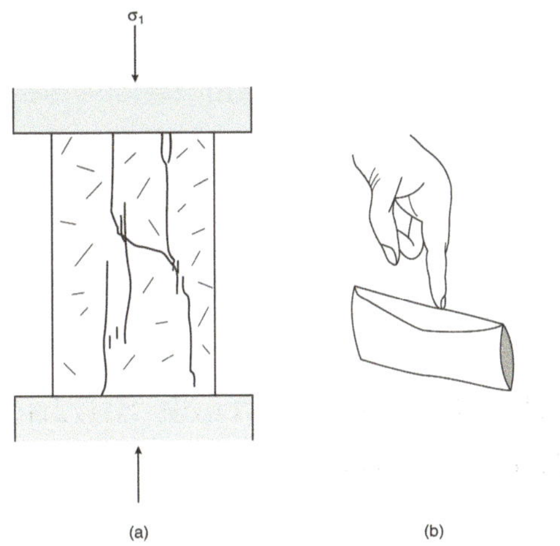

Figure 6.18. (a) A cross section showing a rock cylinder with mesoscopic cracks formed by the process of longitudinal splitting. (b) An "envelope" model of longitudinal splitting. If you push down on the top of an envelope (whose ends have been cut off), the sides of the envelope will move apart.

Source: https://www.geokniga.org/bookfiles/geokniga-es2e-ch06.pdf.

Even in cases where the sides of the envelope were not subjected to a remote tensile stress, they will still pull apart if you apply pressure to their upper edge. When the compressive stress in rocks increases, the tensile stress at the crack tips surpasses the rock's strength, causing the crack to propagate parallel to the direction of the compressive stress. The confining pressure needed in the compressive stress environment shown in Figure 6.18 is very small; however, when adding fluid pressure to the sample's pores and cracks, tensile cracks can also form in a rock cylinder when the remote stress is compressive under higher confining pressure (i.e., the pressure in the pores. A crack may spread as a result of the uniform, outward push of a fluid in a microcrack creating a local tensile stress at the crack tips. This procedure is known as hydraulic fracturing. The fluid pressure drops if no more fluid enters the crack because the crack's volume increases as soon as it starts to grow. When pore pressure falls below the threshold required to generate a significant tensile stress at the crack tip, crack propagation stops and doesn't start up again until pore pressure increases enough. Consequently, pore pressure-induced tensile cracking usually happens in pulses.

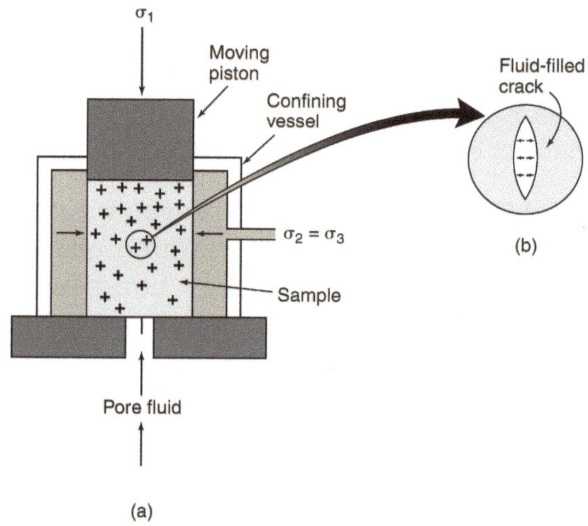

Figure 6.19. (a) Cross-sectional sketch illustrating a rock cylinder in a triaxial loading experiment. Fluid has access to the rock cylinder and fills the cracks. (b) A fluid-filled crack that is being pushed apart from within by pore-fluid pressure.

Source: https://www.geokniga.org/bookfiles/geokniga-es2e-ch06.pdf.

6.4.3. Modes of Crack-Surface Displacement

One more important matter to discuss before we move on from Griffith cracks is the direction in which a single crack grows when it is loaded. We have only been able to investigate cracks that are perpendicular to distant stresses thus far. Materials scientists identify three configurations of crack loading, but what about cracks in other orientations with respect to stress, and how do they propagate? Three different modes of crack-surface displacement are produced by these configurations. Keep in mind that when discussing crack propagation, we are talking about the minuscule movement that starts the crack tip's propagation, not the mesoscopic displacement that can be measured in faults. During Mode I displacement, cracks open slightly perpendicular to the crack surface, creating tensile cracks. These cracks form parallel to the principal stress plane that is perpendicular to the σ3 direction and can expand within their plane without changing orientation. In Mode II displacement (the sliding mode), rock on one side of the crack surface moves slightly parallel to the fracture surface and perpendicular to the fracture front. In Mode III displacement (the tearing mode), rock on one side of the crack slides slightly parallel to the crack but in a direction parallel to the fracture front.

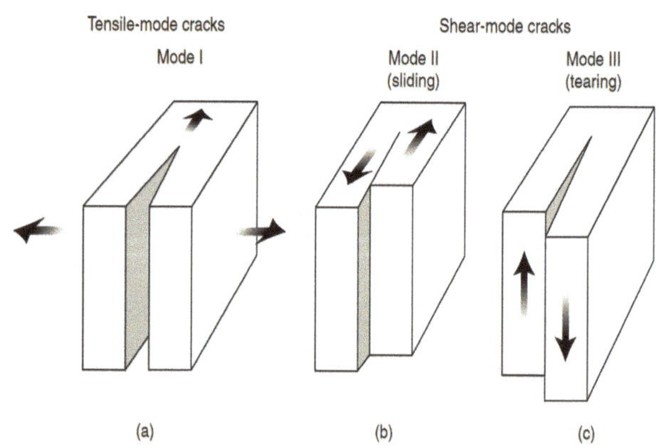

Figure 6.20. Block diagrams illustrating the three modes of crack surface displacement: (a) Mode I, (b) Mode II, (c) Mode III. Mode I is a tensile crack, and Mode II and Mode III are shear-mode cracks.

Source: https://www.geokniga.org/bookfiles/geokniga-es2e-ch06.pdf.

Mode II (and III) cracks are not just microscopic versions of faults, despite the fact that shear-mode cracks resemble mesoscopic faults in appearance. The distinction becomes apparent when we look at how shear-mode cracks spread. Shear-mode cracks do not grow in their plane; instead, as they expand, they instantly curve into the orientation of tensile or Mode I cracks. Wing cracks, which are new tensile cracks that arise from propagating shear-mode cracks, are depicted in Figure 6.21.

6.5. PROCESSES OF BRITTLE FAULTING

A brittle fault is a surface where slip has occurred due to differential stress, rather than plastic deformation mechanisms. This type of faulting occurs when there is a shear stress parallel to the fault plane, typically when σ_1 does not equal σ_3, and the fault surface is not parallel to a principal plane of stress. When faulting occurs, it changes the shape of the rock body where it is located, contributing to regional strain. Unlike ductile strain, which requires measurable deformation in the surrounding rock, brittle faulting occurs as a distinct discontinuity within the rock body.

There are two primary methods of forming a brittle fault. The first involves the shearing rupture of an originally unbroken rock body. The second method involves the shearing reactivation of a preexisting weak surface, such as a joint, bedding surface, or fault, within a rock body. In some cases, the preexisting weak surface may slip before the differential stress is sufficient to cause shearing rupture of the intact rock. Once formed, movement on brittle faults occurs through frictional sliding, the development of fault-parallel veins, or cataclastic flow.

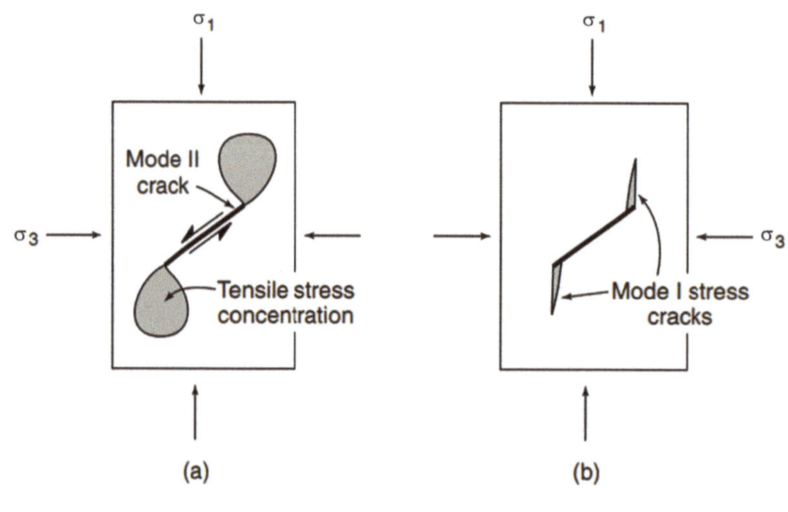

Figure 6.21. Propagating shear-mode crack and the formation of wing cracks. (a) A tensile stress concentration occurs at the ends of a Mode II crack that is being loaded. (b) Mode I wing cracks form in the zones of tensile-crack concentration.

Source: https://www.geokniga.org/bookfiles/geokniga-es2e-ch06.pdf.

6.5.1. Slip by Growth of Fault-Parallel Veins

In the brittle field, not all faults displace through frictional sliding. Some faults have opposing surfaces separated by mineral crystals (such as quartz or calcite) formed from water solutions along the fault during movement. These fault-surface veins can consist of mineral fibers, blocky crystals, or a combination of both.

The formation mechanism of fault-surface veins is poorly understood. They could occasionally form as a result of a crack developing along a weak fault surface due to high fluid pressures. After a fault cracks, one side of the fault moves slightly relative to the other, and vein material precipitates into the crack to seal it. In other situations, vein formation could be the result of the asperities (or steps) gradually dissolving on the fault surface and then ions moving through fluid films to locations with less stress, where mineral precipitation occurs. The formation of an actual discontinuity that causes the rock to lose cohesiveness may not be necessary for this second process to take place. Depending on the strain rate and the presence of water, a fault surface may experience frictional sliding or syn-slip veining. It is likely that the presence of water and slow movement along the fault will increase the frequency of fault-surface veining.

6.5.2. Cataclasis and Cataclastic Flow

Cataclasis is when a fault moves due to a combination of small cracks, frictional sliding of fragments, and the rotation and movement of grains. An analogy for this process is imagining corn being turned into cornmeal as it passes between two grinding stones that slide past each other. When cataclasis affects a wide area of rock, it causes ductile strain and can be referred to as cataclastic flow, as the rock effectively flows over the entire width of the affected band.

Figure 6.22. Photograph of stepped calcite slip fibers on a fault surface; pencil indicates displacement direction.

Source: https://www.geokniga.org/bookfiles/geokniga-es2e-ch06.pdf.

To visualize cataclastic flow, consider how, although the individual grains are undoubtedly solid, the cornmeal we just made behaves remarkably like a fluid when it is poured from one container to another. Often, the term "cataclastic shear zone" refers to movement on a fault that involves the development of a zone where cataclastic flow occurs.

6.6. FORMATION OF SHEAR FRACTURES

Tensile and shear fractures are very different from one another. When the shear stress parallel to the surface is high enough, a rock loses continuity across that surface, which is known as a shear fracture. In laboratory rock cylinders, shear fractures are initiated at an angle of approximately 30° to σ_1 when confining pressure is present ($\sigma_1 > \sigma_2 = \sigma_3$). Friction prevents the fracture from sliding during its formation because, in addition to shear stress, the fracture is also subject to a component of normal stress. The fracture widens and displacement builds up if the shear stress exerted on it keeps rising above the frictional resistance to sliding. Because shear-mode cracks, as we have seen, are unable to grow in their own plane, shear fractures, also known as faults, are therefore more complex than large shear-mode cracks. .

This conceptual distinction is crucial. Therefore, by creating shear ruptures using a rock cylinder under confining pressure during a laboratory triaxial loading experiment, we can obtain insight into the process of shear-fracture formation. Therefore, we first describe such an experiment in order to start our search for an answer to this question.

A rock cylinder is used in a confined-compression triaxial-loading experiment. It is covered in rubber or copper, placed inside a pressure chamber with a confining fluid, and squeezed between two hydraulic pistons. The actual rock in the experiment remains dry. Throughout the experiment, we apply an axial load (σ_1) to the ends of the cylinder by moving the pistons together at a constant rate, and a confining pressure ($\sigma_2 = \sigma_3$) to the sides of the cylinder by raising the pressure in the surrounding fluid. Through a constant σ_3 value and a gradual increase in σ_1, we raise the differential stress ($\sigma_d = \sigma_1 - \sigma_3$). The cylinder's volume (Δ), axial strain (e_a), and change in length (e_a) are all measured in this experiment along with the magnitude of σ_d.

There are four stages in the experiment, as indicated by a graph of σ_d versus ea. In Stage I, we discover that e_a increases along with σ_d and that there is a concave-up curve in the relationship between these two quantities. The relationship between σ_d and e_a in Stage II of the experiment is a straight line with a positive slope. There is a slight decrease in sample volume during Stage I and most of Stage II. The experiment's third stage sees a decrease in the slope of the line that depicts the relationship between σ_d and e_a. The yield strength is the stress at which the slope of the curve changes. We notice a slight increase in volume during the latter part of Stage II and all of Stage III, a phenomenon known as dilatancy. If we were to attach a very sensitive microphone to the sample at this time, we would hear a lot of popping sounds, which are indicative of the formation and growth of microcracks. A shear rupture surface appears at a roughly 30-degree angle to the cylinder axis, and there is a sudden stress drop when σ_d equals the failure stress (σ_f).

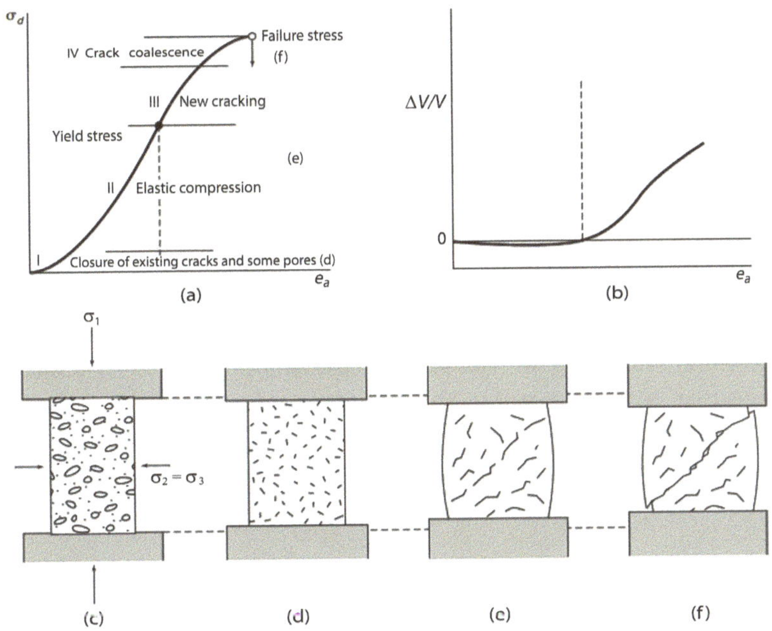

Figure 6.23. Fracture formation. (a) Stress–strain plot (differential stress versus axial shortening) showing the stages (I–IV) in a confined compression experiment. The labels indicate the process that accounts for the slope of the curve. (b) The changes in volume accompanying the axial shortening illustrate the phenomenon of dilatancy; left of the dashed line, the sample volume decreases, whereas to the right of the dashed line the sample volume increases. (c–f) Schematic cross sections showing the behavior of rock cylinders during the successive stages of a confined compression experiment and accompanying stress–strain plot, emphasizing the behavior of Griffith cracks (cracks shown are much larger than real dimensions). (c) Pre-deformation state, showing open Griffith cracks. (d) Compression begins and volume decreases due to crack closure. (e) Crack propagation and dilatancy (volume increase). (f) Merging of cracks along the future throughgoing shear fracture, followed by loss of cohesion of the sample (mesoscopic failure).

Source: https://www.geokniga.org/bookfiles/geokniga-es2e-ch06.pdf.

In this scenario, a stress drop refers to a sudden decrease in the axial stress of the specimen, causing a buildup of large strain at a lower stress level. An analogy to understand this concept is pushing a car stuck in a ditch – initially, you exert a great amount of force until the car is freed from the ditch, but once it is out, you must ease off to avoid falling as the car moves. The moment when the shear rupture forms and the stress drops is known as the failure strength for shear rupture, denoted as σ_d. After failure occurs, the sample loses its integrity, and its frictional resistance on the fracture surface determines its future behavior.

In Stage I, existing open microcracks closed. In Stage II, the sample experienced elastic shortening along the axis, while also expanding slightly in the perpendicular direction due to the Poisson effect. The Poisson effect occurs when a rock undergoing elastic shortening in one direction extends in the direction perpendicular to the shortening direction. The ratio between shortening and extension is known as Poisson's ratio, ν. In Stage III, tensile microcracks started to grow throughout the sample, with wing cracks forming at the tips of shear-mode cracks. This resulted in a slight increase in volume and popping noises. Tensile cracking intensified in Stage III along a narrow band at approximately 30° to the axial stress. Stage IV saw failure as cracks organized to form a continuous surface, causing the rock to lose continuity and allowing for frictional sliding between the two sides. Consequently, the cylinders could move together more easily, leading to a sudden drop in stress.

The scenario outlined above demonstrates that the failure strength for shear fracture is not defined by the stress level at which a single crack propagates, but by the stress level at which numerous small cracks merge to create a complete rupture. It should also be noted that in certain experiments, two ruptures may form, both at approximately 30° to the axial stress. The angle between these conjugate fractures is around 60°, with the acute bisector aligning with the maximum principal stress. As displacement continues, it becomes impossible for both fractures to remain active simultaneously, as movement on one fracture will counteract the other. Therefore, typically only one fracture will develop into a complete fault line.

6.7. PREDICTING INITIATION OF BRITTLE DEFORMATION

It is evident that stress causes brittle structures to form in rock; however, the precise stress conditions that lead to brittle deformation remain unclear. It is obvious that engineers, who need to be able to calculate the amount of stress that a building or bridge can withstand before collapsing, as well as geologists who seek to understand the when, where, and why brittle geologic structures (joints, faults, veins, and dikes) develop in the Earth, benefit from an understanding of the stress state at which brittle deformation begins. In reality, there are three phenomena that are involved in brittle deformation: frictional sliding, shear fracture formation, and tensile crack growth. We look at the stress conditions required for each of these phenomena to manifest in this section.

6.7.1. Tensile Cracking Criteria

A mathematical formula known as a tensile cracking criterion indicates the stress level at which a crack starts to spread. Since preexisting flaws in the rock cause stress concentrations, all tensile cracking criteria are predicated on the idea that macroscopic cracks originate from these flaws (also known as Griffith cracks). One of the earliest researchers to suggest a criterion for tensile cracking was Griffith. He achieved this by examining the energy used in cracking. Griffith proposed modeling a material where a crack originates as a thermodynamic system made up of an elastic plate with an elliptical crack that already exists. The Griffith energy balance can be used to determine the total energy of a system in which a load is applied to the sheet's ends causing it to stretch and the crack to spread.

$$dU_T = dU_s - dW_r + dU_E \qquad (1)$$

where dU_T is the change in total energy of the system, dU_s is the change in surface energy due to growth of the crack (this term arises because crack formation breaks bonds, and energy stored in a broken bond is greater than the energy stored in a satisfied bond), dW_r is the work done by the load in deforming the plate, and dU_E is the change in the strain energy stored in the plate (strain energy is the energy stored by chemical bonds that have been stretched out of their equilibrium length or angle). Griffith pointed out that because the system starts with the load already in place, formation of the crack does not change the total energy of the system. Thus, the change in total energy for an increment of crack growth (dc) equals 0 (in equation form, $dU_T/dc = 0$ for an equilibrium condition). With this point in mind, Equation 1 can be rewritten as

$$dW_r = dU_s + dU_E \qquad (2)$$

In words, Equation 2 means that the work done on a system is divided between creation of new elastic strain energy in the sheet and creation of new crack surface by propagation of the crack. Using this concept, along with several theorems from elasticity theory, Griffith devised a tensile cracking criterion, which we present without derivation:

$$\sigma_t = (2E\gamma/\pi c)^{1/2} \qquad (3)$$

where σ_t = critical remote tensile stress (tensile stress at which the weakest Griffith crack begins to grow), E = Young's modulus, ⊠ = energy used to create new crack surface, and c = half-length of the preexisting crack. Reading this equation, we see that the critical remote tensile stress for a rock sample is proportional to material properties of the sample and the length of the crack. Note that as crack length increases, the value of critical remote tensile stress (σ_t) decreases.

Subsequently, researchers have utilized concepts from the engineering study of linear elastic fracture mechanics to develop tensile cracking criteria. According to this work, the following equation defines conditions at which Mode I cracks propagate:

$$K_I = \sigma_t\, Y(\pi c)^{1/2} \qquad (4)$$

where K_I (pronounced "kay one," where the "one" represents a Mode I crack) is the stress intensity factor, σ_t is the remote tensile stress, Y is a dimensionless number that takes into account the geometry of the crack (e.g., whether it is penny-shaped, blade-shaped, or tunnel-shaped), and c is half of the crack's length. In this analysis, all cracks in the body are assumed to have very large ellipticity; that is, cracks are assumed to be very sharp at their tips. Note that cracks with smaller ellipticity require higher stresses to propagate.

Equation 4 says that the value of K_I increases when σ increases. A crack in the sample begins to grow when K_I attains a value of K_{Ic}, which is the critical stress intensity factor or the fracture toughness (a measure of tensile strength). The fracture toughness is constant for a given material. When K_I reaches K_{Ic}, the value of σ reaches σ_t, where σ_t is the critical remote tensile stress at the instant the crack starts to grow. We can rewrite Equation 6.4 to create an equation that more directly defines the value of σ_t at the instant the crack grows:

$$\sigma_t = K_{Ic}/[Y(\pi c)^{1/2}] \qquad (5)$$

Keep in mind that the amount of stress needed to cause cracking in this equation is influenced by the fracture toughness, crack shape, and crack length. All else being equal, a longer crack usually spreads faster than a shorter crack. Additionally, crack shape plays a role in determining which crack will spread first. As the crack grows, the fracture toughness typically increases, leading to failure of the specimen. This dynamic has an intriguing implication for natural rocks: because crack length is dependent on the grain size of the sample, fine-grained rocks are likely to be stronger than coarse-grained rocks.

Equations 4 and 5 can be applied to Mode II and Mode III cracks as well. When comparing the equations for the three modes of cracking, it is found that, all else being equal, a Mode I crack (perpendicular to σ_3) will typically propagate before a Mode II or Mode III crack. However, factors such as crack shape and length can influence the propagation, causing Mode II or Mode III cracks to sometimes propagate before Mode I cracks in real materials. It is important to note that when Mode II and III cracks propagate, they may either bend and turn into Mode I cracks, or develop wing cracks at their tips, as shear cracks cannot propagate significantly in their own plane.

6.7.2. Shear-Fracture Criteria and Failure Envelopes

An expression that characterizes the stress state at which a shear rupture occurs and splits a sample in half is known as a shear-fracture criterion. Shear-rupture criteria are also frequently referred to as shear-failure criteria because shear-fracture initiation in a laboratory sample invariably results in sample failure, meaning that following rupture the sample can no longer support a load that exceeds the frictional resistance to sliding on the fracture surface.

Charles Coulomb was one of the first to propose a shear-fracture criterion. He suggested that if all the principal stresses are compressive, as is the case in a confined compression experiment, a material fails by the formation of a shear fracture, and that the shear stress parallel to the fracture surface, at the instant of failure, is related to the normal stress by the equation

$$\sigma_s = C + \mu\sigma_n \qquad (6)$$

where C is the cohesion of the rock, a constant that indicates the shear stress required to cause failure if the normal stress across the potential fracture plane equals zero (note that this C is different from the c in Equations 3–5), σ_n is the normal stress across the shear fracture at the moment of failure, and μ is a constant commonly referred to as the coefficient of internal friction. These values are represented by the shear stress parallel to the fracture surface at failure and σ_n, respectively. In the context of shear failure of a solid rock, the name "μ" is essentially meaningless because it originated from studies on the friction between grains in unconsolidated sand and the control that friction has on the slope angles of sand piles. Instead, μ should be understood as a constant of proportionality. The basic idea of Equation 6, sometimes referred to as Coulomb's failure criterion, is that the normal stress across the fracture surface determines the shear stress required to start a shear fracture.

The Coulomb criterion is represented by a straight line on the Mohr diagram, which can be seen by plotting the results of four triaxial loading experiments on confined granite cylinders. Starting with a low confining pressure in the first experiment and increasing the axial load until failure, we plot the Mohr circle representing the critical stress state. As we repeat the experiment with higher confining pressure, the differential stress at failure also increases, resulting in larger Mohr circles shifted to the right. All four circles, when plotted, are tangent to a straight line with a slope of μ (tan φ) and a y-intercept of C, which is the Coulomb criterion. This line can also be represented below the σ_n-axis on the Mohr diagram.

Specifically, 2θ is the angle (usually about 30°) between the σ_3 direction and the plane of shear fracture, and is defined as a line drawn from the center of a Mohr circle to the point of its tangency with the Coulomb criterion. For the range of confining pressures for which the Coulomb criterion is valid, this angle remains constant since the criterion is a straight line.

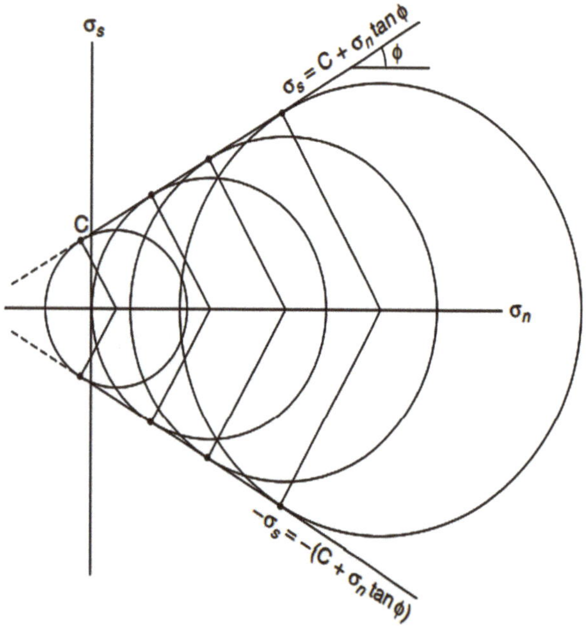

Figure 6.24. Mohr diagram showing a Coulomb failure envelope based on a set of experiments with increasing differential stress. The circles represent differential stress states at the instant of shear failure. The envelope is represented by two straight lines, on which the dots represent failure planes.

Source: https://www.geokniga.org/bookfiles/geokniga-es2e-ch06.pdf.

A graph that shows the normal stress and shear stress magnitudes as a function of the angle between the plane and the σ1 direction makes the rationale for the 30° angle clear. Observe that the maximum shear stress and the minimum normal stress do not occur in the same plane. On a potential failure plane oriented at 45° to σ_1, shear stress reaches its maximum; however, the normal stress across this potential plane is still too great to allow shear fracturing in planes with this orientation. Although it is still relatively high, the shear stress is slightly less across a plane oriented at 30° to σ_1. Shear-fracture formation is, however, encouraged by the significantly lower normal stress across the 30° plane.

As an empirical relation, Coulomb's criterion is not predicated on theoretical concepts or understanding of mechanisms operating at the atomic or crystallographic scales; rather, it is based solely on experimental observation. Unlike the Griffith criterion, this failure criterion does not define the state of stress at which the microcracks that eventually come together to form the shear rupture start to propagate, nor does it relate the stress state at failure to physical parameters.

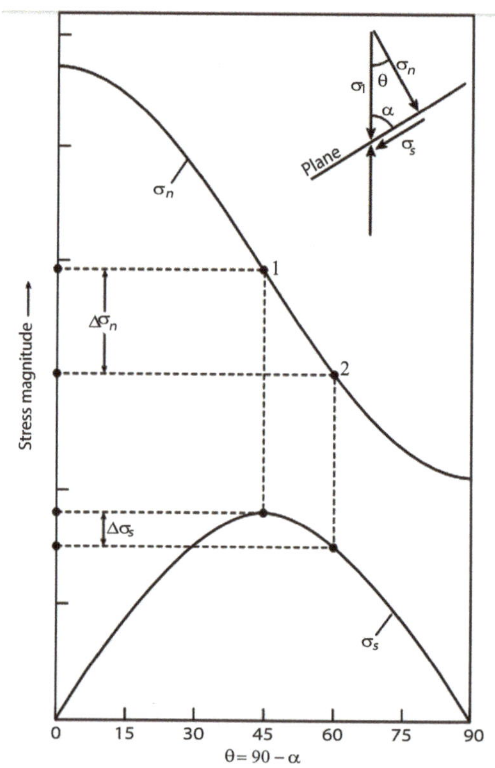

Figure 6.25. The change in magnitudes of the normal and shear components of stress acting on a plane as a function of the angle α between the plane and the σ1 direction; the angle θ = 90 – α is plotted for comparison with other diagrams. At point 1 (α=θ= 45°), shear stress is a maximum, but the normal stress across the plane is quite large. At point 2 (θ = 60°, α

= 30°), the shear stress is still quite high, but the normal stress is much lower.

Source: https://www.geokniga.org/bookfiles/geokniga-es2e-ch06.pdf.

In a triaxial loading experiment, the Coulomb criterion does not predict whether the fractures that form will dip to the left or right with respect to the rock cylinder's axis. As was previously mentioned, it is possible for conjugate shear fractures to develop, one with a left lateral shear sense and one with a right lateral shear sense. The two fractures, which are normally spaced about 60 degrees apart, match the tangency points on the circle that shows the stress condition at failure in relation to the Coulomb failure envelope. Further studies by German engineer Otto Mohr revealed that Coulomb's linear relationship for shear-fracture criteria is only applicable within a specific range of confining pressures. Mohr observed that as the confining pressure decreased, the stress state at failure displayed a more pronounced curve with a steeper slope, while at higher confining pressures, the curve had a shallower slope.

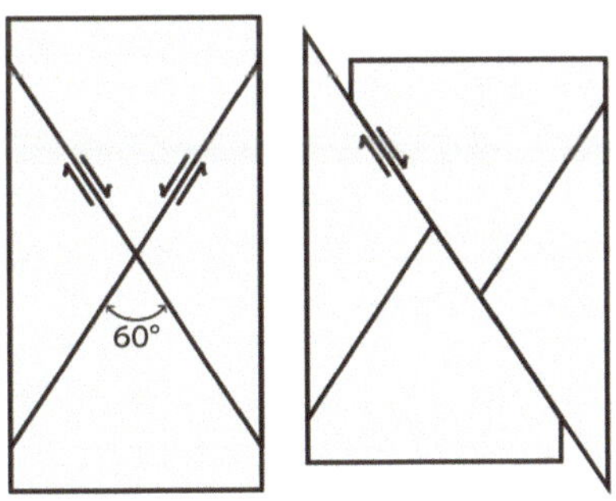

Figure 6.26. Cross-sectional sketch showing how only one of a pair of conjugate shear fractures (a) evolves into a fault with measurable displacement (b).

Source: https://www.geokniga.org/bookfiles/geokniga-es2e-ch06.pdf.

This led Mohr to propose that the failure criteria for shear rupture could be represented by a parabolic curve on its side, known as the Mohr-Coulomb criterion. Unlike Coulomb's empirical straight-line relation, the Mohr-Coulomb criterion acknowledges that the angle between the shear fracture plane and σ_1 varies depending on the stress state. Specifically, at lower confining pressures, the angle is smaller, whereas at higher confining pressures, the angle becomes steeper.

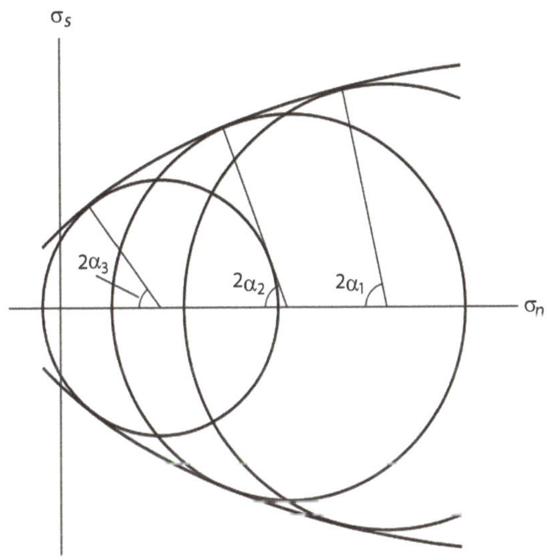

Figure 6.27. Mohr failure envelope. Note that the slope of the envelope steepens toward the σ_s-axis. Therefore, the value of α (the angle between fault and σ_1) is not constant (compare $2\alpha_1$, $2\alpha_2$, and $2\alpha_3$).

Source: https://www.geokniga.org/bookfiles/geokniga-es2e-ch06.pdf.

On the Mohr diagram, the Mohr-Coulomb criterion defines a failure envelope for both positive and negative values of σs. The field on the diagram where stress states are "stable" and the field where stress states are "unstable" are divided by a failure envelope.

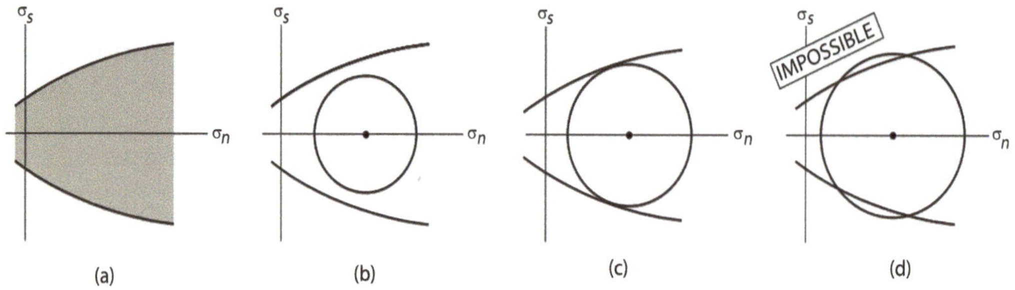

Figure 6.28. (a) A brittle failure envelope as depicted on a Mohr diagram. Within the envelope (shaded area), stress states are stable, but outside the envelope, stress states are unstable. (b) A stress state that is stable, because the Mohr circle, which passes through values for σ_1 and σ_3 and defines the stress state, falls entirely inside the envelope. (c) A stress state at the instant of failure. The Mohr circle touches the envelope. (d) A stress state that is impossible.

Source: https://www.geokniga.org/bookfiles/geokniga-es2e-ch06.pdf.

According to this definition, a sample can withstand a stress state without breaking into pieces. Since the sample will have fractured before an unstable stress state can be reached, it is impossible to reach an unstable stress state. Put differently, a stress state that is fully contained within the envelope of a Mohr circle indicates stability and prevents the sample from developing a shear rupture. A tangent circle to the envelope indicates the stress level at which brittle failure takes place. Stress states delineated by circles that surpass the envelope are inherently unstable and thus unattainable within the specific rock under investigation.

At high confining pressures, samples may exhibit plastic deformation rather than brittle deformation. In these circumstances, the concept of a "failure" envelope is no longer applicable. Instead, we can represent the "yield" envelope on a Mohr diagram by two lines parallel to the σn-axis. This yield criterion, known as the Von Mises criterion, implies that plastic yielding is primarily determined by the yield stress and is not significantly affected by differential stress.

A continuous tensile crack forms on the sample if the tensile stress is high enough. A point, T_0, the tensile strength, along the σ_n-axis to the left of the σ_s-axis can be used to represent the tensile stress required to cause tensile failure.

However, as we've seen, the location of this point is influenced by how big the sample's flaws are. Experiments thus demonstrate that the tensile strength is highly variable, even for the same type of rock, and that it is best represented by a range of points along the σ_n-axis.

There is a debate among geologists about how rocks fail when they are exposed to tensile stresses below their tensile strength. Some believe that a combination of tensile cracks and shear ruptures, known as transitional-tensile fractures or hybrid shear fractures, are responsible for failure in these conditions. This type of failure is represented by the steep portion of the parabolic failure envelope. However, most fracture experts argue that transitional-tensile fractures do not occur naturally, as no experiments have been able to replicate them in the laboratory.

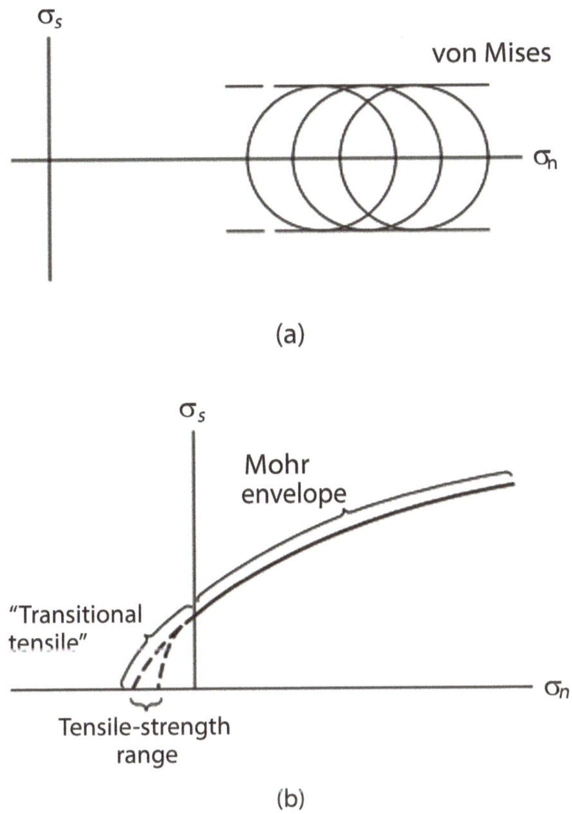

Figure 6.29. (a) Mohr diagram illustrating the Von Mises yield criterion. Note that the criterion is represented by two lines that parallel the σ_n-axis. (b) The extrapolation of a Mohr envelope to its intercept with the σ_n-axis, illustrating the "transitional-tensile" regime, and the tensile strength (T0). Note that the tensile strength has a range of values, because it depends on the dimensions of preexisting flaws in the deforming sample.

Source: https://www.geokniga.org/bookfiles/geokniga-es2e-ch06.pdf.

By considering all of the aforementioned empirical criteria, we are able to create a composite failure envelope that, for a broad range of confining pressures and conditions where one of the principal stresses is tensile, represents the border between stable and unstable stress states. The envelope has the general appearance of a cup on its side cut in half. The curve's various components have labels. The horizontal lines in the diagram represent the Von Mises criteria, which begin at the right side. It's important to keep in mind that the Von Mises section of the envelope pertains to plastic yield rather than brittle failure. The transition between brittle and plastic is best represented by the area of the curve where the lines start to slope. The envelope, which represents Coulomb's criterion for shear rupturing, is made up of two straight, sloping lines to the left of the brittle–plastic transition. Keep in mind that the angle between the shear rupture and the σ_1 direction is independent of the confining pressure when it comes to failure related to the Coulomb criterion. The envelope's slope steepens as it approaches the σs-axis, taking on characteristics of a section of a parabola. Mohr's criterion is represented by this parabolic portion of the curve, and for failure in this region, the degree to which the Mohr circle touches the failure curve to the left determines how much the angle between the fracture and the σ_1 direction decreases. As we have discussed, the existence of supposed transitional tensile fractures formed at a very small angle to σ_1 is still up for debate. The part of the parabolic envelope with steep slopes specifies failure criteria for these fractures. The failure criterion for tensile cracking is represented by the point where the envelope crosses the πn-axis; however, as we have discussed, this criterion really shouldn't be specified by a point, as a material's tensile strength depends on the dimensions of the flaws it contains. Observe that, for a circle tangent to the composite envelope at T_0, $2\theta = 180$ (or, $\alpha = 0$), meaning that the fracture that forms is parallel to σ_1! Additionally, keep in mind that, as long as the magnitude of the differential stress (the diameter of the Mohr circle) is less than or equal to $4T_0$—that is, the circle whose curvature is equal to the apex of the parabola—no unique value of differential stress is required to cause tensile failure.

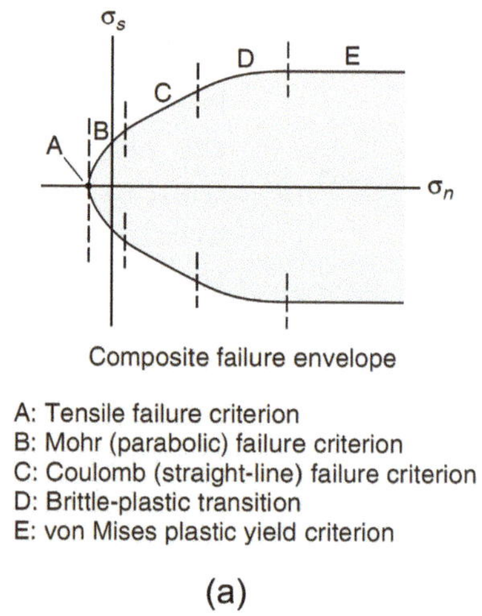

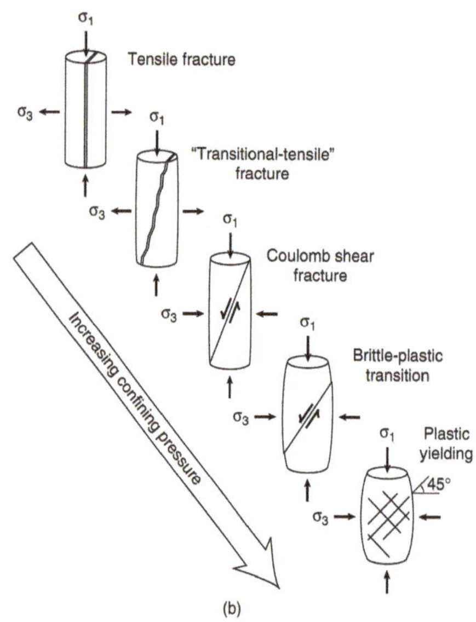

Figure 6.30. (a) A representative composite failure envelope on a Mohr diagram. The different parts of the envelope are labeled, and are discussed in the text. (b) Sketches of the fracture geometries that form during failure. Note that the geometry depends on the part of the failure envelope that represents failure conditions, because the slope of the envelope is not constant.

Source: https://www.geokniga.org/bookfiles/geokniga-es2e-ch06.pdf.

CASE STUDY

BRITTLE INTERFACIAL FAILURE ON A ENI/IAU PWB

Background

Interfacial failures on printed wiring boards using an ENi/IAu surface finish continue to be problem within the electronics industry. This persistent type of assembly quality defect has references in the open literature as early as 1994. The defect level generally is low but there is no definitive root cause. When these catastrophic failures occur on expensive electronic assemblies, they create a major quality and manufacturing crisis.

Failure analysis of the interconnect shows that wetting occurs and a Ni3Sn4 intermetallic layer forms within the solder joint. When an appropriate stress exists during processing or use, a small percentage of the solder joints fracture at the interface between the Ni plating and Ni3Sn4 intermetallic. The low incidence and random nature of the failures suggest that the defects may be due to poor control of the ENi/IAu plating process. Recent hypotheses have targeted the corrosive characteristics of the immersion Au process as the root cause for the weak interfacial strength.

Failure Analysis

When defects occur during new product introduction, particularly with advanced interconnection packages, the initial reaction often is to "try anything" to correct the defect, rather than perform the requisite root cause analysis. Figure 1a shows an Ethernet product with a 10% drop out at assembly during the initial product introduction. Functional testing identifies the defect as an open circuit at pin T1 of a 256 I/O 1.0 mm pitch BGA. These defects raised many issues, because this was the first utilization of 1.0 mm pitch packages on thin (i.e., 0.45mm) PCMCIA substrates at this manufacturing facility. It became imperative to address these questions in a timely, analytical manner to avoid hampering the failure mode analysis and delaying the root cause analysis. Some of the issues and initial suggestions for the cause of the defects include:

- Defective BGA Components: The components may exhibit excessive warpage during surface mount assembly, leading to soldering defects or high residual stresses in the interconnections after assembly. Also, the flexible tape substrate used in the BGA may not provide as good a package interconnection as the more conventional rigid, laminate substrate.
- Defective PWB Substrates: The PWB surface finish gold plating could be too thick, causing gold embrittlement. Also, poor wetting could result from surface contaminants or a nickel oxide layer created by diffusion of the underlying nickel through a porous gold layer.

- Poor Assembly or Manufacturing Processes: The assembly processes used by the contract manufacturer may be "uncontrolled" or "rough." and might not be appropriate for PCMCIA product. The combination of 1.0 mm pitch BGA package and 0.45mm thick substrate may not provide a sufficiently robust interconnection to survive marginal assembly processing and subsequent rough handling

Figure shows cross sections of solder balls in an outer row beneath the 1.0 mm pitch BGA on a defective Ethernet card. These cross sections reveal that the fractures in the corner solder joints are at the PWB pad, not at the BGA flexible tape pad.

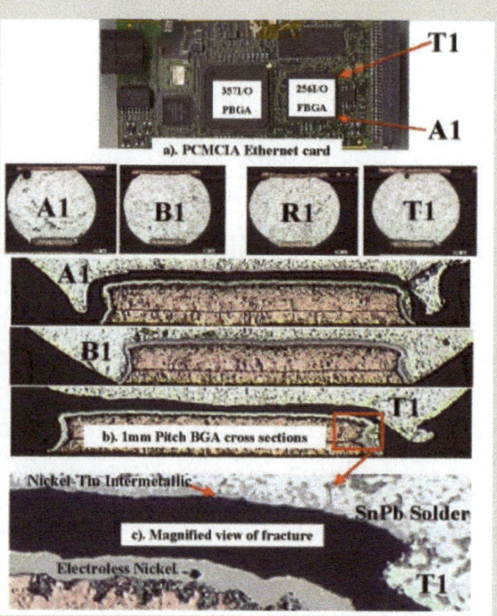

The magnified photomicrograph of the T1 solder joint defect presented in Figure 1c clearly shows a fracture at the PWB pad electroless nickel plating. Comparison of SEM/EDX analysis of non-fractured solder joints to the fractured joints confirms that wetting occurs on the electroless nickel surface and that the fracture path is between the Ni_3Sn_4 intermetallic layer and the plated electroless nickel.

Cross sections and analysis of other BGAs on defective cards reveal that fractures occur on 1.0 mm pitch BGA packages with rigid laminate substrates as well as those with flexible tape substrates. In addition, similar fractures are observed to a lesser extent, on the more frequently used 1.27 mm pitch plastic BGA. With this analysis and the fact that the fractures occur at the PWB pad, the focus of the evaluation shifts to the PWB surface finish.

Further investigation shows that the first production units use PWB substrates from a quick-turn fabricator that outsources the ENi/IAu surface finish plating operation. X-Ray Fluorescence (XRF) measurements of the PWB surface finish indicate a gold thickness of 0.075 µm (3 µin) over 3.75 µm (150 µin) of nickel. Surface analysis by Scanning Auger Microprobe (SAM) reveals no significant contaminates or oxidized nickel at the surface. Auger depth profiling confirms the 0.075 µm (3 µin) gold thickness XRF measurement. The presence of a detectable amount of P (>5 wt. %) in the underlying bulk nickel confirms the plating to be electroless nickel. Consistent with results from earlier published evaluations, optical microscopy, Auger and SEM/EDX do not reveal a root cause for these fractures. Therefore, these defects are assumed to be what is referred to in the industry as brittle nickel-gold fractures.

To verify this assumption and address the quality of the boards and assembly processes, a comparative evaluation of two different sets of PCMCIA Ethernet cards is performed at the contract manufacturing location. One set of Ethernet cards is plated with the ENi/IAu surface finish and is fabricated by a supplier experienced with telecommunication products and the ENi/IAu surface finish. The second set of cards is plated with an immersion silver (IAg) surface finish and is fabricated by a supplier experienced with this alternative surface finish. Approximately 60 PCMCIA Ethernet cards of each surface finish are assembled and a significant number of each are temperature cycled as an accelerated screening test. The ENi/IAu assemblies have approximately a 10% defect rate, while the IAg assemblies have zero defects. Although the failure mode is confirmed to be brittle nickel-gold fractures and the defects are eliminated by using the IAg surface finish, a definitive root cause of the fracture phenomena has not been identified.

CLASS ACTIVITY

This activity extends student understanding of the deformation a rock undergoes as a result of stress, plus several factors that contribute to the behavior. Students work in pairs to explore deformation using Silly Putty™. The student worksheet ties the deformation to the concepts of faulting (earthquakes) and folding, and ultimately concludes with an opportunity for students to design their own deformation experiment. Inclusion of the "Marble Tongs" activity is not required, but reinforces the concept that rock can be elastic.

Students will be able to:

- Demonstrate the three types of stress using Silly Putty™.
- Define strain in a written paragraph.
- Demonstrate brittle, ductile and elastic deformation using common objects.
- Demonstrate that rocks can be elastic with Marble Tongs.
- Describe the cause of an earthquake using the terms stress, strain, elastic rebound, fracture, potential energy, kinetic energy.

SUMMARY

- The brittle regime is where the physical conditions promote brittle deformation mechanisms such as frictional sliding along grain contacts, grain rotation and grain fracture.
- Frictional sliding of grains may be widely distributed throughout a rock volume, but can also be localized to millimeter- to decimeter-wide zones or bands.
- A shear fracture or slip surface is a fracture along which the relative movement is parallel to the fracture.
- Brittle and plastic deformation show different stress–strain curves: the more ductile the deformation, the greater the amount of plastic deformation prior to fracturing.
- Brittle deformation is simply the permanent change that occurs in a solid material due to the growth of fractures and/or due to sliding on fractures once they have formed.
- Longitudinal splitting is similar to tensile cracking except that, in uniaxial compression, the cracks that are not parallel to the σ_1 direction are closed, whereas cracks that are parallel to the compression direction can open up.
- A brittle fault is a surface on which measurable slip has developed without much contribution by plastic deformation mechanisms.
- Cataclasis refers to movement on a fault by a combination of microcracking, frictional sliding of fragments past one another, and rotation and transport of grains.
- A shear fracture is a surface across which a rock loses continuity when the shear stress parallel to the surface is sufficiently large.
- A tensile cracking criterion is a mathematical statement that predicts the stress state in which a crack begins to propagate.

REVIEW QUESTIONS

1. What do you understand by brittle deformation mechanisms?
2. What are the types of fractures?
3. What are the modes of crack-surface displacement?
4. Discuss about the cataclasis and cataclastic flow.
5. Write shot notes on Stress Concentration and Griffith Cracks.

REFERENCES

1. Antonellini, M., & Aydin, A. (1994). Effect of faulting on fluid flow in porous sandstones: Petrophysical properties. *American Association of Petroleum Geologists, 78*(3), 355–377.

2. Atkinson, B. K. (1987). *Fracture Mechanics of Rock*. London: Academic Press.
3. Aydin, A., & Johnson, A. M. (1978). Development of faults as zones of deformation bands and as slip surfaces in sandstones. *Pure and Applied Geophysics, 116*(4–5), 931–942.
4. Bowden, F. P., & Tabor, D. (1950). *The Friction and Lubrication of Solids*. Clarendon Press: Oxford.
5. Brace, W. F., & Bombolakis, E. G. (1963). A note on brittle crack growth in compression. *Journal of Geophysical Research, 68*(12), 3709–3713.
6. Brace, W. F., & Byerlee, J. D. (1966). Stick-slip as a mechanism for earthquakes. *Science, 153*(3739), 990–992.
7. Davis, G. H. (1999). Structural geology of the Colorado Plateau region of southern Utah. *Geological Society of America Special Paper, 342*, 1–157.
8. Engelder, T. (1993). *Stress Regimes in the Lithosphere*. Princeton University Press.
9. Fletcher, R. C., & Pollard, D. D. (1981). Anticrack model for pressure solution surfaces. *Geology, 9*(9), 419–424.
10. Fossen, H., Schultz, R., Shipton, Z., & Mair, K. (2007). Deformation bands in sandstone: A review. *Journal of the Geological Society, London, 164*(4), 755–769.
11. Hancock, P. L. (1985). Brittle microtectonics: Principles and practice. *Journal of Structural Geology, 7*(4), 437–457.
12. Hubbert, M. K., & Rubey, W. W. (1959). Role of pore fluid pressure in the mechanics of overthrust faulting. I: Mechanics of fluid-filled porous solids and its application to overthrust faulting. *Geological Society of America Bulletin, 70*(2), 115–205.
13. Jaeger, J. C., & Cook, N. G. W. (1979). *Fundamentals of Rock Mechanics* (3rd ed.). Chapman and Hall: London.
14. Jamison, W. R. (1989). Fault-fracture strain in Wingate Sandstone. *Journal of Structural Geology, 11*(8), 959–974.
15. Lockner, D. A. (1991). A multiple-crack model of brittle fracture. *Journal of Geophysical Research, 96*(B13), 19623–19642.
16. Mollema, P. N., & Antonellini, M. A. (1996). Compaction bands: A structural analog for anti-mode I cracks in aeolian sandstone. *Tectonophysics, 267*(1–4), 209–228.
17. Narr, W., & Suppe, J. (1991). Joint spacing in sedimentary rocks. *Journal of Structural Geology, 13*(9), 1037–1048.
18. Paterson, M. S. (1978). *Experimental Rock Deformation—The Brittle Field*. Springer-Verlag: New York.
19. Pollard, D. D., & Aydin, A. (1988). Progress in understanding jointing over the past century. *Geological Society of America Bulletin, 100*(8), 1181–1204.
20. Price, N. J. (1966). *Fault and Joint Development in Brittle and Semi-Brittle Rock*. Pergamon Press: London.
21. Rawling, G. C., & Goodwin, L. B. (2003). Cataclasis and particulate flow in faulted, poorly lithified sediments. *Journal of Structural Geology, 25*(3), 317–331.

22. Reches, Z., & Lockner, D. A. (1994). Nucleation and growth of faults in brittle rocks. *Journal of Geophysical Research, 99*(B9), 18159–18172.

23. Scholz, C. H. (1998). Earthquakes and friction laws. *Nature, 391*(6662), 37–42.

24. Scholz, C. H. (2002). *The Mechanics of Earthquakes and Faulting* (2nd ed.). Cambridge University Press: Cambridge.

25. Schultz, R. A. (1996). Relative scale and the strength and deformability of rock masses. *Journal of Structural Geology, 18*(9), 1139–1149.

26. Secor, D. T. (1965). Role of fluid pressure in jointing. *American Journal of Science, 263*(7), 633–646.

27. Segall, P., & Pollard, D. D. (1983). Nucleation and growth of strike slip faults in granite. *Journal of Geophysical Research, 88*(B1), 555–568.

28. Underhill, J. R., & Woodcock, N. H. (1987). Faulting mechanisms in high-porosity sandstones: New Red Sandstone, Arran, Scotland. In M. E. Jones & R. M. F. Preston (Eds.), *Deformation of Sediments and Sedimentary Rocks*. Geological Society Special Publication, *29*, 91–105.

CHAPTER

7 Faults

LEARNING OBJECTIVES

After studying this chapter, you will be able to:

- Focus on fault terminology
- State the definition, parts, and types of fault
- Focus on fault anatomy
- Discuss about displacement distribution
- Identify faults in an oil field setting
- Discuss the birth and growth of faults

KEY TERMS FROM THIS CHAPTER

Dip fault
Fault surface
Graben fault
Horst fault
Parallel fault
Rift fault
Step fault
Earthquakes
Fault zone
Hanging wall
Normal fault
Reverse fault
Seismic hazards
Strike fault

INTRODUCTION

Massive rock slabs in the lithosphere of Earth make up the crust. the African, Antarctic, Eurasian, Indo-Australian, North American, Pacific, and South American tectonic plates are the seven major tectonic plates that make up Earth. Numerous minor tectonic plates exist in addition to these major tectonic plates. The tectonic plates that make up the Earth's crust come together and arrange themselves like a big puzzle. The convection currents beneath the surface of the Earth drive these tectonic plates to move slowly and continuously. The tectonic plates are moved by convection currents, which can cause them to spin and move past or toward one another. A plate boundary is the location where two tectonic plates converge. A particular kind of fault forms at a plate boundary based on the movement of the plates. A fault is a break in the rocks that comprise the crust of the Earth. Normal faults, reverse faults, and strike-slip faults are the three categories into which scientists divide faults.

Layered rock sequences are disturbed by faults, which cause "faults" or "defects" to be introduced into the main stratigraphic framework. Faults are very interesting structures that have fascinated structural geologists as much as they have frustrated stratigraphers and miners, despite posing challenges to stratigraphers and geologists mapping rocks in the field or interpreting seismic data. Because of the significance of faults to the petroleum industry, we have a lot more knowledge about them now than we did a few decades ago. Active faults are strongly linked to earthquakes and seismic hazards, and faults also pose difficulties for tunnel operations and waste repositories. With examples and applications pertinent to the petroleum industry, we will concentrate on fault geometry, fault anatomy, and the evolution of faults and fault populations in this chapter.

7.1. FAULT TERMINOLOGY

Large amounts of strain in the upper crust can be accommodated by faults, which are far more complex compound features. Different geologists use the term "fault" in different contexts. A fault is any surface or narrow zone that has visible shear displacement along the zone, according to a straightforward and conventional definition.

Some geologists use the terms interchangeably because this definition is nearly the same as that of a shear fracture. Microfaults are the term used by geologists to describe shear fractures that have offsets on the millimeter to centimeter scale. Most geologists, however, would only use the term "failure" to describe more composite structures with offsets of at least a meter, reserving the term "shear fracture" for smaller structures.

Another problem with a fault is its thickness. When faults are closely examined, it becomes clear that they are made up of fault rock material and subsidiary brittle structures, giving them a defined thickness. Faults are commonly represented as planes and surfaces in oral and written communication as well as sketches. On the other hand, the thickness is typically orders of magnitude smaller than the fault length and much smaller than the offset. The degree of observation, the goals, and the requirement for accuracy all play a major role in determining whether a fault is better categorized as a zone or a surface.

Faults are typically complex zones of deformation that include deformation bands, subsidiary fractures, and several slip surfaces. This is especially evident when taking into account large faults that have offsets on the kilometer scale. On a map or seismic line, these faults may appear as single faults, but upon closer inspection in the field, it becomes clear that they are actually composed of several smaller faults. Stated differently, the significance of the scale dependency is noteworthy for descriptive structural geologists. A fault is a tabular volume of rock made up of a central slip surface or core formed by intense shearing and a surrounding volume of rock affected by more gentle brittle deformation genetically and spatially related to the fault. This has led most geologists to define a fault as a volume of brittlely deformed rock that is relatively thin in one dimension.

Fault can also refer to deformation mechanisms (plastic or brittle). Both brittle discontinuities and ductile shear zones dominated by plastic deformation are loosely referred to as faults in this informal sense. When talking about significant faults on seismic or geologic sections that pierce most or all of the crust, this is occasionally implied. If the deformation mechanism needs to be specified, the term "brittle fault" (as opposed to "ductile shear zone") may be used. The term "brittle fault" is redundant because, in most cases, geologists implicitly limit the term to slip or shear discontinuities dominated by brittle deformation mechanisms: "A fault is a discontinuity with wall-parallel displacement dominated by brittle deformation mechanisms."

A discontinuity characterized by wall-parallel displacement and brittle deformation mechanisms is called a fault. In this case, discontinuity mainly refers to layers, i.e., faults split apart and discontinue rock layers. Faults, however, also signify discontinuities in displacement and mechanics. The displacement field rapidly varies across faults in both map view and cross-section, as shown in Figure 7.1. Therefore, a kinematic definition can be added: A fault is a discontinuity in the velocity or displacement field associated with deformation. This definition is especially helpful for experimental work and GPS-monitoring of active faults.

Shear fractures are not the same as faults because a simple shear fracture cannot grow into a larger structure in its own plane. On the other hand, faults can expand through the formation of a multifaceted process zone containing a large number of tiny fractures, some of which join to form the fault slip surface and the remainder of which are abandoned.

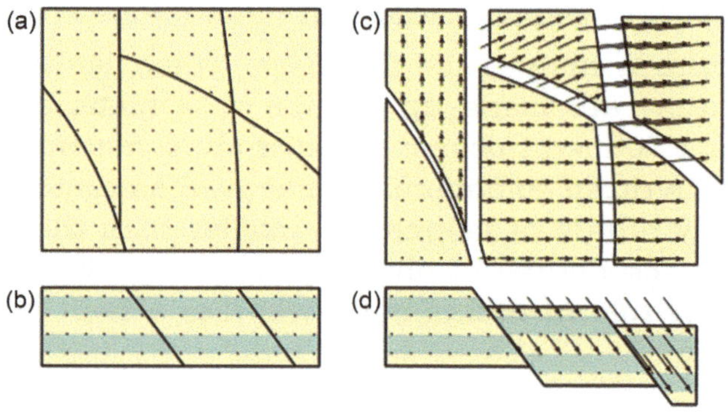

Figure 7.1. Faults appear as discontinuities on velocity or displacement field maps and profiles. The left blocks in the undeformed map (a) and profile (b) are fixed during the deformation. The result is abrupt changes in the displacement field (arrows) across faults.

Source: Haakon Fossen (2010). Structural Geology. Cambridge University Press. ISBN: 978-0-521-51664-8.

7.1.1. Geometry of Faults

The hanging wall and the underlying footwall are divided by non-vertical faults (Figure 7.2). A typical fault occurs when the hanging wall is lowered or downthrown in relation to the footwall. A reverse fault occurs in the opposite situation, where the hanging wall is upthrown in relation to the footwall. When there is lateral movement, i.e., if the fault is strike-slip in the horizontal plane. The Latin terms sinister and dexter, which signify left and right, respectively, can be used to describe strike-slip faults that are either sinistral (left-lateral) or dextral (right-lateral).

The whole range from vertical to horizontal faults is found in naturally deformed rocks, though some fault dip ranges are more common than others. For example, strike-slip faults usually occur as steep faults, and reverse faults typically have lower dips than normal faults. Low-angle faults have a dip angle of less than thirty degrees, whereas steep faults have a dip angle of more than sixty degrees. Thrust faults are low-angle reverse faults that move tens or hundreds of kilometers in either direction.

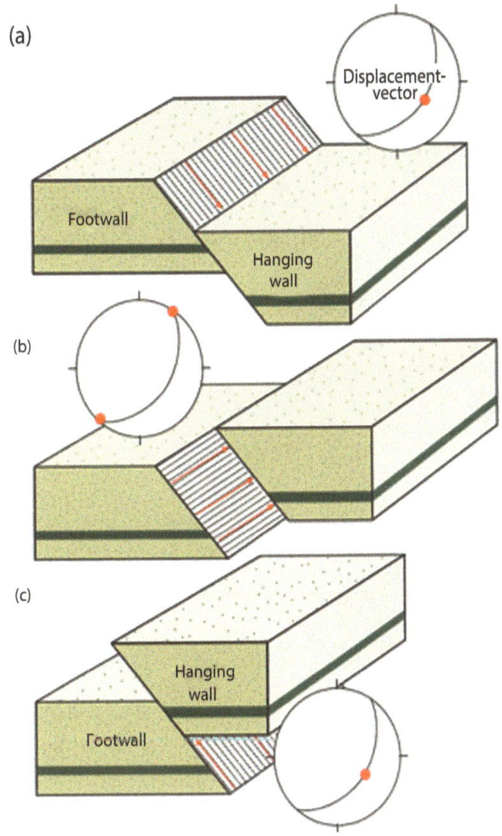

Figure 7.2. Normal (a), strike-slip (sinistral) (b) and reverse (c) faults. These are end-members of a continuous spectrum of oblique faults. The stereonets show the fault plane (great circle) and the displacement vector (red point).

Source: Haakon Fossen (2010). Structural Geology. Cambridge University Press. ISBN: 978-0-521-51664-8.

Listric faults are defined as downward-flattening faults (Figure 7.3), whereas downward-steepening faults are occasionally referred to as antilistric faults. Originally from thrust fault terminology, the terms ramps and flats refer to alternating steep and subhorizontal sections of any fault surface. For instance, a fault with ramp-flat-ramp geometry would change from steep to flat and back again.

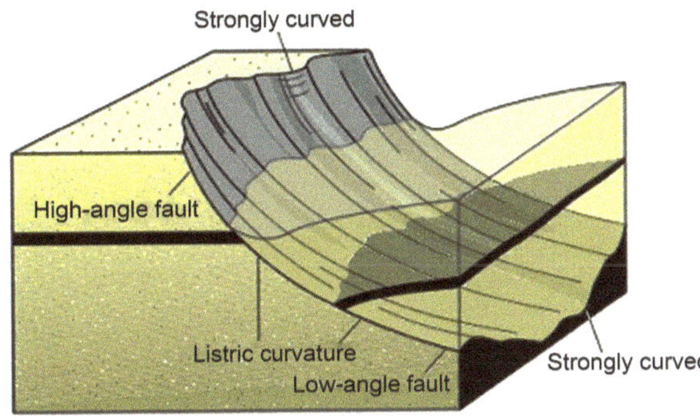

Figure 7.3. Listric normal fault showing very irregular curvature in the sections perpendicular to the slip direction. These irregularities can be thought of as large grooves or corrugations along which the hanging wall can slide.

Source: Haakon Fossen (2010). Structural Geology. Cambridge University Press. ISBN: 978-0-521-51664-8.

The section perpendicular to the fault slip direction is where irregularities are most prevalent. This translates to curved fault traces in map view for normal and reverse faults, as the faults of the extensional oil field in Figure 7.4 demonstrate. As long as the axes of the irregularities align with the slip vector, there is no conflict during fault slippage due to the irregularities in this section. If the slip direction exhibits irregularities as well, the hanging wall and/or footwall will need to undergo deformation. For instance, a hanging-wall rollover is usually caused by a listric normal fault.

Any shape of fault can exist perpendicular to the slip direction; however, non-linearity in the slip direction causes space problems that can cause strain on the hanging or footwall.

Figure 7.4. The main faults in the North Sea Gullfaks oil field show high degree of curvature in map view and straight traces in the vertical sections (main slip direction). Red lines represent some of the well paths in this field. From Fossen and Hesthammer (2000).

Source: Haakon Fossen (2010). Structural Geology. Cambridge University Press. ISBN: 978-0-521-51664-8.

Traditionally, a group of subparallel faults or slip surfaces that are near enough to one another to define a zone have been referred to as fault zones. The width of the zone varies according to the scale of observation; for example, when studying large-scale faults like the San Andreas Fault, it can range from centimeters or meters in the field to the order of a kilometer or more. These days, the term "fault zone" is also applied erratically to the core of the fault, which encompasses the surrounding deformation zone and the core itself, or to the central portion of the fault where most or all of the original rock structures have been destroyed. Any usage of the term "fault zone" needs to be clarified because it is still somewhat confusing and is widely used in the literature related to petroleum today.

A graben is a downthrown block formed by two distinct normal faults dipping toward one another (Figure 7.5). A typical fault that dips apart from one another produces an upthrown block known as a horst. Minor faults that may be synthetic or antagonistic are linked to the largest faults in a faulted area, known as master faults. A synthetic fault dips in the same direction as the master fault, whereas an antithetic fault dips in the opposite direction (Figure 7.5). These are relative statements that only make sense when specific larger-scale faults are connected to minor faults.

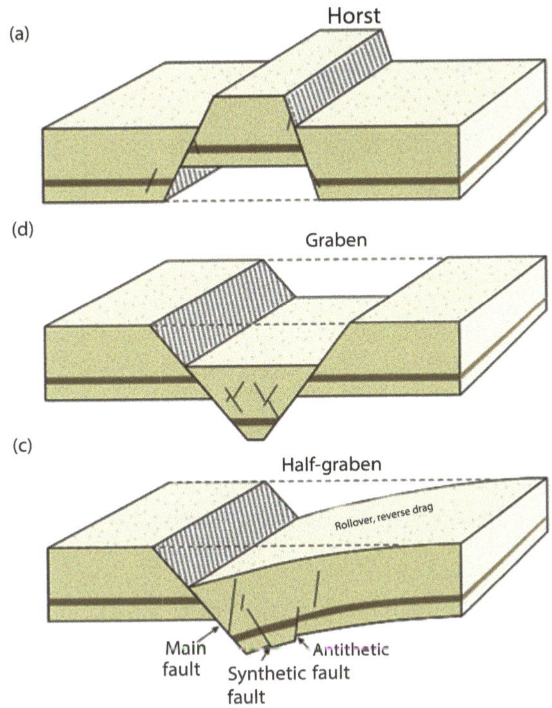

Figure 7.5. A horst (a), symmetric graben (b) and asymmetric graben (c), also known as a half-graben. Antithetic and synthetic faults are shown.

Source: Haakon Fossen (2010). Structural Geology. Cambridge University Press. ISBN: 978-0-521-51664-8.

7.1.2. Displacement, Slip, and Separation

The local displacement vector or net slip direction is indicated by the vector that joins two points that were connected before faulting (Figure 7.6). Normal and reverse faults have displacement vectors in the dip direction, whereas strike-slip faults ideally have a horizontal slip direction. On the majority of faults, the total slip that we see is typically the result of multiple increments, or earthquakes, each with a unique displacement or slip vector. It's possible that the slip directions varied for each individual slip event. This brings us back to the distinction between deformation history and deformation sensu stricto, which only considers the relationship between the undeformed and deformed states. We could search for indicators of the slip history in the field, like multiple striations.

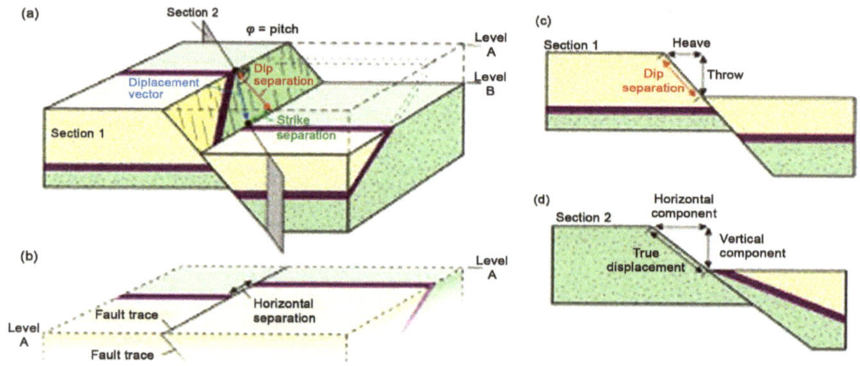

Figure 7.6. Illustration of a normal fault affecting a tilted layer. The fault is a normal fault with a dextral strike-slip component (a), but appears as a sinistral fault in map view (b, which is the horizontal section at level A). (c) and (d) show profiles perpendicular to fault strike (c) and in the (true) displacement direction (d).

Source: Haakon Fossen (2010). Structural Geology. Cambridge University Press. ISBN: 978-0-521-51664-8.

The displacement field, also known as the slip field, on the surface is given by a set of displacement vectors over the slip surface. Information about the direction, sense, and amount of slip is provided to the field geologist by striae, kinematic indicators, and layer offsets.

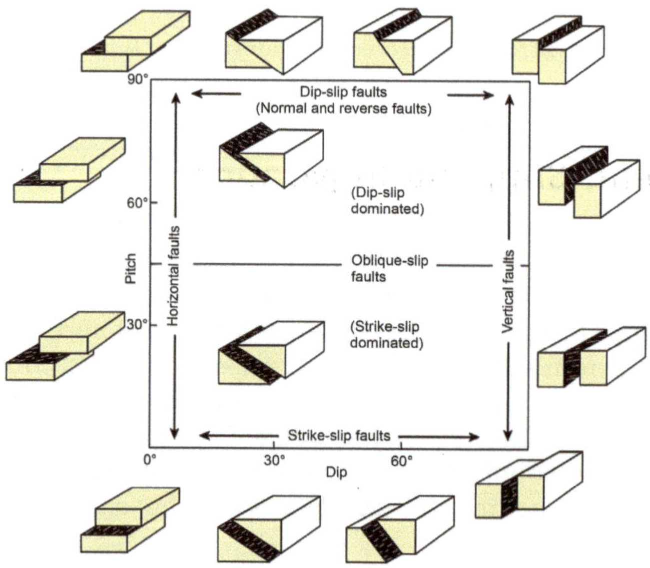

Figure 7.7. Classification of faults based on the dip of the fault plane and the pitch, which is the angle between the slip direction (displacement vector) and the strike. Based on Angelier (1994).

Source: Haakon Fossen (2010). Structural Geology. Cambridge University Press. ISBN: 978-0-521-51664-8.

Since the net slip vector is oblique, many faults exhibit some deviation from true dip-slip and strike-slip displacement. We refer to these types of faults as oblique-slip faults (Figure 7.7). Pitch, also known as rake, is the angle formed by the slip vector (striation) and the slip surface strike. It indicates the degree of obliquity.

The offset shown on any section through the faulted volume, whether it be an outcrop or a seismic section, could deceive us if we don't know the actual displacement vector (Figure 7.6b). The (apparent) separation is the apparent displacement that is seen on a section or plane. The layers seen on a horizontal exposure or map are said to be separated horizontally (Figure 7.6b), whereas the layers seen in a vertical section are said to be separated diply (Figure 7.6c). The dip separation in a vertical section can be broken down into a vertical and a horizontal separation. Keep in mind that this horizontal separation is not the same as what Figure 7.6b depicts. The terms "heave" (horizontal component) and "throw" (vertical component) are more frequently used to describe these two separations that are shown in a vertical section (Figure 7.6c). The true displacement or total slip on the fault is only visible in the section that contains the true displacement vector (Figure 7.6d).

When a fault occurs in a layered sequence, it will split each surface (the stratigraphic interface) into three dimensions, resulting in the appearance of two fault cutoff lines (Figure 7.8). A map of the faulted surface will display an open area between the two cutoff lines if the fault is non-vertical and the displacement vector is not parallel to the layering. Both the fault dip and the fault's dip separation are related to the width of the unconstrained open space. Moreover, the aperture mirrors the horizontal division, or heave, observed on the fault's vertical segments (Figure 7.8).

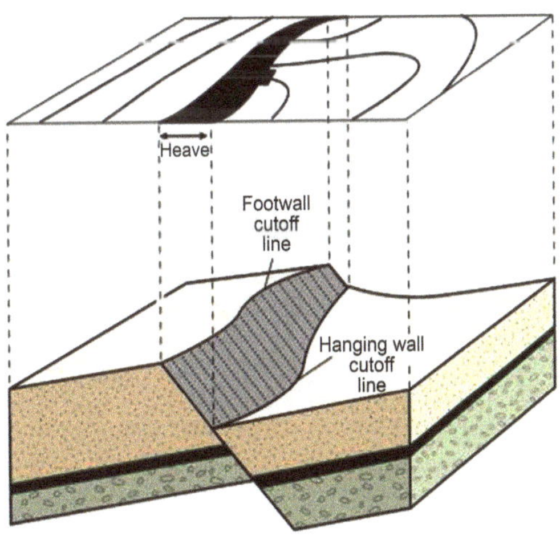

Figure 7.8. The relationship between a single fault, a mapped surface and its two fault cutoff lines. Such structure contour maps are used extensively in the oil industry where they are mainly based on seismic reflection data.

Source: Haakon Fossen (2010). Structural Geology. Cambridge University Press. ISBN: 978-0-521-51664-8.

7.1.3. Stratigraphic Separation

At the fault cut, where the wellbore crosses the fault, drilling through a fault causes either a repeated section or a missing section. It is straightforward for vertical wells: reverse faults result in repeated stratigraphy in the well, whereas normal faults omit stratigraphy (Figure 7.9a). Stratigraphic repetition is observed across normal faults for deviated wells, such as well G in Figure 7.9b, where the well bore plunge is less than the fault dip. Stratigraphic separation is the general term for the stratigraphic section that is absent or repeated in wells drilled through faults. If the strata are horizontal, stratigraphic separation—a measurement of fault displacement obtained from wells in subsurface oil fields—is equal to the fault throw. Since most faulted strata are not horizontal, it is necessary to compute or construct the throw.

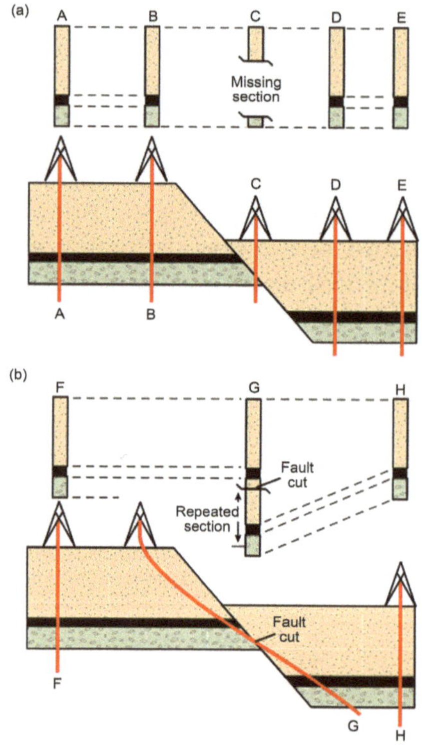

Figure 7.9. (a) Missing section in vertical wells (well C) always indicates normal faults (assuming constant stratigraphy). (b) Repeated section (normally associated with reverse faults) occurs where the normal fault is steeper than the intersecting wellbore (well G).

Source: Haakon Fossen (2010). Structural Geology. Cambridge University Press. ISBN: 978-0-521-51664-8.

7.2. DEFINITION, PARTS, AND TYPES OF FAULT

Let's examine the flaws now. They are defined as the natural fissures where the opposing walls have moved relative to one another. A fault introduces "faults" or "defects" into the stratigraphic framework by severing rock sequences. When geologists are mapping rocks in the field, they present difficulties. Because of the importance of faults in the petroleum industry, we now know a great deal more about them than we did a few decades ago. Faults also cause issues when choosing a site for civil engineering purposes.

A fault is a fracture along which the two rock blocks are visibly displaced. The rock adjacent to it is moved by it. These rock blocks' displacements are nearly parallel to the fracture surface. Faults are found in many mountain belts, continents, and oceans, and they come in all sizes similar to folds. Faults in rock are frequently the result of tectonic stresses, which can be tangential, compressional, tensional, or of any combination of these types.

7.2.1. Definition of Faults

Faults are the cracks where one block has moved in relation to other blocks. The range of this movement is contingent upon the type and strength of the stresses and resistance provided by the rocks, and can range from a few centimeters to several kilometers.

7.2.2. Importance of Faults in Geological Studies

Faults are significant in geological studies because:

1. Faults may bring material from the deeper level to or near the surface of the Earth, and thus they provide us knowledge of the subsurface.
2. Many mineralized solution or petroleum rise up along the fault. Faults sometime trap petroleum from migration and loss.
3. Faults may develop beautiful landscapes which enhance geotourism.
4. They may cause earthquakes.
5. Fault locations are very important in mining.

7.2.3. Parts of a Fault

The following are important from the subject point of view:

- *Fault Plane*: A plane along which the rupture has actually taken place or where one block is moved with respect to other is known as 'Fault Plane.' It may be noted that such a plane is generally formed along the line of least resistance.
- *Hanging wall and Footwall*: The upper block or, in other words, the block above the fault plane is called 'Hanging wall.' The block below the fault plane or, in other words, beneath the fault plane is called the Footwall.
- *Hade*: It is the inclination of the fault plane that is vertical.
- *Throw*: It is the vertical displacement between the Hanging wall and Footwall.
- *Heave*: It is the horizontal displacement between the Hanging wall and Footwall.

7.2.4. Types of Faults

The direction of slip along the fault plane and the angle, or dip, that the fault plane makes with the Earth's surface are the primary classification criteria for faults. Faults can be classified as follows based on the direction of slip.

- *strike-slip*, where the offset is predominantly horizontal, parallel to the fault trace (Allaby, Michael (2015));
- *dip-slip*, offset is predominantly vertical and/or perpendicular to the fault trace; or
- *oblique-slip*, combining strike-slip and dip-slip.

Depending upon the inclination of the fault number of types of faults are recognized.

7.2.4.1. Normal Fault

A fault in which Hanging wall (HW) has apparently come down with respect to the Footwall (FW) is termed as 'Normal Fault' (Oskin, Michael E. (2019)).

7.2.4.2. Reverse Fault

A fault in which hanging wall has apparently gone up with respect to the Footwall is termed as 'Reverse Fault.'

The only difference between the Normal Fault and Reverse Fault is that, in Normal Fault the Hanging wall is downward with respect to the Footwall whereas in a Reverse Fault the apparent movement of the Hanging wall is upwards with respect to the Footwall.

7.2.4.3. Thrust Fault

A fault which is a very small angle of hade (i.e., the inclination of fault plane with the vertical plane is very small) and the Hanging wall that apparently goes up with respect to the Footwall is called 'Thrust Fault.'

7.2.4.4. Vertical Fault

A fault in which the fault plane is vertical (having an angle of hade up to 5 degrees) and either of the walls has moved upwards or downwards.

7.2.4.5. Horst Fault

Horst (German, Horst = upthrow) Fault is one in which wedge shaped block has gone up with respect to the side blocks.

7.2.4.6. Graben Fault or Rift Fault or Trench Fault

Graben (German, Graden= Trench) Fault is one in which wedge shaped block has down with respect to side block.

7.2.4.7. Dip Fault or Transverse Fault

A Dip Fault whose strike is parallel to the dip of the strata is called 'Transverse Fault.'

7.2.4.8. Strike Fault

A Strike Fault is one whose strike is parallel to the strike of the strata.

7.2.4.9. Parallel Fault

A series of faults running more or less parallel to one another and all handing in the same direction is called 'Parallel Fault.'

7.2.4.10. Step Fault

The term 'Step Fault' is applied to that parallel fault where downthrown of all is in the same direction and it gives a step-like arrangement.

7.2.5. Causes of Faulting

We may categorize the causes of folding and faulting into tectonic and non-tectonic categories.

- *Tectonic Causes*: The tectonic reasons developed due to compressive forces that may include orogenic (mountain building) forces and endogentic forces which are generated from within the Earth.
- *Non-tectonic Causes*: The non-tectonic reasons which may cause local faulting may be due to hillside creep, collapse or gravity sliding, erosion of certain bed, glacial ice push, solution, differential compaction and contemporaneous or synsedimentary deformations.

7.2.6. Field Evidence of Faulting

Field evidence of faulting can be divided into two groups.

They are:

- *Lithological Evidence:* This evidence includes:
    i. *Slickensides*: The movements of one wall against another along fault results in polishing and grooving of one or both surfaces. These are known as 'Slickensides.' The direction of the movement is indicated by the trend of the striations or grooves.
    ii. *Fault Breccia and Gouge*: Along some faults the rocks are found highly fractured or even crushed to angular fragments. Such crushed rocks are called 'Fault Breccia.' When the dislocation forces are very severe, as is frequently the case in thrusting, the rock may be ground to fine clay like powder called 'Gouge.'
    iii. *Drag*: Drag is the minor folding of strata along the walls of a fault. It is caused by fault displacement.
    iv. *Dislocations*: The displacement of beds, igneous dykes; veins, etc. along a fault may be seen in either plan or section.
    v. *Repetition or Omission of Beds*: The repetition and omission of beds often establish the fault.

vi. *Abrupt Termination of Structures:* An abrupt termination of structures such as folds, beds or dykes along a common line or zone suggests faulting.

vii. *Silicification and Mineralization*: Action water while percolation through a fault zone may deposit fine-grained Quartz causing Silicification. Many mineral deposits have also been localized along faults.

- *Physiographic Evidence:* This evidence includes Fault Scrap, Fault Line Scrap and Fault Control of Streams:

  i. **Fault Scrap:** An actual surface of fault displacement may stand up unmodified by erosion as in escarpment or cliff. It is called a 'Fault Scrap.' In this case the escarpment faces towards the down throw side.

  ii. **Fait Line Scrap:** Fault frequently brings together resistant and non-resistant rocks. The resistant rock will stand out prominently as ridge along a fault zone. Such ridges that will generally face the upthrow side of the faults are called 'Fault Scrap Side.'

  iii. **Fault Control or Streams:** Streams may be guided in the direction and course of their flow by faulting such stream, which may follow a straight line or make approximately right angle turns.

7.2.7. Effects of Faults on Outcrop

1. **Effect of Dip Fault:** The effects of Dip Fault is to cause lateral displacement in the outcrops. The amount of displacement becomes less with increase in the dip of rocks and in vertical strata the displacement of outcrops will be nil.
2. **Effect of Strike Fault:** The effects of Strike Fault are either to cause a repetition of the outcrops of the beds or to eliminate the outcrops of some of the beds altogether. Repetition of the outcrops occur when a Strike Fault Hades in the opposite direction to the dip of strata.

7.2.8. Engineering Considerations of Faults

1. Strictly speaking on site should be selected on a fault for any major project because movements along the existing fault plane is much easier than any other planes.
2. For major projects like a dam, tunnel, etc., a site that is highly faulted should be avoided because the engineer may have to face much troubles sooner or later. The structure constructed on fault may collapse at any moment even due to slight disturbance.
3. If the project is of a scattered nature like electric or telegraphic poles the work can be carried out without much of a risk.
4. Safety factor building modification should bear the initial shocks of an earthquake of low intensity.
5. Faults are responsible for lakes, swamps and marshy places.
6. Fault traces often provide potential springheads.
7. Fault zones often form oil traps.

7.3. FAULT ANATOMY

Typically, faults are depicted as single, uniformly thick lines on seismic or geologic sections. Faults are rarely simple surfaces or zones of constant thickness, though. As a matter of fact, the majority of faults are actually intricate structures made up of several unpredictable structural components. It is difficult to describe a fault in a straightforward and universal way due to the differences in expression that exist both along and between faults. The fault damage zone, which is the surrounding volume of brittlely deformed wallrock, should generally be distinguished from the central fault core, also known as the slip surface, as shown in Figure 7.10.

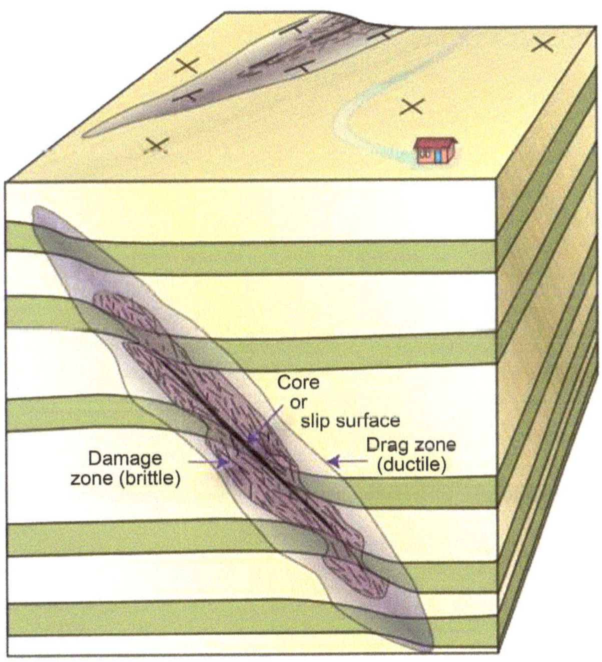

Figure 7.10. Simplified anatomy of a fault.

Source: Haakon Fossen (2010). Structural Geology. Cambridge University Press. ISBN: 978-0-521-51664-8.

The fault core can range from a simple slip surface with a cataclastic zone that is less than millimeters thick through a zone that has multiple slip surfaces to an intensely sheared zone that is up to several meters wide and contains only the remains of the main rock structures. When clay minerals have formed at the expense of feldspar and other primary minerals, the fault core of crystalline rocks can consist of nearly noncohesive fault gouges. In other situations, the fault core is made up of hard, flinty cataclasites, especially when the fault originated in the lower region of the brittle upper crust. Fault cores also contain a variety of cohesive and non-cohesive breccias. Extreme frictional conditions can cause localized, transient melting of crystalline rocks, resulting in the glassy fault rock known as pseudotachylyte.

Fault cores in soft, sedimentary rocks are usually composed of non-cohered, smeared-out layers. Sometimes soft layers like silt and clay smeared out to a continuous membrane can significantly reduce the ability of fluids to cross the fault if the membrane is continuous in three dimensions. Even within the same lithology, there are significant variances in the thickness of the fault core, which generally increases positively with fault throw.

A density of brittle deformation structures higher than the background level characterizes the damage zone. It is present in the tip zone and on all sides of the fault core, enveloping the fault core. Deformation bands, shear fractures, tensile fractures, and stylolites are among the structures that can be found in the damage zone. Figure 7.11 provides an example of how these small-scale structures, such as deformation bands, only appear in close proximity to the fault core, in this case defining a footwall damage zone width of about 15 meters.

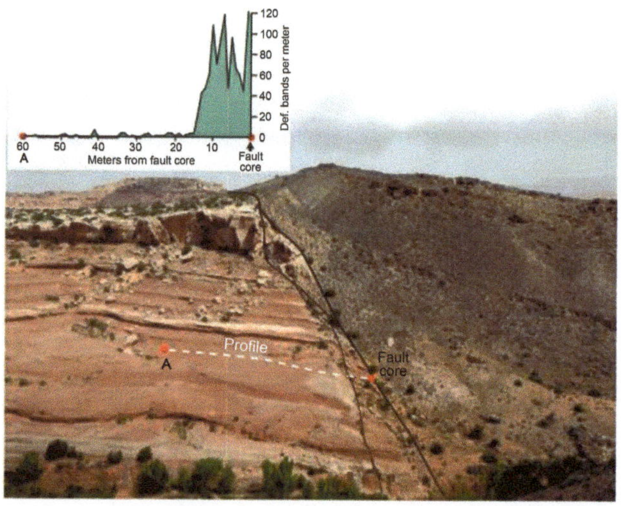

Figure 7.11. Damage zone in the footwall to a normal fault with 150–200 m throw. The footwall damage zone is characterized by a frequency diagram with data collected along the profile line. A fault lens is seen in the upper part of the fault. Entrada Sandstone near Moab, Utah.

Source: Haakon Fossen (2010). Structural Geology. Cambridge University Press. ISBN: 978-0-521-51664-8.

As with the fault core, there is a positive correlation between fault displacement and damage zone thickness; however, the width of the damage zone can vary from layer to layer (Figure 7.12a). Straight lines in logarithmic diagrams, like the one in Figure 7.12, denote a constant relationship between the two plotted parameters, which is why fault analysis uses them so frequently. In particular, the distance between adjacent lines in this figure represents one order of magnitude, and for data that plot along one of the straight lines in this figure, the ratio between fault displacement D and damage zone thickness DT is the same for any fault size. For the faults with displacements up to 100 meters, at least, the fault displacement is near or slightly larger than the damage zone thickness, as indicated by the majority of the data in Figure 7.12a, which plot around or above the line D = DT. This diagram can be used to estimate throw from damage zone width or vice versa, but there is a significant amount of uncertainty due to the wide range of data (more than two orders of magnitude).

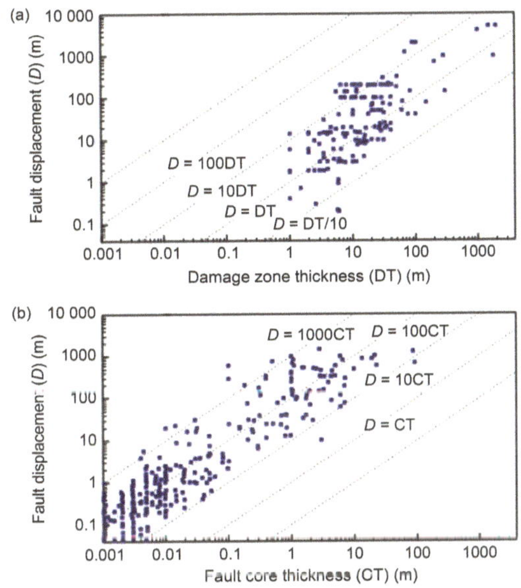

Figure 7.12. (a) Damage zone thickness (DT) (one side of the fault) plotted against displacement (D) for faults in siliciclastic sedimentary rocks. (b) Similar plot for fault core thickness (CT). Note logarithmic axes. Data from several sources.

Source: Haakon Fossen (2010). Structural Geology. Cambridge University Press. ISBN: 978-0-521-51664-8.

Fault displacement and fault core thickness (CT) have a comparable relationship (Figure 7.12b). The straight lines D = 1000CT and D = 10CT place constraints on this relationship, indicating that for faults with displacements up to 100 meters, the fault core is statistically approximately 1/100 of the fault displacement.

Especially in faulted sedimentary rocks, layers frequently deflect (fold) around faults. Drag is the traditional term for this behavior, and it should only be used in a descriptive or geometric sense. The drag zone may exist entirely or it may be greater or smaller than the damage zone. The damage zone and the drag zone are different in that the former is limited by definition to brittle deformation, while the latter is an expression of strain related to ductile faults. Both of them are a component of the entire strain zone connected to faults. Generally speaking, soft rocks generate greater drag than stiff rocks.

7.4. DISPLACEMENT DISTRIBUTION

Displacement variations can occasionally be mapped in both the horizontal and vertical directions along a field fault. As seen in Figure 7.13, faults typically exhibit a maximum displacement in the middle of the fault trace, which progressively decreases toward the tips in both directions. The displacement profile can be elliptic, bell-shaped, or linear in shape. There are two types of displacement profiles: those with a broad, central portion of relatively constant displacement (plateau type) and those with a well-defined central maximum (peak type). Illustrations are displayed in Figure 7.14.

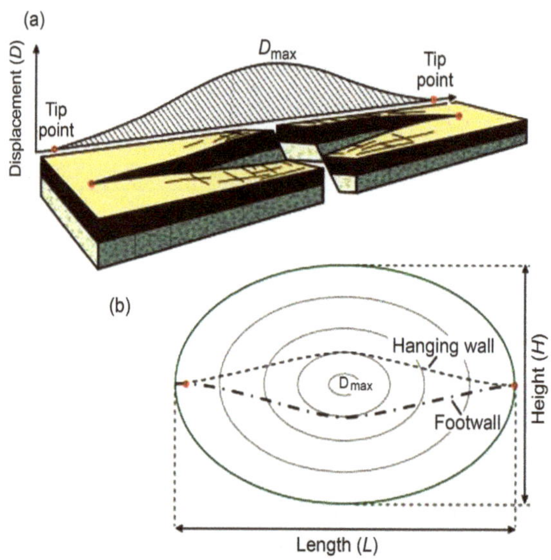

Figure 7.13. (a) Schematic illustration of an ideal, isolated fault. The displacement profile indicates maximum displacement near the center. (b) The fault plane with displacement contours. Stippled lines are the hanging-wall and footwall cutoff lines and the distance between them indicates the dip separation.

Source: Haakon Fossen (2010). Structural Geology. Cambridge University Press. ISBN: 978-0-521-51664-8.

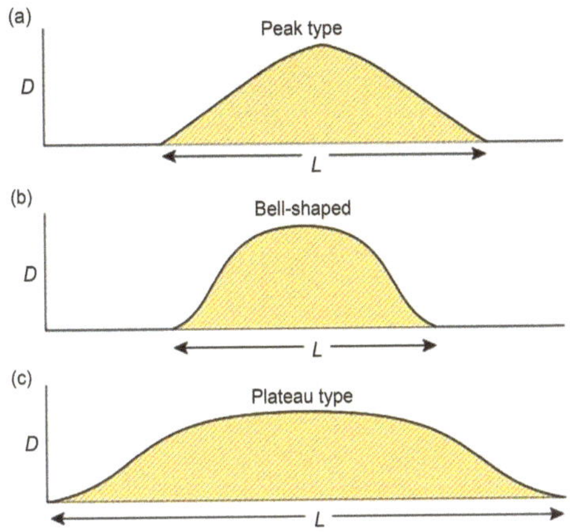

Figure 7.14. Types of displacement (D) profiles along faults.

Source: Haakon Fossen (2010). Structural Geology. Cambridge University Press. ISBN: 978-0-521-51664-8.

In most cases, displacement from the tip line toward a central point increases gradually in single faults. To get a clear picture of the displacement distribution on the fault surface, it could be challenging to gather enough displacement data from a single fault. However, contour displacement on several faults has been made possible by coalmine data sets and high-quality 3-D seismic data. An illustration of these studies is Figure 7.15, which illustrates the general pattern that the displacement is typically largest in the center of a single, isolated fault and gradually decreases toward the tip line. This conclusion is in line with the field observations that were discussed in the previous paragraph. Therefore, the idealized model depicted in Figure 7.16, which features an isolated fault with elliptical displacement contours and an elliptical tip line, is supported by these observations. When dealing with extension fractures, where the displacement vectors are perpendicular to the fracture, an elliptical model that is geometrically similar to one can be used. It is important to note that an isolated fault in an isotropic medium is intended to be described by the simple elliptical fault model. Generally, fault growth is complicated by mechanical layering and fault interaction, leading to deviations from this basic model.

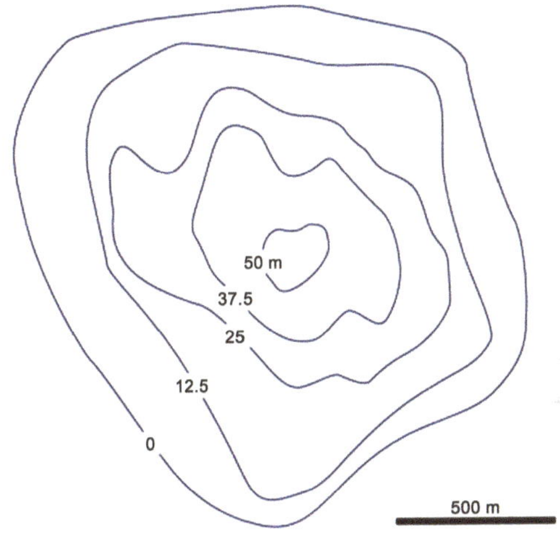

Figure 7.15. Displacement contours for a fault interpreted from high-resolution seismic data from the Gulf of Mexico. Modified from Childs et al., (2003).

Source: Haakon Fossen (2010). Structural Geology. Cambridge University Press. ISBN: 978-0-521-51664-8.

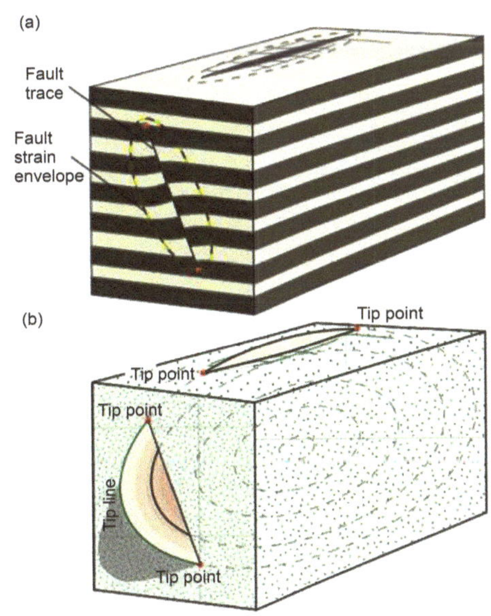

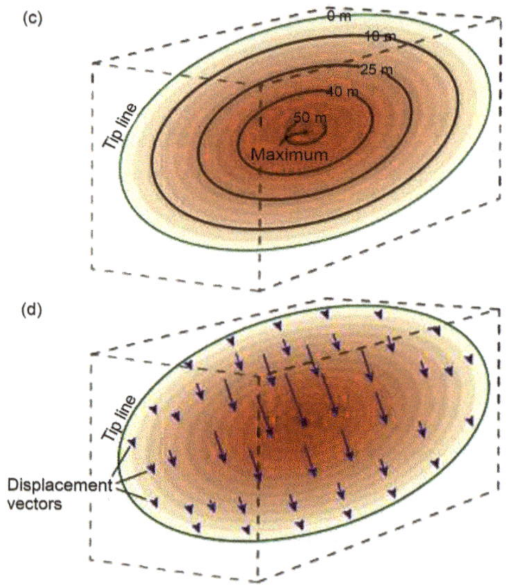

Figure 7.16. Geometric aspects of an isolated fault (elliptical fault model). (a) The fault trace is the intersection between the fault surface and an arbitrary surface (outcrop, seismic line). The endpoints of the fault trace are called tip points (b). They lie on the tip line, which is the zero-displacement line that outlines the fault (c). Displacement increases towards the center of the fault. This can be expressed in terms of displacement contours (c) or displacement vectors (d).

Source: Haakon Fossen (2010). Structural Geology. Cambridge University Press. ISBN: 978-0-521-51664-8.

7.5. IDENTIFYING FAULTS IN AN OIL FIELD SETTING

In the process of discovering and producing petroleum, it is essential to gather and accurately interpret data regarding faults. Here, we'll examine a few data sources that offer essential details regarding faults in an extensional oil field; the same ideas hold true for contractional and strike-slip regimes.

7.5.1. Seismic Data

Faults in the subsurface are most commonly located and mapped through the interpretation of seismic data. Fault identification is contingent upon the existence of seismic reflectors that can be mapped. As seen in Figure 7.17, discontinuous reflectors identify the locations of faults, and the correlation of seismic reflectors across the fault is used to estimate the dip separation.

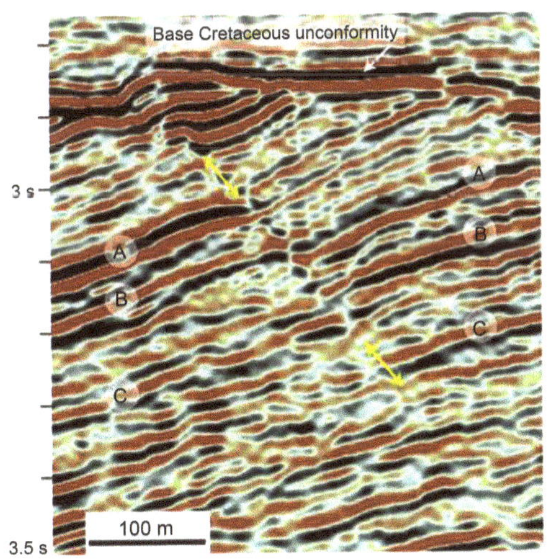

Figure 7.17. A fault imaged in seismic data (arrows). There are no clear reflections from the fault itself. The dip separation is identified by the discontinuity of reflectors (cutoffs). Three-dimensional seismic from the Visund Field, North Sea.

Source: Haakon Fossen (2010). Structural Geology. Cambridge University Press. ISBN: 978-0-521-51664-8.

Even on high-quality 3-D seismic data sets, it is typically difficult or impossible to identify faults with throw less than 15–20 meters. Seismic data have a resolution limit that varies from data set to data set. Subseismic faults are generally defined as faults that are below seismic resolution. Seismic data alone may not be sufficient to resolve fault-related complications like fault lenses and fault branching. Well data, when available, is used to limit the interpretation of the seismic data.

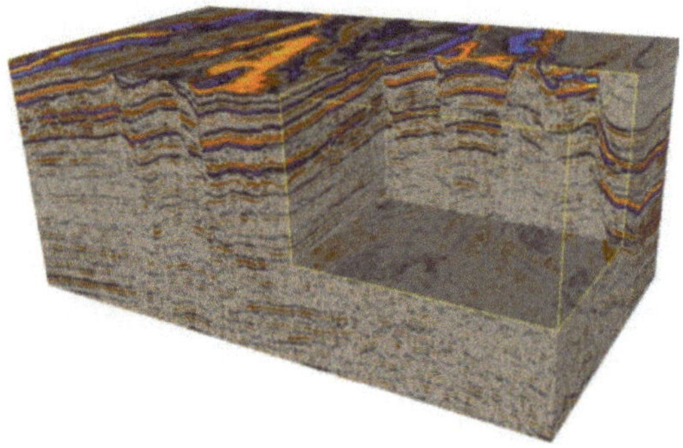

Figure 7.18. Faults as they may appear in a 3-D seismic data cube. Data from the Barents Sea.

Source: Haakon Fossen (2010). Structural Geology. Cambridge University Press. ISBN: 978-0-521-51664-8.

A 3-D data set is a cube of data that can be studied and interpreted in any direction, including horizontal sections (time slices), as shown in Figure 7.18. This includes faults and reflectors. As a result, the 3-D geometry of faults and fault populations can be understood.

Regional 2-D lines are more often used to map regional fault arrays and basin-scale faults. Certain 2-D lines are deep seismic lines that image the upper mantle and the deepest regions of the crust. Large faults that pierce the upper crust and occasionally the entire crust as they enter deep shear zones are visible along deep seismic lines.

7.5.2. Fault Cut and Well Log Correlation

Stratigraphic correlation is usually utilized to identify faults along the wellbore. Stratigraphic sections are repeated in reverse faults and omitted in normal faults, respectively. This kind of fault identification is based on knowledge of the stratigraphy from nearby wells.

The number of wells, the separation between them, the sedimentary facies variations, the well's orientation, and other factors all affect how big faults are that can be identified using this method. Characteristic signatures found in the well logs are also necessary for identifying the fault cut, or the location of the fault inside the wellbore. For stratigraphic well correlations, standard logs like resistivity logs, density logs, gamma-ray logs, and neutron logs are utilized instead of cores, which are typically unavailable. An example of how faults with as little as 6 meters of stratigraphic separation can be found in areas of the North Sea Brent Group with a high well density is provided by Figure 7.19. In this instance, core inspection verified the fault.

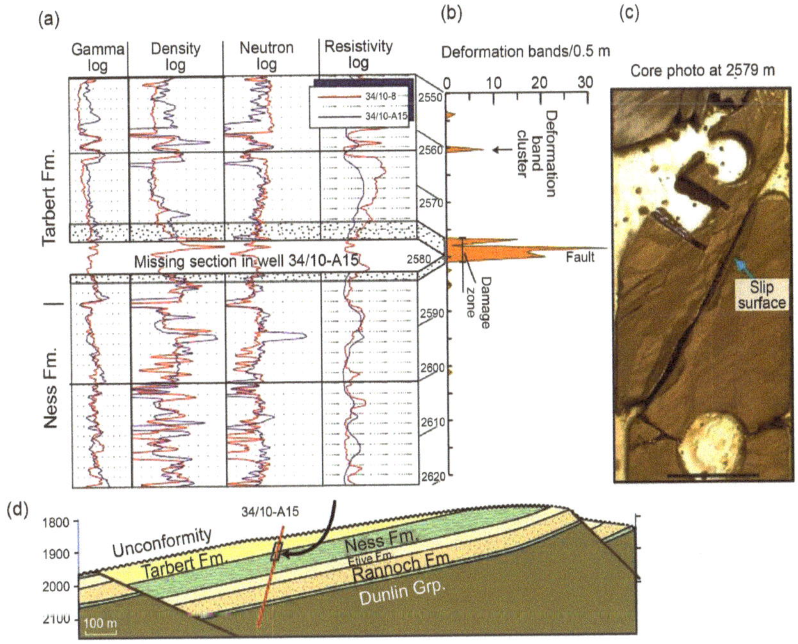

Figure 7.19. Fault separation detected by log correlation with a neighboring well. The complete log in well 8 is shown in red, and the correlation gives a missing section of 6 m in well A15. The damage zone (orange) was estimated from core inspection to be a few meters wide. Based on Fossen and Hestshammer (2000).

Source: Haakon Fossen (2010). Structural Geology. Cambridge University Press. ISBN: 978-0-521-51664-8.

7.5.3. Dipmeter Data and Borehole Images

With a dipmeter tool, microresistivity is continuously measured along the wellbore using three or more (typically sixteen) electrodes. To fit a plane, the responses from the various electrodes are correlated around the borehole at short depth intervals. The planes can be fractures or deformation bands, but they are usually bedding or lamination.

In dipmeter diagrams, orientations—typically indicated by dip and dip azimuth—are plotted. As seen in Figure 7.20, breaking up dip azimuth and dip into separate plots and condensing the vertical scale can be helpful for structural analysis. There are then a minimum of three methods available for identifying faults. Abrupt changes in dip or dip azimuth are indicative of the first. These happen when a fault divides two blocks with differently oriented layering; this is a fairly typical occurrence in many faults. Figure 7.20 shows an example in the area between the 17- and 31-meter faults.

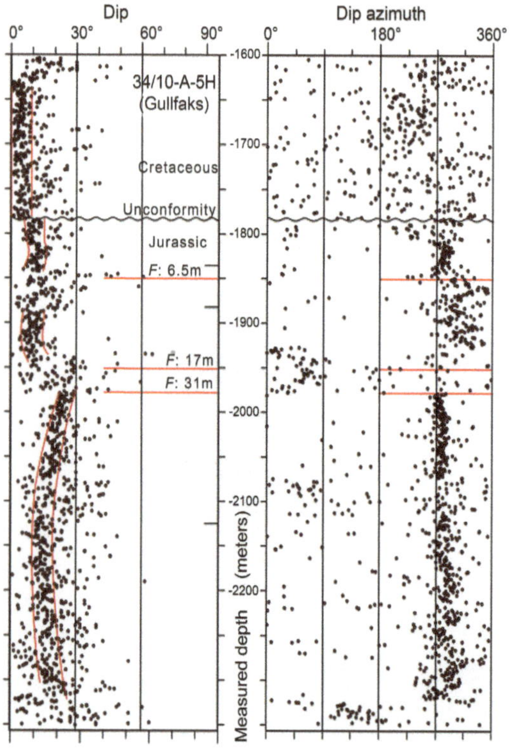

Figure 7.20. Dipmeter data from the Gullfaks Field, North Sea, where dip and dip azimuth (dip direction) are plotted against the depth measured along the wellbore. Faults identified by stratigraphic correlation (missing section) are indicated.

Source: Haakon Fossen (2010). Structural Geology. Cambridge University Press. ISBN: 978-0-521-51664-8.

The existence of local intervals with fast but progressive changes in dip and/or dip azimuth is another distinguishing feature. Cusps are the term for these anomalies (Figure 7.21). In many cases, cusp-shaped dip patterns indicate drag related to the fault. The example presented in Figure 7.21 has a 9-meter missing section, according to stratigraphic log correlation.

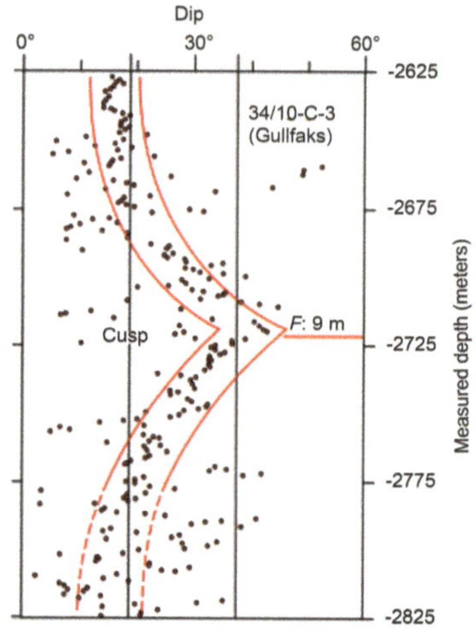

Figure 7.21. Dipmeter data (dip against depth) showing a classic cusp geometry related to drag around a minor fault. Data from the Gullfaks Field, North Sea.

Source: Haakon Fossen (2010). Structural Geology. Cambridge University Press. ISBN: 978-0-521-51664-8.

The emergence of anomalous orientations associated with fractures or deformation bands in the damage zone or away from the main slip surface itself is the third characteristic. These could be the high dip values at the fault locations in Figure 7.20.

A (nearly) continuous microresistivity image of the borehole is now typically produced using resistivity data from more advanced instruments, like the FMI (Formation MicroImager). With the use of several hundred electrodes, this instrument detects microresistivity. The end effect is a continuous wall image that resembles a real photograph of the rock. These photos are examined at workstations where structural and bedding characteristics are deciphered.

7.5.4. Drill Core Information

Faults are rarely represented in the drill core material, and only a small portion of the drilled section in a reservoir will be cored. Drillers are hesitant to cut cores across faults due to the possibility of pressure issues and jamming. Moreover, certain cored fault rocks could be so incoherent that they crumble and create rubble zones. On the other hand, damage zones and successfully cored faults provide important information. Permeability tests and microscopic investigations are possible with these samples. It is also possible to estimate the length and make-up of the damage zone, as well as occasionally the fault core. An illustration of a core through a fault with a 6-meter missing section can be found in Figure 7.22. Visible in the damage zone are the deformation bands and the central slip surface, which is a very thin core.

Figure 7.22. One-meter-long core section across a minor (6 m missing section) fault in the Gullfaks Field. Holes are from plugs sampled for permeability analysis.

Source: Haakon Fossen (2010). Structural Geology. Cambridge University Press. ISBN: 978-0-521-51664-8.

If the core is oriented, the orientation of the faults and fractures can be measured. Most of the time, it isn't, so its orientation needs to be rebuilt using dipmeter or seismic data to determine the bedding orientation. Only when the bedding is non-horizontal is this possible.

7.6. THE BIRTH AND GROWTH OF FAULTS

Tensile fractures, hybrid fractures, and micro-fractures all form and connect during the intricate process of a fault's birth and growth. These fractures form ahead of the fracture tip zone in the frontal process zone, also known as the frictional breakdown zone. Because of the irregularity of the fault's initial surface, the walls grind and microfracture, eventually forming a thin core of brecciated or crushed rock. A well-defined core of extreme deformation encircled by a zone of less severe fracturing damage is produced as the fault widens and new fractures appear in the walls adjacent to the fault core.

7.6.1. Fault Formation in Non-Porous Rocks

Small shear fractures in rocks with low or no porosity somehow give rise to faults. However, since shear fractures are unable to expand in their own planes, this cannot occur directly from a single shear fracture. Rather, they will twist and create wing-like cracks or other related cracks that are tensioned. Research demonstrates that before fracture initiation or propagation, there is a period of intense microfracturing. Favorably oriented microfractures link to form the main fracture, which grows once the density of microfractures reaches a critical point. The frictional breakdown zone, also known as the process zone, is the region of microfractures (and mesofractures) that lies ahead of the fracture tip zone.

Many tiny shear fractures, tensile fractures, and hybrid fractures must form and connect in order for a fault to develop. Because of the uneven nature of the early fault surface, the walls are being ground down and are microfractured. Usually, a thin core of crushed or brecciated rock forms. New fissures appear in the walls adjacent to the fault core during fault growth. Because of this, the majority of faults feature a clearly defined core with significant cataclastic deformation and a damage zone around it with less severe fracturing.

The complex process of fault formation and growth involves a frontal process zone where microfractures originate and eventually unite.

Because natural rocks are not isotropic, faults frequently form along the rock's preexisting mesoscopic weaknesses. Dikes and layer interfaces are examples of such weaknesses, but joints—as well as, of course, preexisting faults—are the structures most likely to become faults. Joints are typically very flimsy planar structures with negligible cohesiveness. Because joints are extension structures with the freedom to expand in their own plane, they can also form surfaces that are tens of meters or longer in length and/or height. Joint faults that result from joint faulting carry over some of the characteristics of the original joints. When a fault develops as a result of frictional sliding on a single, large joint, it usually starts out as a sharp slip surface with little to no damage zone and no fault core. But if slip builds up, the fault expands beyond the joint and connects to nearby joints near its tip zone. After then, the fault core may expand and the damage zone thickens.

Joint reactivation fault formation results in a smaller damage zone, less off-fault damage, and possibly a lower displacement gradient along the fault. It also requires less stress.

7.6.2. Fault Formation in Porous Rocks

A slightly different course is taken by fault growth in extremely porous rocks and sediments. Grain reorganization is made possible by pore space. Sandstone will reorganize its grains through rotation and frictional grain boundary sliding (translation) during deformation if the grains are only loosely cemented together. Grain breaks can also occur internally in other situations. Either way, the deformation is probably going to concentrate into small areas or bands to create structures called deformation bands.

Experimental and numerical work, along with field observations, demonstrate that deformation occurs sequentially as new deformation bands form next to the original one (Figure 7.23). This implies that eventually, it will be simpler to start a new band next to the current one than to continue cutting the original band. A deformation band zone is the end result, and strain hardening is frequently used to explain this development. The loss of porosity in the band is assumed to be connected to strain hardening, which is most noticeable in areas where grains are crushed (cataclastic bands). The process zone in non-porous rocks weakens the host rock and increases porosity by the formation of cracks. This is the difference between process zones in highly porous and non-porous rocks. The deformation bands in the process zone often harden the rock and decrease porosity in high-porosity rocks.

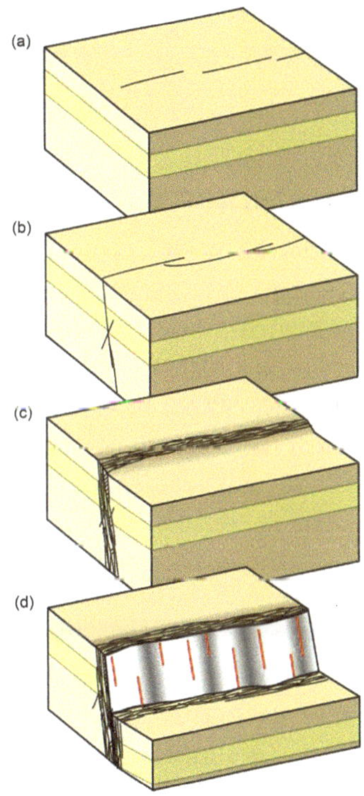

Figure 7.23. The general model for fault formation in porous sandstone, as proposed by Aydin and Johnson (1978). (a) Individual deformation bands, (b) linking of bands, (c) the formation of a deformation band zone, and (d) faulting of the zone.

Source: Haakon Fossen (2010). Structural Geology. Cambridge University Press. ISBN: 978-0-521-51664-8.

Porosity is sufficiently decreased in the deformation band zone once a certain number of deformation bands have accumulated for a slip surface to begin forming and expanding. Small patches serve as the nucleus of slip surfaces, which spread, connect, and eventually form continuous slip surfaces. Slip surfaces are weak mechanical structures that can accumulate meters or more of slip very quickly. A thin (millimeter-thick) zone of ultracataclasite is frequently linked to through-going slip surfaces; this zone could be regarded as the local fault core.

In highly porous rocks, precursory deformation band zones are where faults originate.

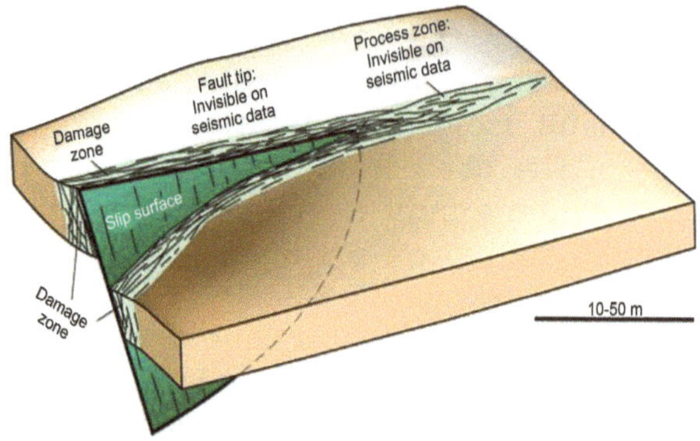

Figure 7.24. A fault is contained within a damage zone, which means that there is a (process) zone ahead of the tip where the rock is "processed" prior to fault propagation. The process zone may potentially contribute to compartmentalization of petroleum reservoirs. From Fossen et al., (2007).

Source: Haakon Fossen (2010). Structural Geology. Cambridge University Press. ISBN: 978-0-521-51664-8.

7.6.3. The Damage Zone

Our comprehension of the damage zone may be affected by the development of deformation bands and/or regular fractures before a through-going slip surface forms. The surrounding area of already-existing structures will become the damage zone as soon as the slip surface (fault) forms. When a fault is established, it expands and leaves behind a zone that becomes the initial fault damage zone, with the process zone ahead of the fault tip moving ahead of the fault tip (Figure 7.24). This zone in a porous rock is probably made up of bands of deformation. As illustrated in Figure 7.23, faults in porous rocks originate from the faulting of a deformation band zone, which is why the length of the deformation band process zone is typically longer than that of the process zone found in many non-porous rocks. This is especially true in the case of cataclastic deformation bands, where the process zone may extend several hundred meters.

The damage zone should be marginally older than the corresponding slip surface if its structures form ahead of a propagating fault tip. This assumption would lead to the conclusion that a damage zone's width and strain are independent of fault displacement. Despite the fact that the fault (slip surface) is the weakest area of the rock and is therefore prone to reactivation without causing additional sidewall damage, empirical data indicates that this is not the case. The reason is that faults don't expand inside a perfect plane or have perfectly planar structures. Because the rocks in which faults form are both heterogeneous and anisotropic, they exhibit irregularities on a variety of scales. Faults can bend, for instance, when they connect with other faults or when they come into contact with a different lithologic layer (Figure 7.25). An illustration of how damage can occur near a fault bend—in this case, together with a gentle fault-bend fold—can be found in Figure 7.26.

The structures within the damage zone develop before, during, and after the slip surface (fault) locally forms.

There may be times when displacement builds up without the wall rocks deforming if the fault is momentarily or locally planar and smooth, i.e., without the damage zone expanding further. However, wall damage can also happen during fault growth at fault bends or linkage locations, which causes a local widening of the damage zone. The damage zone may eventually become inactive again as the fault finds a more planar path through zones of complications. As a result, damage zones may grow locally and temporally (Figure 7.27), which adds to the scatter observed in Figure 7.12.

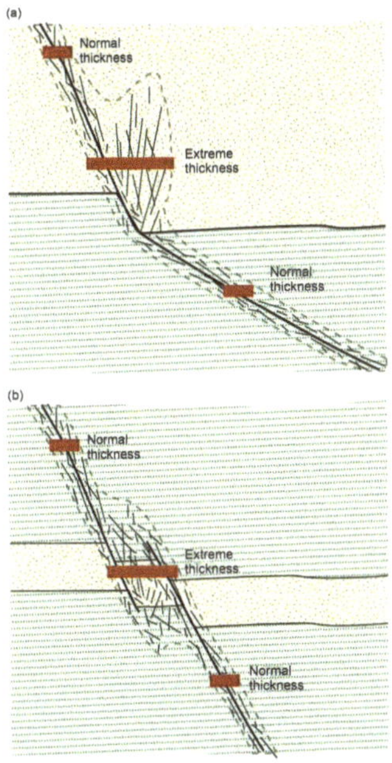

Figure 7.25. Variations in the thickness of damage zones related to (a) change in dip and (b) linkage. In these situations, minor structures are added to the damage zone until the fault cuts through the complex zone along a straighter path.

Source: Haakon Fossen (2010). Structural Geology. Cambridge University Press. ISBN: 978-0-521-51664-8.

Figure 7.26. Hanging-wall rollover (fault-bend fold) related to bend in the main fault. The damage zone is unusually wide due to the complications posed by the fault bend. Synthetic and antithetic shear bands are separated by color. Matulla Formation, Wadi Matulla, Sinai. The offset of the fault exceeds the 4 m cliff height.

Source: Haakon Fossen (2010). Structural Geology. Cambridge University Press. ISBN: 978-0-521-51664-8.

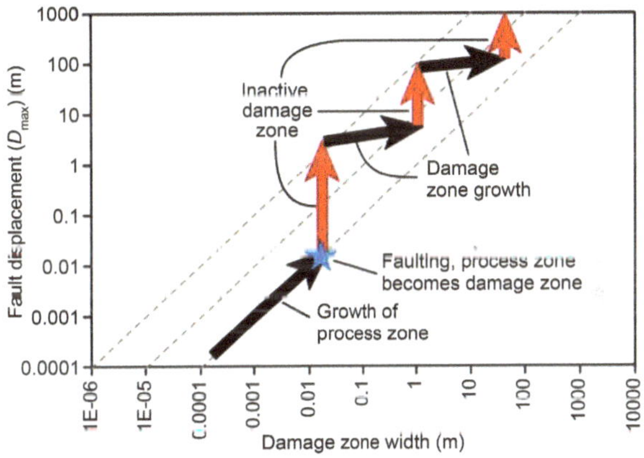

Figure 7.27. Schematic illustration of how a damage zone can grow periodically. The first stage is the growth of the process zone. Once the fault forms, the process zone becomes the damage zone and slip occurs smoothly for a while (red arrow) until complications lock the fault, which causes renewed growth of the damage zone. This repeats itself as the fault grows. The result is considerable scatter of fault data in fault displacement–damage zone width diagrams.

Source: Haakon Fossen (2010). Structural Geology. Cambridge University Press. ISBN: 978-0-521-51664-8.

7.6.4. The Ductile Drag Zone

The best way to characterize drag is as any systematic shift in the orientation of markers or layers next to a fault that indicates the change in deflection is genetically related to the fault. Zones that are a few meters or tens of meters wide are typically referred to as drag zones. Nonetheless, several hundred meters can be found extending into the hanging wall of continental rift basins along hanging-wall synclines associated with normal faults. Large listric faults are linked to large-scale rollovers, also known as reverse drag, which can extend up to several kilometers on the hanging-wall side.

Drag is the folding of layers around a fault caused by brittle deformation mechanisms; this process is directly associated with the fault's genesis and/or expansion.

Figure 7.28. Drag of layers of siltstone and sandstone along vertical fault located along the left margin of the picture. Colorado National Monument, USA.

Source: Haakon Fossen (2010). Structural Geology. Cambridge University Press. ISBN: 978-0-521-51664-8.

Drag is typically observed in faulted sedimentary sequences, where layers are soft enough to deform ductilely in the upper, brittle part of the crust. Drag is typically restricted to a zone a few meters wide along the fault (Figure 7.28), but it can also be large enough to be imaged on seismic data, as shown by the dipmeter data in Figure 7.29.

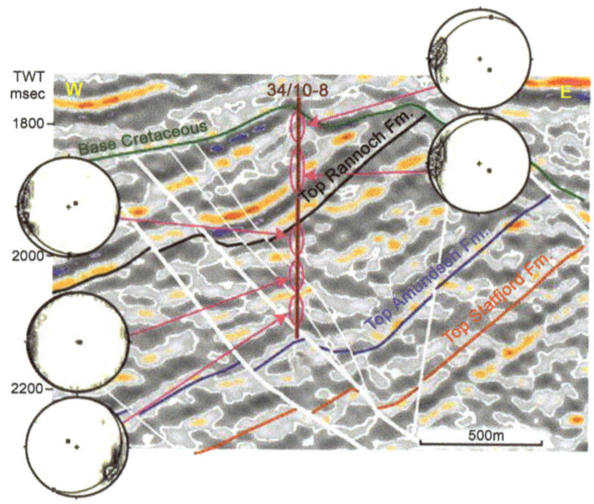

Figure 7.29. Dipmeter data collected along a vertical North Sea well. Stereonets show dip azimuth of layering in selected intervals. A change from west- to east-dipping layers is consistent with the normal drag portrayed by the seismic reflectors. Also note that the drag zone widens upward, consistent with the trishear model.

Source: Haakon Fossen (2010). Structural Geology. Cambridge University Press. ISBN: 978-0-521-51664-8.

Within any tectonic regime, drag can form. The angle between the fault's slip vector and the layering must not be too small, according to kinematic requirements. Since sedimentary rocks typically have subhorizontal layering, drag is more frequently developed along normal and reverse faults and less frequently along strike-slip faults. Along strikeslip faults, folds also form in subhorizontal layers; however, these folds are not drag folds. Therefore, we might want to include one more feature of drag folds:

- The axes of drag folds make a high angle to the displacement vector of the fault.
- For dip-slip faults and subhorizontal layers that means subhorizontal fold axes.

Normal drag and reverse drag are the two geometrically distinct types of drag. The shear zone-like geometry where layers flex toward parallelism with the fault is known as normal drag. The geometry displayed in Figures 7.28–7.32 is this one. Since normal drag involves displacement, the discrete fault displacement plus the ductile normal drag add up to the total offset. For the rollover structures that typically occur on the hanging-wall side of listric normal faults, reverse drag is employed. The layers in this instance are concave in the direction of the slip. Depending on the specific geometry of the fault, there can be both normal and reverse drag along faults.

Drag was once believed to be the outcome of friction along the fault during fault growth, but the term is now used to refer to layer bending that occurs before fault formation. A lot of drag instances seem to fit the latter model. This model is comparable to the damage zone development model in that layers are distorted following the fault tip in both scenarios. The distinction lies in the fact that drag folding is malleable up to a specific scale, typically equivalent to a hand sample.

A drag fold's geometry provides insight into its formation. Simple shear can be used to model a drag fold with layers that have the same geometry along the fault and drag zone dip isogons that are always the same width and run parallel to the fault trace. We have a straightforward shear zone with a central discontinuity in this instance.

In other situations, a different kinematic model needs to be used because the drag zone is expanding upward. The trishear model is a well-liked model. According to this model, strain is dispersed in an active deformation zone that is fan- or triangle-shaped and located ahead of the fault tip (Figure 7.30). As the fault spreads, this zone moves through the rock; once the fault has sliced through the layers, no more folding takes place. Each case is different in terms of the width of the triangular deformation zone, but the drag zone always widens upsection. This model appears to function especially well in areas where basement faults that have been reactivated grow into sedimentary strata above. Forced folds, as the fold structures are commonly called, are found in numerous examples in the Rocky Mountain foreland in Wyoming and Colorado, as well as in the uplifts on the Colorado Plateau.

Figure 7.30. Trishear modeling of normal drag development. In this fault propagation fold model the drag zone widens upward.

Source: Haakon Fossen (2010). Structural Geology. Cambridge University Press. ISBN: 978-0-521-51664-8.

Fault propagation folds are folds that develop in front of a propagating fault tip. Therefore, a large number of drag folds are faulted fault propagation folds. Drag, however, can also develop or intensify within the walls of an already-existing fault. Similar to the damage zone, fault drag can arise as a result of fault linkage, locking of the fault at fault bends, and other issues that raise the friction along faults. Figure 7.31 shows the impact of non-planar fault geometry and the evolution of normal drag between two overlapping fault segments. The latter mechanism has the capacity to drag material down to the point where a smear of rotated layer, usually shale or clay, forms along the fault.

Drag can form both ahead of the fault tip and in the walls of an active fault.

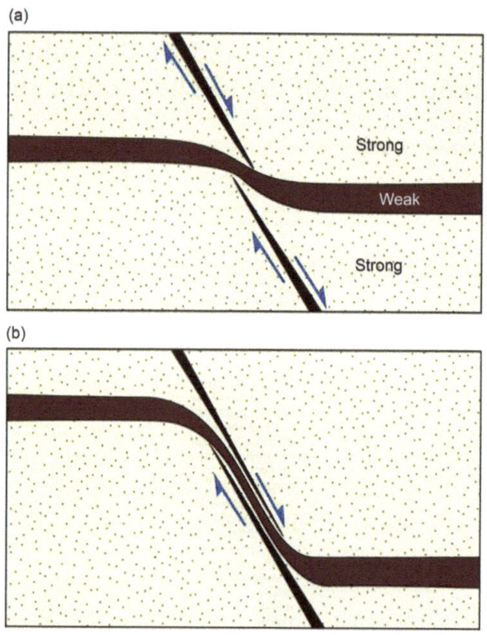

Figure 7.31. Normal dragging of mechanically weak layer (e.g., clay) between two overlapping fault segments.

Source: Haakon Fossen (2010). Structural Geology. Cambridge University Press. ISBN: 978-0-521-51664-8.

7.6.5. Drag, Deformation Mechanisms, and the Damage Zone

Granular flow can cause drag, especially in sediments with low lithification. With the exception of the modification of sedimentary structures or the rotation of layering, granular flow leaves little to no trace of deformation. Grain fractures in consolidated sedimentary rocks can lead to distributed cataclastic flow as a result. Drag folding results in less localized deformation and generally lower strain, but the mechanisms are the same as those that work in the various deformation bands. However, as Figure 7.32 illustrates, there is a strain gradient in the direction of the fault.

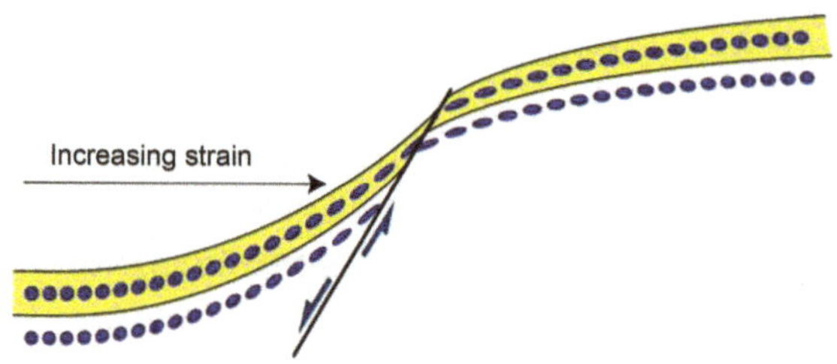

Figure 7.32. Strain ellipses portraying the relation between layer orientation and strain. Generated by means of the trishear program FaultFold. (Almendinger, 2003).

Source: Haakon Fossen (2010). Structural Geology. Cambridge University Press. ISBN: 978-0-521-51664-8.

Drag zones can have deformation bands or fractures. In these situations, as Figure 7.33 illustrates, the density of fractures or deformation bands increases toward the fault. We are in the damage zone when mesoscopically mappable fractures or deformation bands appear. Although the opposite circumstance can also happen, the drag zone is typically wider than the damage zone where drag folds are well developed.

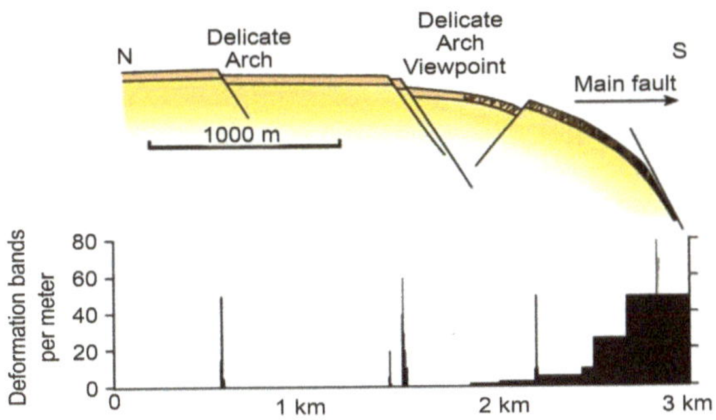

Figure 7.33. Folding of layers (rollover or reverse drag) adjacent to a major fault (not shown) accommodated by deformation band formation. Note the relationship between bed rotation and density of deformation bands. Example from Arches National Park, based on Antonellini and Aydin (1994).

Source: Haakon Fossen (2010). Structural Geology. Cambridge University Press. ISBN: 978-0-521-51664-8.

Certain faults exhibit drag-like folding of the layering, akin to that depicted in Figure 7.28, especially in metamorphic rocks. Upon closer inspection, many of these "drag" folds are found to be shear zones surrounding faults that could have formed in a variety of ways, most commonly in the zone of brittle–plastic transition, and are therefore controlled by plastic deformation mechanisms. Even though there can be a lot of similarities, we usually do not think of such plastic fold structures as drag folds.

7.6.6. Fault Growth and Seismicity

Once a fault surface is identified, it will stand for a structure that is weak mechanically and is prone to failure during subsequent stress builds. There are two ways that faults spread. The most prevalent type is known as stick-slip, in which there are intervals of no slip in between extremely abrupt seismic slip events (Figure 7.34). Between slip events, stress accumulates until it surpasses the frictional resistance of the fault. According to this model, the magnitude of each earthquake that results from a slip event is determined by the energy released during the stress drop. This relates to strain in terms of how much elastic strain is released as the fault moves.

Stable sliding, also known as seismic slip, is another way that faults can accumulate slip. During stable sliding, displacement should ideally accumulate at a constant rate (Figure 7.34c). According to certain lab tests, slip requires a force that increases gradually. This phenomenon, which is associated with damage to the slip surface during deformation, is known as slip hardening (Figure 7.34d).

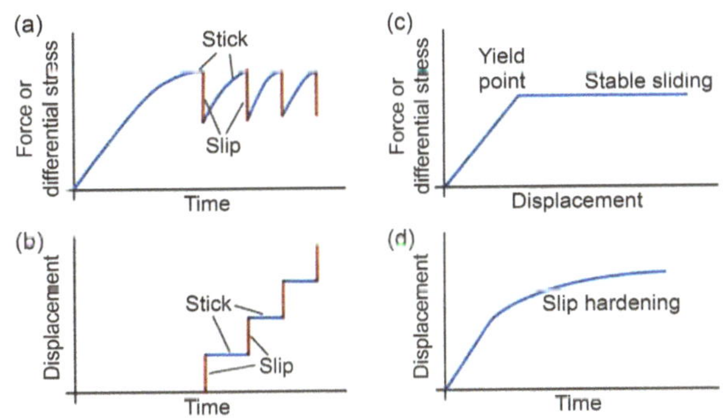

Figure 7.34. Idealized graphs illustrating the difference between stick-slip and stable sliding: (a) and (b) stick-slip graphs, (c) ideal stable sliding and (d) stable sliding with slip hardening.

Source: Haakon Fossen (2010). Structural Geology. Cambridge University Press. ISBN: 978-0-521-51664-8.

The accumulation of fault displacement due to sudden slip events or gradual accumulation is determined by several factors. According to rock experiments, stable sliding is more likely to occur in the uppermost portion of the brittle crust than it is deeper down, where there is a small normal stress across the fault. Even at depths of several kilometers, low-angle faults along over-pressured layers in a sedimentary sequence are likely to experience stable sliding because overpressure lowers the effective normal stress across the fault.

Lithology is another significant factor: compared to low-porosity crystalline rocks, porous sediments and sedimentary rocks are more likely to deform due to stable sliding. Specifically, in low-porosity quartz-rich siliceous rocks, stick-slip is preferred, whereas clay facilitates stable sliding. Similar to claystone, clay-bearing incohesive fault gouge in the fault core encourages stable sliding in thick, continuous zones. Gouges' (or rather, the slip surfaces along gouge zones') propensity to serve as channels for fluid flow may contribute to their capacity to slide steadily.

High temperatures introduce plastic deformation mechanisms near the brittle–plastic transition, which also facilitate stable sliding. In the plastic regime (heats over 300°C for granitic rocks), stick-slip deformation plays a minor role.

In summary, we could say that low normal stresses, weak and unconsolidated fault cores (gouges), and, at least in sedimentary basins, weak and porous rocks in general are expected to cause earthquake rarities in the upper kilometer or two of the crust. Until the brittle–plastic transition is reached, a lot of seismic or stickslip activity should be expected below this depth. The seismogenic zone (Figure 7.35) corresponds precisely to what the seismic data show.

Stick-slip and stable sliding are typically observed on a single fault within the seismogenic zone. Although the bulk of the total displacement accumulated over time may have been caused by seismic events, there is slow, aseismic "creep" that is discovered in between seismic events. Additionally, minor postseismic adjustments—which might or might not be seismic—are required. Seismic events are defined as abrupt releases of energy and accumulations of displacement caused by earthquakes. Seismic refers to the progressive accumulation of displacement without the production of earthquakes.

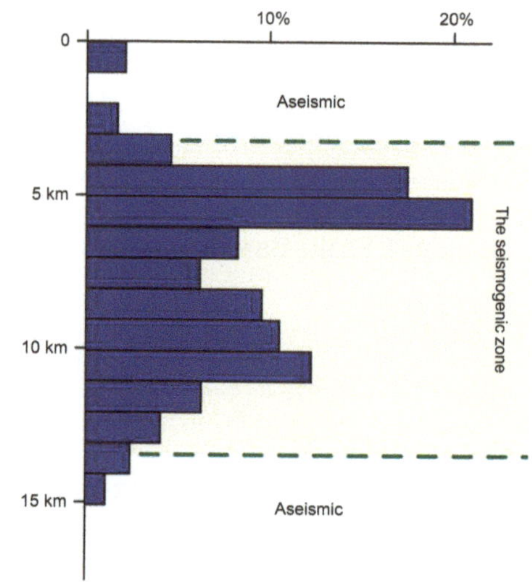

Figure 7.35. Distribution of 630 earthquakes in the crust beneath Parkfield, California. The distribution is characteristic for the continental crust away from subduction zones. Data from Marone and Scholz (1988).

Source: Haakon Fossen (2010). Structural Geology. Cambridge University Press. ISBN: 978-0-521-51664-8.

Seismicity and seismic slip behavior are frequently used to discuss fault slip and displacement accumulation. It's critical to understand that an earthquake is unlikely to cause more displacement than a few meters. A 15–20 km long fault that is activated by an earthquake with a magnitude of 6.5–6.9 adds no more than one meter of maximum displacement to the fault. The fact that only the biggest earthquakes can produce offsets of 10–15 m. has very significant implications:

a fault with a displacement of a kilometer must have resulted from hundreds of smaller earthquakes.Remember that depending on the local displacement rate, the accumulation of such displacements could take millions or even thousands of years. When offset sedimentary layers are dated and their displacements are measured, throw rates for faults can be determined. Regarding major faults in tectonically active areas, average displacement rates of about 1–10 mm/year have been reported.

Large faults typically only extend along a small section of the entire fault surface. Therefore, the sum of the displacements contributed by individual slip events (earthquakes) determines the total displacement distribution for a large fault (Figure 7.36). While it is evident that individual slip events result in displacement contours that are roughly elliptical and resemble those in Figure 7.15, it is more challenging to forecast or comprehend the finite displacement distribution that arises from numerous slip events, such as earthquakes. The characteristic earthquake model makes the assumption that the slip distribution and rupture length of each slip event are the same as those of the others. But for every slip event, the displacement maximum's location is altered. While the uniform slip model assumes that the slip at a given point is the same in every slip event (the area varies), the variable slip model predicts that both the amount of slip and the rupture length vary from event to event. We won't delve into these models' specifics here; suffice it to say that displacement accumulation leads to a displacement maximum close to the fault's center, which progressively tapers off toward the tips.

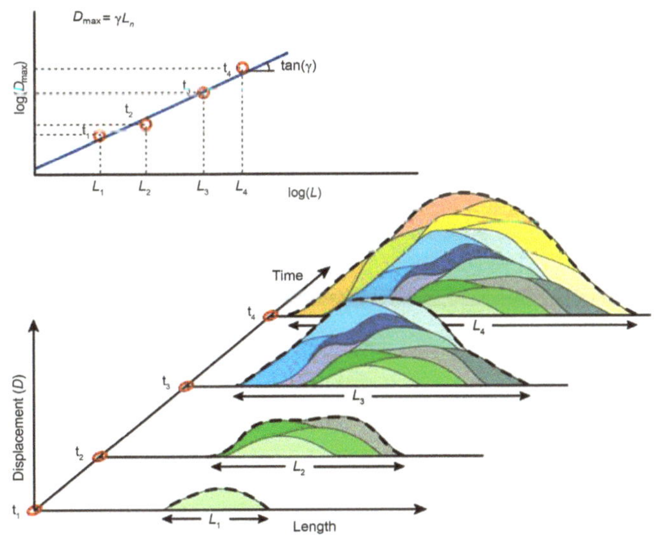

Figure 7.36. Schematic illustration of displacement accumulation through repeated slip events (earthquakes). Each event results in up to a few meters of displacement. In this model a bell-shaped cumulative displacement profile emerges which resembles that of a single slip event. The result of this model is a straight line in a logarithmic length– displacement diagram.

Source: Haakon Fossen (2010). Structural Geology. Cambridge University Press. ISBN: 978-0-521-51664-8.

ROLE MODEL

ARNOLD ESCHER VON DER LINTH

Arnold Escher von der Linth (8 June 1807 in Zürich – 12 July 1872) was a Swiss geologist, the son of Hans Conrad Escher von der Linth (1767–1823).

He made the first ascent of the Lauteraarhorn on 8 August 1842 together with Pierre Jean Édouard Desor and Christian Girard, and guides Melchior Bannholzer and Jakob Leuthold.

He studied geology and other sciences in Geneva, where one of his teachers was Nicolas Theodore de Saussure, and in Berlin as a student of Leopold von Buch and Alexander von Humboldt. In 1856 he became professor of geology at the *École Polytechnique* in Zürich and established the Geological Institute there. His researches led him to be regarded as one of the founders of Swiss geology.

With Bernhard Studer, he was the first to systematically explore the geology of the Swiss Alps and its neighboring regions (eastern Switzerland, Vorarlberg, Tyrol, Piedmont and Lombardy). Also with Studer, he produced a highly acclaimed geological map of Switzerland (1853).

In particular, his scientific liaison with the Scottish geologist Roderick Murchison (1792–1871) made him a contributor to the discovery of the Silurian system, and the first systematic description of sedimentary rocks and their index fossils.

He was the author of *Geologische Bemerkungen über das nordliche Vorarlberg und einige angrenzenden Gegenden* (Geological observations on the northern Vorarlberg and some adjacent areas), published at Zürich in 1853.

CLASS ACTIVITY

Student with your clay block, slice an inclined fault plane. With this fault plane, make either a normal or reverse fault scenario. Draw and color a block diagram of the faulted clay block, and label the hanging wall and the footwall. Be sure to use the colors you defined in your legend above, indicate the sense of movement with arrows, and title your design "normal fault" or "reverse fault."

SUMMARY

- A fault is a tabular volume of rock consisting of a central slip surface or core, formed by intense shearing, and a surrounding volume of rock that has been affected by more gentle brittle deformation spatially and genetically related to the fault.
- Faults are fractures along which movement of one block with respect to others has taken place. This movement may vary from a few centimeters to many kilometers depending on the nature and magnitude of the stresses and resistance offered by the rocks.
- Interpretation of seismic data is the most common way of identifying and mapping faults in the subsurface. Identification of faults depends on the presence of mappable seismic reflectors. Discontinuous reflectors indicate fault locations, and the dip separation is estimated by correlation of seismic reflectors across the fault
- Faults along the wellbore are typically identified by means of stratigraphic correlation. Reverse and normal faults cause repetition and omission of stratigraphic sections, respectively. Knowledge of the stratigraphy from other wells in the area forms the basis for this type of fault identification.
- Microresistivity is measured continuously along the wellbore by the three or more (usually 16) electrodes of a dipmeter tool. The responses from the different electrodes are correlated around the borehole in narrow depth intervals to fit a plane. The planes are generally bedding or lamination, but may also represent deformation bands or fractures.
- Fault formation and growth is a complicated process involving a frontal process zone where microfractures form and eventually connect.
- Drag is folding of layers around a fault by means of brittle deformation mechanisms, directly related to the formation and/or growth of the fault.

REVIEW QUESTIONS

1. What is a Fault?
2. List the types of faults
3. Distinguish between normal fault and reverse fault.
4. Mention the causes of faulting.
5. Why would the damage zone grow during faulting?

REFERENCES

1. Allaby, M. (2015). Strike-Slip Fault. In *A Dictionary of Geology and Earth Sciences* (4th ed.). Oxford University Press.

2. Aydin, A. (2000). Fractures, faults, and hydrocarbon entrapment, migration, and flow. *Marine and Petroleum Geology, 17*(7), 797–814.
3. Barnett, J. A. M., Mortimer, J., Rippon, J. H., Walsh, J. J., & Watterson, J. (1987). Displacement geometry in the volume containing a single normal fault. *American Association of Petroleum Geologists Bulletin, 71*(8), 925–937.
4. Benedicto, A., Schultz, R. A., & Soliva, R. (2003). Layer thickness and the shape of faults. *Geophysical Research Letters, 30*(23), 2076.
5. Caine, J. S., Evans, J. P., & Forster, C. B. (1996). Fault zone architecture and permeability structure. *Geology, 24*(11), 1025–1028.
6. Cerveny, K., Davies, R., Dudley, G., Kaufman, P., Knipe, R. J., & Krantz, B. (2004). Reducing uncertainty with fault-seal analysis. *Oilfield Review, 16*(2), 38–51.
7. Childs, C., Manzocchi, T., Walsh, J. J., Bonson, C. G., Nicol, A., & Schöpfer, P. L. (2009). A geometric model of fault zone and fault rock thickness variations. *Journal of Structural Geology, 31*(2), 117–127.
8. Childs, C., Watterson, J., & Walsh, J. J. (1995). Fault overlap zones within developing normal fault systems. *Journal of the Geological Society, 152*(3), 535–549.
9. Ferill, D. A., & Morris, A. P. (2001). Displacement gradient and deformation in normal fault systems. *Journal of Structural Geology, 23*(4), 619–638.
10. Hatcher, R. D. Jr. (1990). *Structural Geology Principles, Concepts, and Problems*. Merrill Publishing Company.
11. Hull, J. (1988). Thickness–displacement relationships for deformation zones. *Journal of Structural Geology, 4*(4), 431–435.
12. Kim, Y. S., Peacock, D. C. P., & Sanderson, D. J. (2004). Fault damage zones. *Journal of Structural Geology, 26*(3), 503–517.
13. Morewood, N. C., & Roberts, G. P. (2002). Surface observations of active normal fault propagation: Implications for growth. *Journal of the Geological Society, 159*(3), 263–272.
14. Oskin, M. E. (2019). *Normal Faults*. LibreTexts. Retrieved from: https://libretexts.org (accessed on 18 July 2024).
15. Peacock, D. C. P., & Sanderson, D. J. (1994). Geometry and development of relay ramps in normal fault systems. *American Association of Petroleum Geologists Bulletin, 78*(2), 147–165.
16. Roberts, G. P., & Michetti, A. M. (2004). Spatial and temporal variations in growth rates along active normal fault systems: An example from the Lazio–Abruzzo Apennines, central Italy. *Journal of Structural Geology, 26*(2), 339–376.
17. Shipton, Z. K., & Cowie, P. (2003). A conceptual model for the origin of fault damage zone structures in high-porosity sandstone. *Journal of Structural Geology, 25*(3), 333–344.
18. Walsh, J. J., & Watterson, J. (1989). Displacement gradients on fault surfaces. *Journal of Structural Geology, 11*(3), 307–316.

19. Walsh, J. J., & Watterson, J. (1991). Geometric and kinematic coherence and scale effects in normal fault systems. In A. M. Roberts, G. Yielding, & B. Freeman (Eds.), *The Geometry of Normal Faults*. Special Publication 56 (pp. 193–203). Geological Society.

20. Wibberley, C. A. J., Yielding, G., & Di Toro, G. (2008). Recent advances in the understanding of fault zone internal structure: A review. In C. A. J. Wibberley, W. Kurz, J. Imber, R. E. Holdsworth, & C. Collettini (Eds.), *The Internal Structure of Fault Zones: Implications for Mechanical and Fluid-Flow Properties*. Special Publication 299 (pp. 5–33). Geological Society.

21. Yielding, G., Walsh, J., & Watterson, J. (1992). The prediction of small-scale faulting in reservoirs. *First Break, 10*(10), 449–460.

CHAPTER

8

Kinematics and Paleostress in the Brittle Regime

LEARNING OBJECTIVES

After studying this chapter, you will be able to:

- Understand the kinematic criteria
- Discuss on stress from faults
- Describe the kinematic approach to fault slip data
- Define the contractional and extensional structures

KEY TERMS FROM THIS CHAPTER

Block rotation
Conjugate fault
Fault slip data
Finite displacement vector
Mineral growth
Slip lines
Strike slip faulting
Brittle deformation
Fault kinematic analysis
Fault surface
Kinematic method
Paleostress
Stress ratio
Strike-slip faults

INTRODUCTION

Fault kinematic analysis often involves measuring mesoscale faults and fault striae to quantify brittle deformation. This analysis focuses on determining the reduced stress tensor, which includes the directions of principal stresses ($\sigma_1 > \sigma_2 > \sigma_3$) and the stress ratio $R = (\sigma_2 - \sigma_3/\sigma_1 - \sigma_3)$ at a specific location. While this analysis provides information about the orientation and shape of the stress ellipsoid, it does not provide details about the exact values of stress.

There are three basic assumptions of fault kinematic analysis which have to be verified if fault kinematic data are interpreted in terms of stress:

1. the stress tensor is symmetric, i.e., deformation is coaxial, pure shear;
2. deformation is homogeneous;
3. faulting is consistent with the Mohr-Coulomb yield criterion (Coulomb, 1773), i.e., faults develop parallel to σ_2 and with an material-inherent fracture angle (the "angle of internal friction") to σ_1

It is important to remember that most of these assumptions are not realized in nature when analyzing fault kinematic data. At a small (outcrop) scale, pure shear and homogeneous deformation (assumptions 1 and 2) might be a realistic simplification, but most likely not at the regional scale (e.g., Pollard et al., 1993, Homberg et al., 1996, Jiang and Williams, 1999, Tikoff and Wojtal, 1999, Gapais et al., 2000). Consequently, inferred stress tensors reflect a local state of stress.

Assumption 3 is limited to isotropic rocks exclusively. In these rocks, faults can form parallel to σ_2 with fracture angles ranging from 0° (pure tensional fractures) to 30° (shear fractures) and σ_1 (Griffith, 1921, Coulomb, 1773). Experimentally obtained fracture angles range between 0° and 40° and are mostly around 30°, which might therefore be a reasonable approximation in most natural cases (e.g., Handin et al., 1963; Byerlee, 1978). On the other hand, pre-existing discontinuities in anisotropic rocks, such as foliations, bedding planes, and fractures, may become reactivated regardless of how they are oriented with respect to σ_1. Even angles between 45 and 90° are observed if the applied differential stress is low and the pore fluid pressures are high enough to surpass lithostatic pressures (e.g., Secor, 1965; Sibson, 1985).

From the above follows that fault kinematic data actually should not be interpreted as stress-related. Instead, an interpretation of the derived tensors as reflecting the local finite strain (P, B, T) is proposed. In this work, stress-related terms (stress tensor, principal stresses, stress ratio) are used consequently when dealing with the results of fault kinematic analysis to facilitate correlation with other publications. Interpretation will follow, however, purely kinematic ideas.

8.1. KINEMATIC CRITERIA

The actual finite displacement vector on a fault surface can be determined by connecting a point in the hanging wall to a nearby point in the footwall, such as faulted fold hinges or other distinct linear features intersecting the fault surface. Unfortunately, these points are rarely found, so researchers often focus on correlating layers or seismic reflectors across the fault instead. When the fault surface is visible, lineations can be used to estimate the orientation and length of the displacement vector, with the assumption that the lineation represents the direction of displacement.

However, it is possible that the lineation only shows the most recent deformation event, while earlier events may have been obscured. Therefore, it is important to carefully look for multiple, overlapping lineations when collecting fault slip data in the field. Correlating stratigraphy or marker horizons across faults is not feasible in other situations. Afterwards, we don't even know how big the fault is (though the fault core and damage zone's width may provide some indication), nor do we know if the fault moved in a sinistral or dextral direction or in reverse. The fault surface's lineation is helpful, but finding kinematic criteria is still necessary to ascertain the sense of slip. There are these kinds of standards, though most of them are vague. For kinematic fault analysis, therefore, it is necessary to combine as many kinematic criteria as possible.

8.1.1. Mineral Growth and Stylolites

Where there are geometric irregularities, useful kinematic structures may form. Fault surfaces are never perfectly planar structures. Local contraction is induced by irregularities, such as the contractional bend shown in Figure 8.1, and local contraction is induced by stylolites (Figure 8.2). and there might be "cleavage." Irregularities oriented differently, like the left bend in Figure 8.1, can lead to voids where mineral growth occurs expanding and opening. With a high degree of confidence, the sense of slip may be revealed through an analysis of the relationship between the geometry of the fault surface and the occurrence of contractional versus extensional structures.

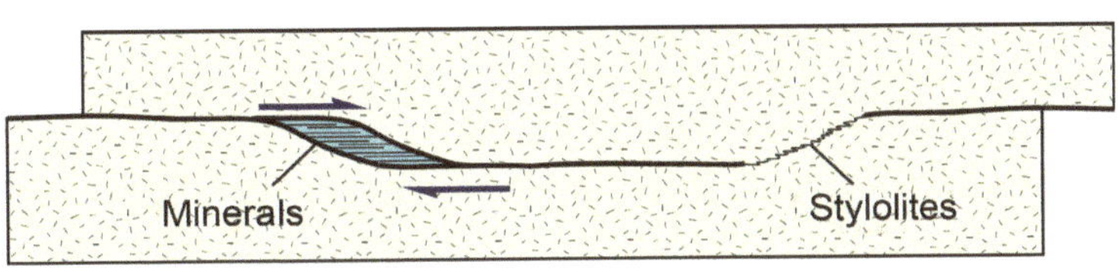

Figure 8.1. Irregularities along a fault create steps where mineral growth or shortening structures (stylolites) can form. The localization of such structures relative to the local fault geometry gives reliable information about the sense of slip.

Source: https://elearning.uniroma1.it/pluginfile.php/624958/mod_resource/content/0/Haakon-Fossen-Structural_Geology_Part%201.pdf.

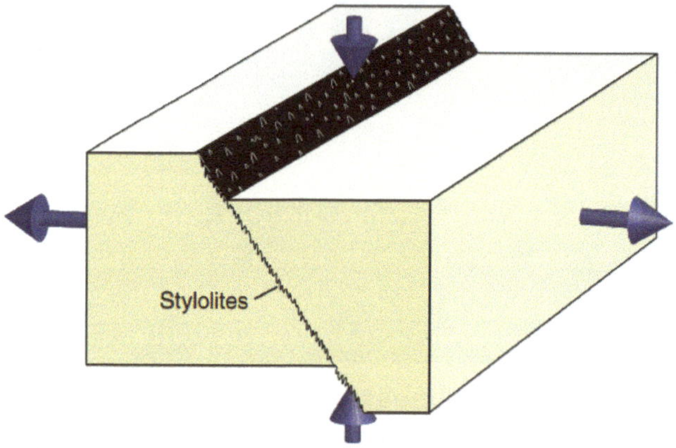

Figure 8.2. Faults with a component of shortening across the fault surface can result in pressure solution and stylolite formation in limestone and marble. Linear stylolitic structures sometimes form, called slickolites. Slickolites form a lineation that parallels the movement direction quite precisely.

Source: https://elearning.uniroma1.it/pluginfile.php/624958/mod_resource/content/0/Haakon-Fossen-Structural_Geology_Part%201.pdf.

8.1.2. Subsidiary Fractures

Tiny cracks that form along a fault or slip surface can provide valuable information about the direction of movement on the fault. These cracks, known by different names based on their orientations and movements, are shown in Figure 8.3e in a section perpendicular to the main slip surface (M) and parallel to the direction of movement. T-fractures, a common type of small extension fractures in this setting, are often mineralized with quartz or carbonates and do not show striations. The orientation of T-fractures in relation to the main slip surface characterizes the direction of movement on the fault. Typically, T-fractures dip around 45 degrees in the direction of movement with respect to a horizontally oriented M-surface (Figure 8.3 a, c).

Sometimes, a group of shear fractures known as P-fractures can be observed dipping in the opposite direction while M remains horizontal. These fractures form low angles to M and are kinematically associated with low-angle "reverse" or "thrust faults." In this scenario, Riedel shear fractures or R-fractures represent low-angle normal "faults," while R'-fractures correspond to antithetic reverse faults that form a high angle to M. Riedel fractures are more common than R' and P fractures, but they all exist and their local kinematics and orientation with respect to M indicate the direction of movement on the main structure. T-structures are considered the most reliable because they can be easily distinguished from other types of shear fractures.

By using the T, P, and R criteria—where the letters stand for the dominant subordinate fracture element of the structure—the French geologist Jean-Pierre Petit was able to distinguish between the different types of structures that can be seen along fault surfaces. The extension fractures (T) that cross the striated fault slip surface (M) make up the T-criteria. As seen in Figure 8.3a, T and M in the cross-section tend to form an acute angle of intersection pointing toward the slip direction. In Figure 8.3b, the intersections can also be represented as curved structures pointing in the direction of slip. These formations bear a resemblance to glacial chatter marks, which are created when ice flows break off chunks of relatively massive bedrock, like granite or quartzite.

P-criteria can occur in conjunction with T-fractures and are primarily dominated by P-fractures (Figure 8.3c). Situations where the main slip surface's undulations create a systematic pattern with striations on the side facing the opposite wall's movement (contractional side) and no striations on the lee (extensional) side (Figure 8.3d) can also be considered P-criteria. P-surfaces can be polished and striated, and they are distinguished by their low angle to Mdot. The acute angle between R and M is the basis for R-criteria, the most widely used kind of kinematic criteria (Figure 8.3f, g). These linear structures on the slip surface, which can be straight (Figure 8.3f) or curved (Figure 8.3g) and show a high (almost 90o) angle to the striae on M, are commonly called chatter marks. Small offset faults might not have developed a through-going slip or M-surface; in this scenario, en-echelon arranged R-fractures (Figure 8.3h) and occasionally R'-fractures could define the fault. The striations are then poorly developed and the R-fractures are grouped closely.

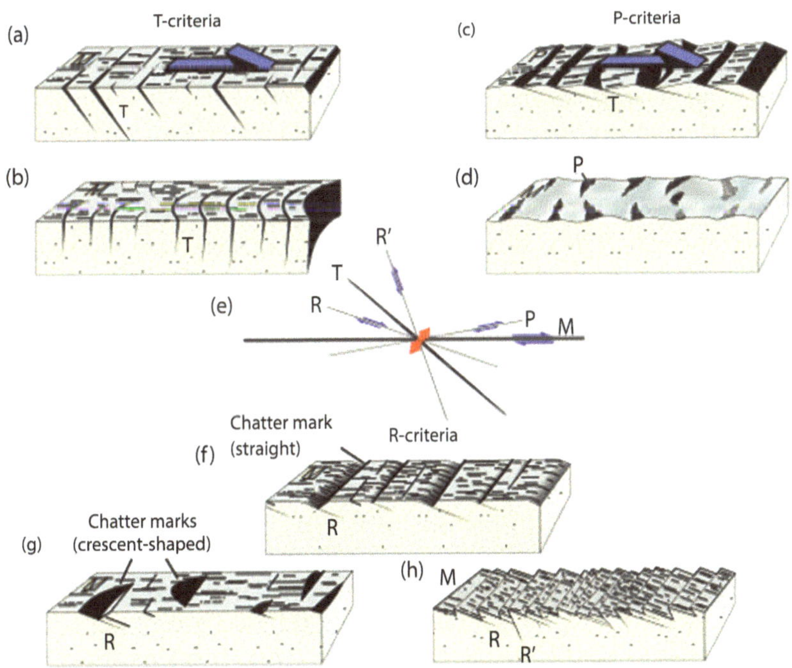

Figure 8.3. Kinematic criteria along a dextral fault with subordinate fractures or irregularities. The general nomenclature for fractures in a shear system is used (R, R' , P, T and M fractures). R, Riedel shears; P, shear fractures; T, extension fractures; M, average slip surface (fault). Identification of the type of subordinate fracture helps interpret the movement on the fault. Based on Petit (1987).

Source: https://elearning.uniroma1.it/pluginfile.php/624958/mod_resource/content/0/Haakon-Fossen-Structural_Geology_Part%201.pdf.

8.1.3. Strike Slip Faulting

An approximate faulting sequence in experimental strike-slip fault zones has been revealed by models employing analogue materials like sand and clay. A set of relatively short R-shears arranged in en echelon arrays typically precedes fault zones, coinciding with minor, conjugate R'-shears. P-shears start to form as further deformation occurs, and R' shears connect propagating and overlapping R-shears. Then, a continuous but irregular fault zone made up primarily of alternating R-shear and P-shear segments, synthetic to the sense of movement on the main fault, is created by the linkage of R-, R'-, and P-shears, as well as the imbrication and duplexing of the resulting rhomb-shaped blocks. The P-shear linkages are in restraining orientations for further displacements because of the variations in orientation of these segments with respect to the overall slip direction. In order to create a more planar, continuous fault zone, additional strike slip faulting may require the modification of these restraining sections through both adhesive and abrasive Y-shear wears. Although shear surfaces can occur and slip together in theory, analog experiments reveal that they typically develop in distinct locations and at distinct times. Only a portion of the segments or splay faults are ever active. Faults may lie dormant for extended periods of time. This causes faults and fault segments to reactivate in a complex and repetitive manner, creating complex stratigraphy and structure.

8.1.4. Relationship Between Folds and Strike-Slip Faults

8.1.4.1. Passive en echelon Folds

Folds related to wrench fault systems are usually not cylindrical, but rather doubly plunging, short, and have steeply dipping axial planes. They are positioned in such a way that the highs and lows of consecutive folds are aligned along lines that form an acute angle with the nearly parallel fold axes. These folds are stepped, overlapping, and are organized in an en echelon pattern. By considering the axial planes as approximately perpendicular to the direction of compression, their distribution can help determine the potential strike-slip fault they are associated with. These types of folds are commonly found above strike-slip faults in the underlying rock, which have not yet ruptured the overlying material. The en echelon folding patterns provide insight into the direction of movement.

8.1.4.2. Local Strike-Slip Faults Associated With Other Structures

Local strike-slip faults can be found in the hanging wall blocks of low-angle faults, and they can accommodate varying degrees of displacement along the fault or between the nearby autochthonous and allochthonous rocks. In deformed areas that are cut off from lower levels, they are typical (e.g., in fold-and-thrust belts found in the frontier. On a décollement separating the deformed cover from the basement below, local strike slip faults in this instance die out downward.

8.1.4.3. Transfer Fault

Transfer faults are strike-slip faults that move objects from one similar non-coplanar structure to another (e.g., in transitions between two thrust or normal faults). It is remarkably parallel to the direction of extension or compression in the region. The lateral ramp, another name for the transfer fault, ends on these two additional structures. It is a localized, passive fracture that results from active faulting on transfer-related faults. Transfer faults can therefore exhibit apparent offsets that are contrary to the true movement sense as well as reversing their strike-slip sense in space and time.

8.1.4.4. Tear Fault

A tear fault is a relatively small strike-slip fault that accommodates differential displacement between two adjacent belt segments and crosses the strike of a contractional or extensional belt. Tear faults are therefore parallel to the thrust or normal fault movement direction, and they are typically found in low-angle fault hanging walls. When folding is involved, fold axes typically come to an end in the presence of tear faults.

8.1.5. Large-Scale Analysis of Strike-Slip Systems

The mature stages of organic belts are characterized by strike-slip tectonics. For instance, Asia's escape tectonics didn't start until much later in the collision between Asia and India. Similarly, long after the Variscan collision, Northwest Europe underwent lateral extrusion away from the North-African indentor.

8.1.5.1. General Features

Steeply dipping transcurrent fault zones and shear zones absorb the mechanical effects of stresses generated during the frictional move. Characteristically, deformed rocks have steep foliations and sub-horizontal lineation.

8.1.5.2. Transform Boundaries

Transform faults are strike-slip faults that occur at plate boundaries and run parallel to the direction of the adjacent plates' relative motion. They consist of transfer faults, where plates slide past one another but at the fault's end, the direction of the motion between the plates changes. For instance, trenches to trenches or divergent and convergent plate boundaries may be connected by the transform fault. At some point, the strike slip movement is converted into the corresponding convergence or divergence, marking the end of such transform faults (Figure 8.4).

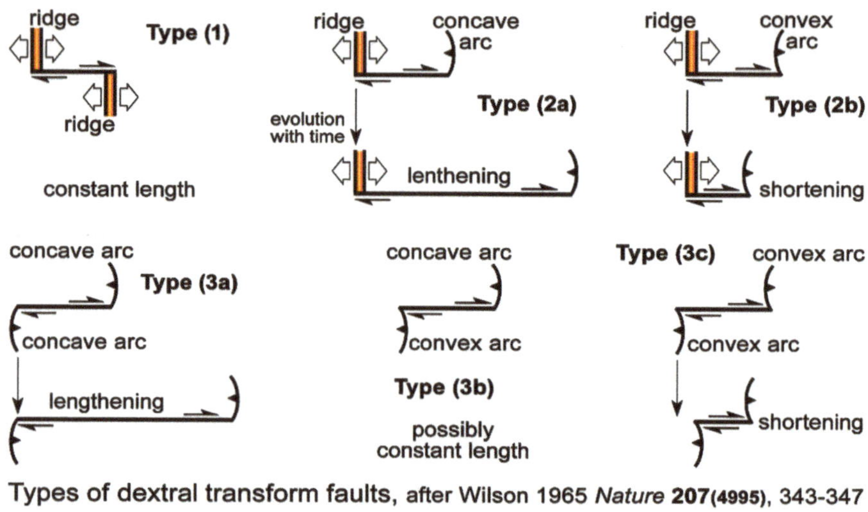

Figure 8.4. Types of dextral transform faults.

Source: https://www.files.ethz.ch/structuralgeology/jpb/files/english/5wrench.pdf.

Transform faults of ideal type (1) are the most prevalent and are found nearly exclusively in coastal areas. Ridge-to-Ridge (R-R) transforms are dynamically stable; shifts in the ridge axis translate into shifts in continental margins, and the transform fault's length never changes. Other transform kinds change over time, though perhaps (3b). The linked spreading and subduction zones' relative velocities determine the displacement rate along the transform fault. In particular, if trench migration is required due to slab roll back, cases (2a) and (3a) are likely to become longer. On the other hand, and for the same reason, the length of cases (2b) and (3c) will generally decrease. The San Andreas Fault, which divides the Pacific Plate from the American Plate to the west, is one of the most well-known examples.

8.1.5.3. Ridge–Ridge Transform Faults: Strike Slip in Oceanic Setting

Divergent plate boundaries always have transform faults that interrupt their smooth, continuous trace. These faults, known as ridge transform faults or fracture zones, are notable for offsetting the ocean ridges to account for variations in spreading rates on either side of a ridge or between adjacent segments.

- Origin: Major transform faults, such as the Romanche transform zones that divide the African and South American plates in the Central Atlantic, are frequently inherited from the continental rifting stage, when transfer faults connected two independent rifts or compensated uneven extension. These faults propagate continuously from the passive margins oceanward to segment the ridge. The majority of transform faults enable orthogonal spreading to continue by forming as a remnant of the original plate boundary. Transform faults in oceanic crust typically follow tiny circles on the surface of the earth, assuming they are pure strike-slip faults. They can be used to determine the rotational pole of divergent plates that are spreading apart on a sphere.

- Morphology: Fracture zones are very obvious, long linear bathymetric depressions with a sharp topography. Ridge-Ridge transform faults have the following characteristics:
- They are nearly parallel to the direction of relative motion of the plates on either side, the spreading direction of the ocean ridge. They connect two offset segments of the ridge, which are nearly perpendicular to the spreading direction. The divergent motion away from the ridge is "transformed" to a transcurrent motion along such a fault.
- They terminate the ridges abruptly. They also are plate boundaries and serve as zones of strike slip accommodation between opposite-travelling domains of seafloor.
- Equal displacement along their length.
- Transform faults can accommodate unlimited amounts of displacement, which may even exceed the length of the fault.
- Contrarily to ordinary strike-slip faults, adjacent and parallel transform faults may show opposite senses of relative displacement. Sense of displacement can be opposite to what it seems to be from the apparent offset of the oceanic ridge.
- The ridge and the transform faults are contemporaneous. Compared to the ridges, earthquake activity is significantly higher along transform faults (energy release 100 times greater). However, the faults are only active between the offset segments of the ridge due to the relative motion between the plates. Beyond this point, the plates on either side of the fracture can be thought of as being linked together because they are moving in the same direction and at the same rate. Rarely does seismic activity occur along the fracture zones that extend past the offset ridge segments. This indicates that the apparent offset of the ridges along the transform fault does not always increase or decrease when the system is active; rather, its geometry can remain stable for an extended period of time.
- Transform faults have rocks that display more shearing and metamorphism compared to the typical oceanic crust. The presence of large serpentinite blocks indicates that ultra-basic rocks are inserted into fracture zones. Additionally, there is some alkali basalt volcanism and hydrothermal activity occurring in the area.
- Oceanic crusts of varying ages are juxtaposed by transform faults. Where oceanic lithospheres of various ages are juxtaposed along oceanic transform faults, 2–4 km large vertical offsets may form as a result of temperature differences. The square root of the age determines the sea floor's depth, which is correlated with cooling and density. Different ages of oceanic crusts have experienced varying degrees of subsidence. As the oceanic lithosphere moves away from the ridge, thermal contraction causes dip-slip movement along fossilized transform fault lines, leading to seismic activity. This process is expected to create a scarp across the fault zone, with the older side on the lower end. This initial scarp may remain preserved for over 100 million years, with some suggesting that

the two sides could eventually fuse together and sink while maintaining their relative heights (Figure 8.5).

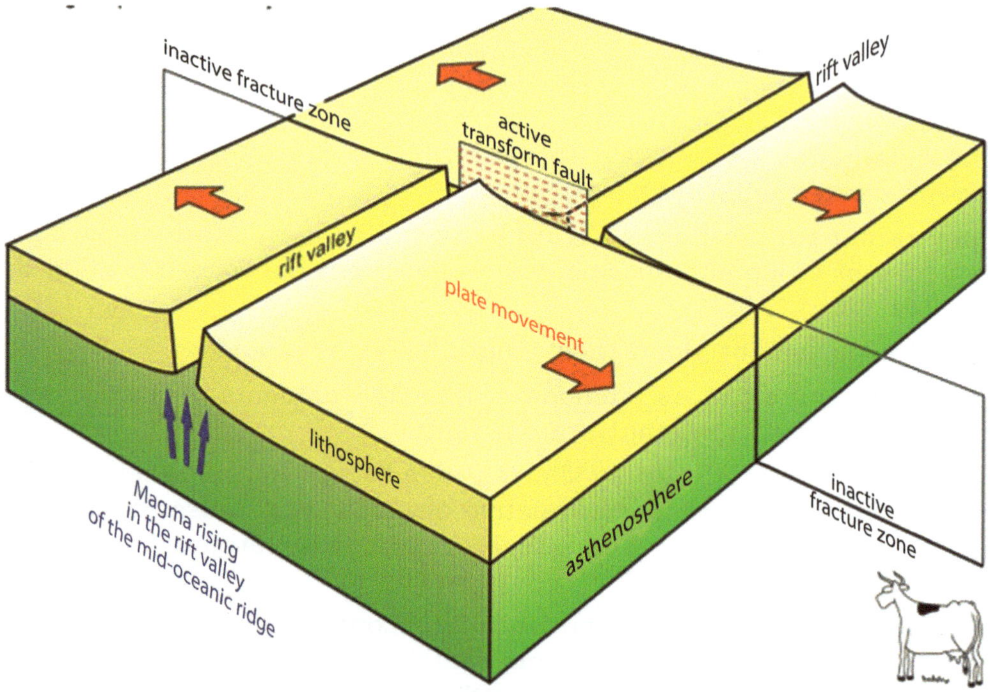

Figure 8.5. Divergent plate boundary with a transform fault.

Source: https://www.files.ethz.ch/structuralgeology/jpb/files/english/5wrench.pdf.

3. Associated Features: Transform faults can be associated with transverse ridges, which are isolated mountains with a height of approximately 6km on one or both sides of the fault. These ridges cannot be easily explained by strike-slip or volcanic activity alone, and instead require elements of compression or tension across the fault, often resulting from small changes in spreading directions. In some cases, these ridges can rise above sea level to form islands, such as St. Peter-Paul which is made up of mantle peridotites. In instances where there is an extension component across a transform fault, it may adjust its trajectory to align with the spreading direction, forming what is known as a leaky transform. Examples of leaky transforms include the Gulf of California and the Andaman Sea, characterized by segments joined by short lengths of spreading ridges.

8.1.5.4. Ridge-Trench Transform

Connecting spreading to convergent plate boundaries (South America) are the longest transform faults. A subtype would connect a belt of mountains to a ridge. This is the case for the Middle-East Fault, which is responsible for the formation of the Red Sea and connects the N-Syria-E-Turkey collision mountains to the Red Sea spreading center.

8.1.5.5. Trench-Trench Transform

Trench-trench transform faults are not common, with the Alpine fault in New Zealand being a notable exception. These faults cause a change in the direction of subduction across them. A specific type of transform fault found on continents connects two convergent plate boundaries, typically linking two mountain ranges. One well-known example is the fault system that connects the N-Pakistan Himalayas to the Zagros mountains in Iran. Another subtype of these transform faults links a trench to a mountain range, with the Java Trench and East Himalayas being an example. Deformation and metamorphism are more significant in continental settings compared to oceanic settings. Trench-trench transform faults can connect segments with different rates of convergence, leading to an increase in strike length over time.

8.1.5.6. Continental Transform Faults

Oceanic transform faults and transform faults that sever continents are comparable. There is little seismicity. The most well-known instances are the Levant Fault Zone, which includes the Dead Sea Rift, the Anatolian Fault in Turkey, and the Alpine Fault in New Zealand. In eastern Turkey, the spreading motion of the Red Sea is transformed into a continental collision by the Levant Fault Zone, which borders the Arabian and African plates in the Middle East.

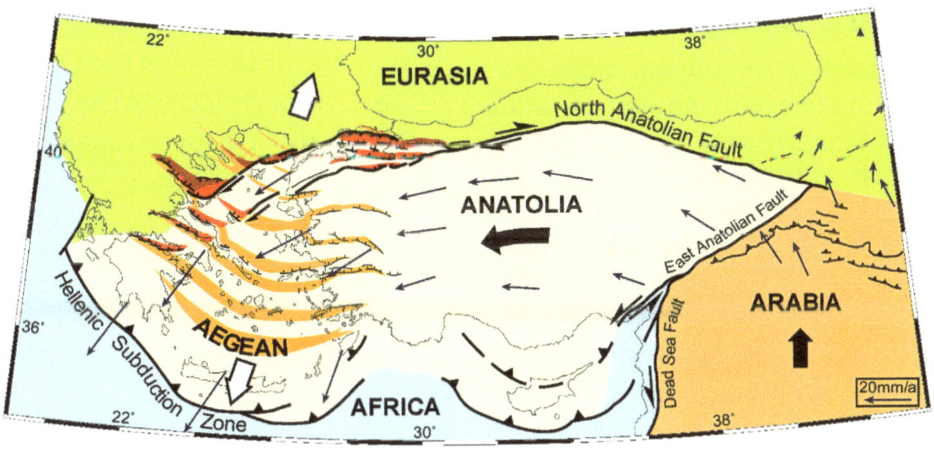

Figure 8.6. Continental extrusion of the Anatolia Aegean block between the North and East Anatolian strike slip fault.

Source: https://www.files.ethz.ch/structuralgeology/jpb/files/english/5wrench.pdf.

8.1.5.7. Oblique Subduction

Whether in an oblique continental collision or subduction, the slip vector oblique to a plate margin is divided into two parts: a strike-slip part that causes transcurrent motion on strike-slip faults parallel to the margin, and a dip-slip, orthogonal part that causes pure underthrusting in the subduction zone. Geographically distinct regions of dip-slip and strike-slip faulting are frequently found, and only by observing a very large area can an accurate kinematic model be inferred. Indirect subduction occurs frequently. Major strike-slip fault zones that run parallel to the trench form in the oblique subduction zones' hanging wall (e.g., the Sumatra Semangko Fault). Trench-related strike-slip faults should form when the vector-trench angle is less than 35–55° with low slab dips and high friction coefficients, according to an analysis of the energy balance between the obliquity of the movement vector to the margin, slab dip, and friction coefficient. Understanding the partitioning of fault motions is crucial to comprehending the lateral tectonic transport of large and small blocks as well as fore-arc domains.

8.1.6. Tectonic Escape

It might be necessary to shorten more than one mountain system can absorb in order for two colliding continents to continue converge. Once shortening is supported by continental extrusion, deformation spreads throughout the continents. Along significant strike-slip (transcurrent) faults, there are significant lateral movements (the escape) of continental blocks as a result of this within-plate deformation. While wrenching is supported by diffuse ductile flow in deeper levels, such massive, isolated transcurrent faults are limited to the upper crust. Crustal thickening is not necessary in a convergent setting when crustal blocks are lateral extruded along such faults.

The concept of tectonic escape has been developed to interpret active tectonics of Asia. Since collision India has penetrated about 2000 km into Asia and the present day rate of convergence of about 5 cm yr^{-1}. To understand this amount of shortening of the Asian continental plate, an analogy was derived from the theory of indentation developed by geotechnical engineers to determine the stability of foundations, cuts, embankments and tunnels. Lithospheric plates are here considered to be thin layers suffering negligible, ie, no vertical strain. The theory predicts mathematically, for simple shapes of both, plastic medium and die (the indentor), the configuration of lines of failure developed on indentation in two dimensions (i.e plane strain), the slip lines.

The problem of India, which has a limited width, pushing against the vast Asia is idealized and compared to the action of a wall on the ground. The horizontal ground and underground (Asia) is a homogeneous halfspace in which the stress field consists of three regions of homogeneous stress states:

- The weight of the wall applies a uniform stress over the width of the wall on the horizontal surface;
- On both sides of the wall the uniform stress applied on this horizontal surface is zero (in reality, the weight of air).

The vertical components of stress in the three regions increase together with depth z according to ρgz , with ρ the density of the ground. The two vertical lines projected from the wall sides down into the half-space represent discontinuities of vertical stress components. The horizontal components of stress, however, must be equal in the three regions in order to preserve horizontal equilibrium.

The Mohr-Coulomb shear criteria describe the behavior of the half-space. The principal stresses in this configuration are the vertical and horizontal components. Two Mohr circles, each of which represents the homogeneous stress states previously discussed, can be used to discuss the yield conditions. Failure happens when the Mohr circles cross the failure envelope, and continuity is expressed if the two circles are tangent at a single common principal stress. Because the wall's surface limits are stepped, the stress field is more intricate. If so, the discontinuity lines are defined angles with the horizontal that open downward from the wall boundaries in a symmetrical manner. The normal and shear stress components along the wedge boundaries need to be continuous from one domain to the next in order to maintain equilibrium. If the two neighboring Mohr circles pass through a point, then the point where the normal and shear stresses are equal for two adjacent domains can be found using the Mohr representation, which plots the orientation of the wedge boundaries. This process is symmetrical for both sides of the wall and determines the states of stress in the three adjacent regions when the three circles are tangential to the Coulomb envelope (Figure 8.7).

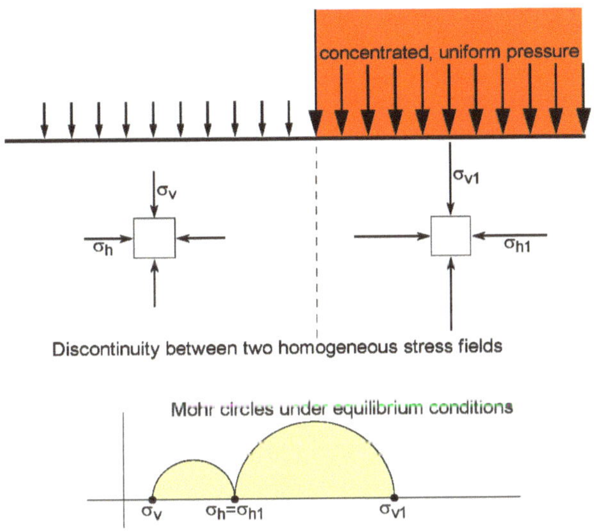

Figure 8.7. Stresses in a half space with a stepped load on its surface.

Source: https://www.files.ethz.ch/structuralgeology/jpb/files/english/5wrench.pdf.

Please note that the primary stress directions in the intermediate wedges are neither horizontal nor vertical. The wedges on either side of the wall form a rigid, head-down triangle with a top base as wide as the wall. This triangle, which is unable to deform, moves downwards vertically due to the pressure from the wall. Surrounding this triangle on both sides, and symmetrically mirrored along the center line, are two zones that tend to move horizontally apart to accommodate the downward motion of the rigid triangle. Meanwhile, sections of the ground on either side of the wall will move upwards to allow for the lateral displacement of these zones. These lateral and upward movements occur along shear surfaces.

8.1.6.1. Slip Lines

The mathematical solution identifies two types of curved slip lines known as α- and β-lines. These lines create a perpendicular grid in perfectly plastic materials with no frictional resistance. In this scenario, the angle bisected by the α and β directions aligns with the direction of the highest principal stress. Additionally, α and β lines correspond to paths of maximum shear stress, resembling faults with dextral and sinistral strike-slip motion in geological terms respectively. In materials that have a non-zero friction φ, the α- and β-lines are no longer orthogonal but must intersect at an angle of $\left[(\pi/2)-\phi\right]$ (Figure 8.8).

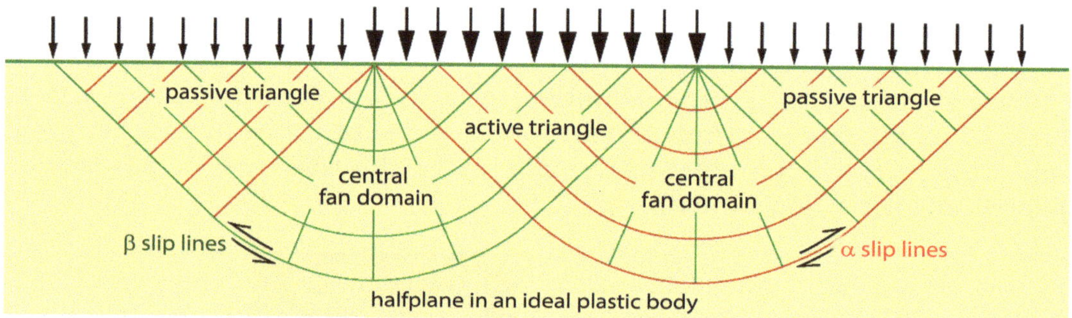

Figure 8.8. Slip Lines 2 dimensions produced by an indenter in an ideal plastic body.

Source: https://www.files.ethz.ch/structuralgeology/jpb/files/english/5wrench.pdf.

8.1.6.2. Application

Analog modeling has helped to further solidify the idea. A block of plasticine that has been distorted by a stiff indenter moving forward steadily is part of the apparatus. Extruded blocks are guided in their sideways translation and rotation by faults on the α- and β-slip lines. The width of the indented block and the indenter width determine the block sizes. Deformation in one-sided plane indentation experiments is asymmetric. Given the geometrical correspondence between Asian tectonic features and slip line model predictions, faulting could be the predominant mechanism causing distributed deformation of the continental lithosphere. Intracontinental deformation and plate tectonics are reconciled by fault propagation in response to indentation on a plate boundary.

Tectonic escape shows the following characteristics:

- The "escaping" blocks are bounded by two dominant strike-slip systems, which have the same or opposite sense of movement.
- The movement increases in the direction of escape.

The process can happen on a large scale, or on a small scale along uneven continental margins colliding with indenting promontories.

8.1.6.3. Block Rotation

Paleomagnetic declination measurements have demonstrated that crustal blocks and bounding faults in wrench systems have frequently rotated. In essence, the rotation axis is vertical. As in the domino model of normal faults, block rotation occurs in accordance with the bulk sense of shear, clockwise in dextral wrench zones and counterclockwise in sinistral wrench zones (Figure 8.9).

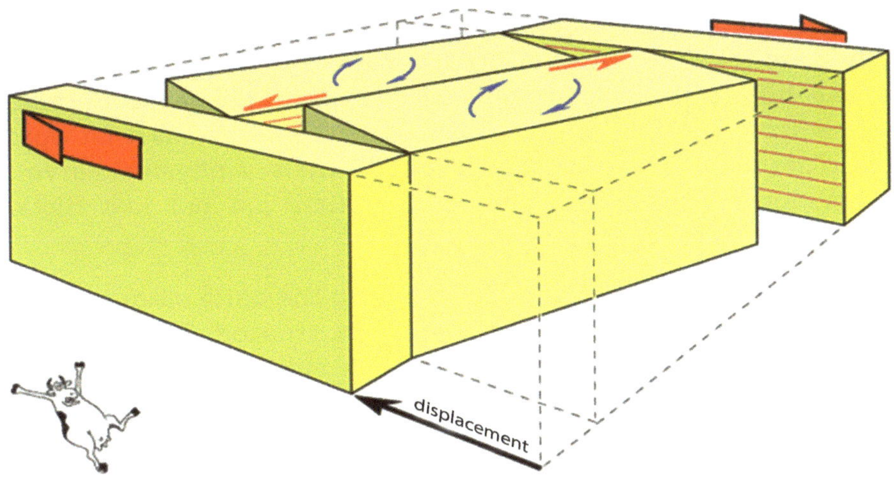

Figure 8.9. Clockwise block rotation in a dextral wrench zones.

Source: https://www.files.ethz.ch/structuralgeology/jpb/files/english/5wrench.pdf

8.1.7. Ploughing, Mineral Growth, and Slickensides

A fault surface may have asperities or relatively hard objects (stones, pebbles, or strong mineral grains) on one side that mechanically plow grooves or striations into the opposite wall. On the lee side, there is a crescent-shaped opening that is usually filled with material from the opposite wall, while material in front of the object is being pushed aside. Sometimes a ridge (or groove, depending on your perspective) is found ahead of the object, as shown in Figure 8.10. These lineations are sometimes called ridge-in-groove lineations.

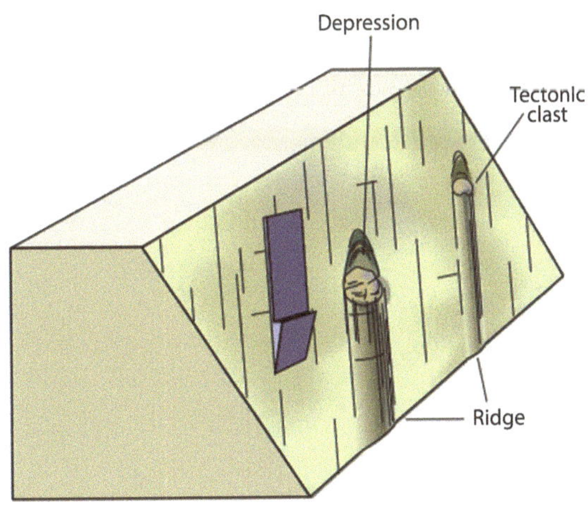

Figure 8.10. Pointed asperities or tectonic clasts sitting on fault surfaces may form depressions on the lee side and ridges on the opposite side. Such lineations are ridge-in-groove lineations.

Source: https://elearning.uniroma1.it/pluginfile.php/624958/mod_resource/content/0/Haakon-Fossen-Structural_Geology_Part%201.pdf.

The opening in the shape of a crescent quickly transforms into a groove whose length perfectly matches the movement of the hard object in relation to the wall as a whole. That is, at the very least, our preferred scenario. However, numerous instances exist where offsets on the centimeter scale result in striations that span decimeters or even meters. Therefore, more mechanisms than merely physical carving must be at play. One explanation for some striations is that they are not just frictional grooves; they are also corrugations.

These are linear structures that developed during the early stages of fracture growth. As slip builds up, they could become polished and striated. Long and well-developed striae may result on slip surfaces with offsets as small as cm or even mm. Therefore, the amount of slip on the surface is not always reflected in small-scale fault corrugation structures. Usually found on polished slip surfaces known as "slickensides," but not exclusively, striations are also referred to as "slickenlines" on these surfaces.

Minerals may crystallize on the lee side of asperities (irregularities) in addition to frictional striations, exposing the sense of slip. The orientation of the fibers tends to be near the slip direction when minerals crystallize as fibers. Minerals that precipitate on fault slip surfaces are frequently found to be impacted by renewed slip. As a result, slickenlines frequently form on deformed mineral fill along fault surfaces.

8.2. STRESS FROM FAULTS

Data collected for fault observations in paleostress analysis typically includes the local strike and dip of the fault surface, the orientation of the lineation (usually indicated by rake or pitch), and the direction of movement. By analyzing such data for a group of faults, valuable insights can be gained not only about strain but also about the orientation and relative magnitudes of the principal stresses (σ_1- σ_2 and σ_3) and the shape of the stress ellipsoid. Paleostress methods are based on a number of assumptions, including that the faults in question were formed under the same stress conditions, that the rocks are relatively uniform, that the strain is minimal, and that there has been limited rotation of structures since their formation.

8.2.1. Conjugate Sets of Faults

Plane strain expressed in terms of conjugate fault systems is a basic example of deformation. A conjugate system consists of two sets of oppositely dipping faults with lineations perpendicular to the line of fault intersection, as illustrated in Figure 8.11 for the three major tectonic regimes. The two sets have complementary senses of slip, and there should always be a constant angle between them.

The angle between the two sets should be compatible with the mechanical characteristics of the rock (and the pressure of the pore fluid) at the moment of deformation, according to the Coulomb fracture criterion, which states that faults or shear fractures form at an angle to $\sigma 1$ that is controlled by the internal friction of the rock. More significantly, for normal, reverse, and strike-slip faults, conjugate sets should form symmetrically around the principal stresses (Figure 8.11). The slip direction or lineation would be in the strike direction (strike-slip faults) or in the dip direction (reverse and normal faults), according to the Andersonian theory of faulting. Therefore, the orientations of the principal stresses are revealed by conjugate faults and their striations.

In certain instances, the main stresses may not align exactly with the horizontal or vertical directions as expected by Anderson. This divergence could suggest that the area has experienced tilting since the deformation occurred, or that the stress field was rotated or refracted within this particular volume of rock.

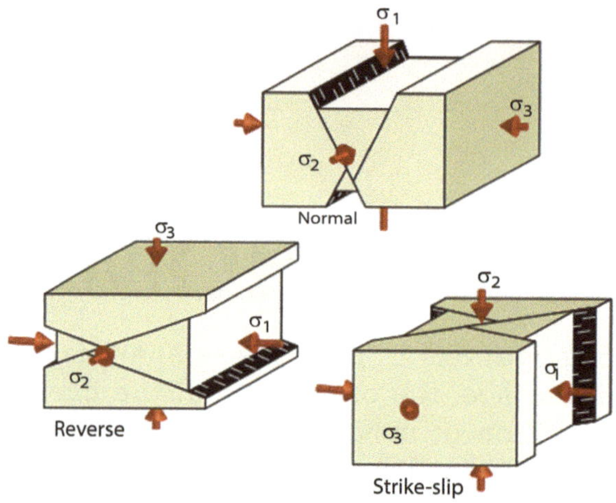

Figure 8.11. Conjugate shear fractures and their relation to stress according to Anderson's conditions for planar strain. The acute angle is bisected by σ_1 and the two fractures intersect along σ_2.

Source: https://elearning.uniroma1.it/pluginfile.php/624958/mod_resource/content/0/Haakon-Fossen-Structural_Geology_Part%201.pdf.

Two sets of conjugate faults are indicative of plane strain (see Figure 8.12a, b), and the strain axes can be determined by plotting these fault sets on a stereonet (see Figure 8.12c). Assuming isotropic rheology, a consistently oriented stress field, and minimal rotation of fault blocks, the orientation of the principal stress axes can be inferred. By observing conjugate fault sets in rock mechanics experiments, we can draw a connection between their angular relationship and the principal stresses.

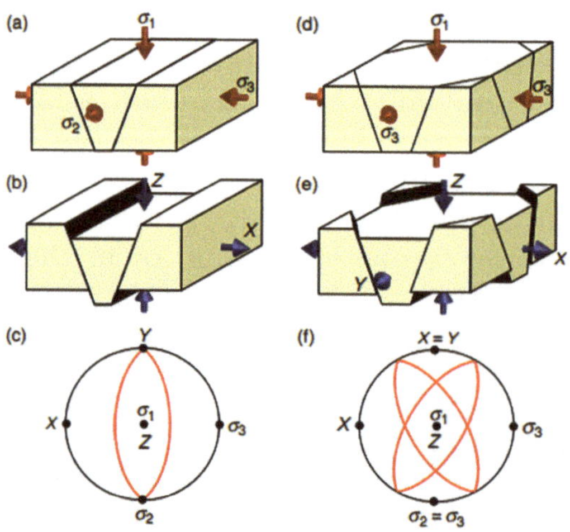

Figure 8.12. A single pair of dip-slip conjugate fault sets (a–c) form in plane strain, while multiple sets imply 3-D strain (d–f). The pattern shown in (e) is called orthorhombic from its high degree of symmetry. In these idealized cases the principal stresses (red arrows) and strains are parallel.

Source: https://elearning.uniroma1.it/pluginfile.php/624958/mod_resource/content/0/Haakon-Fossen-Structural_Geology_Part%201.pdf.

Fault systems may contain more than two conjugate sets in other situations. Figure 8.12d,e depicts a straightforward scenario in which two genetically related pairs of fractures coexist. The symmetry elements of this arrangement give rise to the term orthorhombic arrangement. Once more, strain axes can be found with ease using stereoplots, and stress axes can be deduced (Figure 8.12f).

8.2.2. Complex Fault Populations

Rocks that have undergone one stage of brittle deformation are known to contain conjugate fault sets. Fracture patterns are typically more complicated than simple conjugate patterns, and they frequently exhibit evidence of repeated activation under varying stress conditions. It is obvious that some faults will slip while others will not, where a weak foliation was reactivated for a certain range of orientations. However, what would be the response of such a fracture population to a new stress field, if it were complex in the sense that it contained elements with a wide range of orientations? We now have a population of weak elements to examine, and we also wish to determine the degree and direction of slip. While it's not easy to predict, some significant claims can be made.

The most obvious assumption that can be made is that faults or fractures oriented perpendicular to one of the three principal stresses will not slip. This is obvious since there is no shear component in this situation ($\sigma_s = 0$). For any other orientation $\sigma_s \neq 0$ and frictional sliding (slip) will occur if σ_s exceeds the frictional resistance against slip.

It is reasonable to assume that slip on a surface occurs in the direction of maximum resolved shear stress. If the largest shear stress is in the dip direction, normal or reverse faults will result. If the maximum shear stress vector is horizontal, strike-slip faulting occurs. Other cases lead to oblique-slip movements. This hypothesis assumes planar faults, rigid fault blocks, negligible block rotations, and activation of faults during a single phase of deformation under uniform stress. While idealized, empirical observations and numerical modeling suggest that the Wallace-Bott hypothesis is generally accurate, despite local perturbations caused by intersecting faults. By measuring fault orientation, lineation orientation, and sense of slip, we can use stress inversion techniques to calculate the orientation and relative size of the stress axes.

8.2.3. Paleostress from Fault Slip Inversion

Although it is often difficult to determine the exact values of the principal stresses, we can estimate their relative magnitudes, or the shape of the stress ellipsoid, by analyzing fault population data. We use the stress ratio for this estimation

$$\phi = (\sigma_2 - \sigma_3)/(\sigma_1 - \sigma_3) \quad (1)$$

where $0 \leq \phi \leq 1$. The ratio ϕ is also called R (note that R is defined as 1-ϕ by some authors). ϕ = 0 for a prolate stress ellipsoid, where $\sigma_2 = \sigma_3$ (uniaxial compression). ϕ = 1 implies that $\sigma_1 = \sigma_2$, and the stress ellipsoid is oblate (uniaxial tension). We use the ratio f to express the stress tensor, which in the coordinate system defined by the principal stresses is

$$\begin{bmatrix} \sigma_1 & 0 & 0 \\ 0 & \sigma_2 & 0 \\ 0 & 0 & \sigma_3 \end{bmatrix} \quad (2)$$

The orientation and shape of the stress ellipsoid can be preserved by adding an isotropic stress component to the stress tensor and multiplying the components by a constant. The reduced stress tensor is the stress tensor that includes details about the orientation and form of the stress ellipsoid but not the absolute magnitude of the principal stresses. This tensor is here expressed in terms of two constants k and l, where: $k = l/(\sigma_1 - \sigma_3)$ and $l = -\sigma_3$

$$\begin{bmatrix} \sigma_1 + l & 0 & 0 \\ 0 & \sigma_2 + l & 0 \\ 0 & 0 & \sigma_3 + l \end{bmatrix} \begin{bmatrix} k & 0 & 0 \\ 0 & k & 0 \\ 0 & 0 & k \end{bmatrix} = \begin{bmatrix} 1 & 0 & 0 \\ 0 & \phi & 0 \\ 0 & 0 & 0 \end{bmatrix} \quad (3)$$

Adding the constant l to each principal stress is the same as adding an isotropic stress. Furthermore, multiplying the axes by the constant k means contracting or inflating the stress ellipsoid while maintaining its shape.

While the full stress tensor contains six unknowns, the reduced tensor contains only four unknowns, represented by ϕ and the orientation of the principal stresses. In other words, the reduced stress tensor gives us the orientation and shape of the stress ellipsoid. While we have discussed how stress influences slip on fractures (by causing slip in the direction of the maximum resolved shear stress), the inverse problem where the (reduced) stress tensor is calculated from slip data is more relevant to paleostress analysis. In order to determine the reduced stress tensor, we require information from a minimum of four distinct fault surfaces, as it possesses four unknowns. We are looking for a tensor that fits all of the data. Nevertheless, due to rotations, local stress deflections, and measurement errors, etc. we are trying to find the tensor that fits the fault slip data the best. As a result, we typically gather data from ten to twenty slip surfaces, if not more. This results in far more equations than unknowns, and a statistical model that reduces errors is then used. A stress inversion program—many are available—performs the calculations. The findings are displayed as dimensionless Mohr diagrams, which depict the relative sizes of the principal stresses, and stereoplots, which display the orientations of the principal stress axes (Figure 8.13).

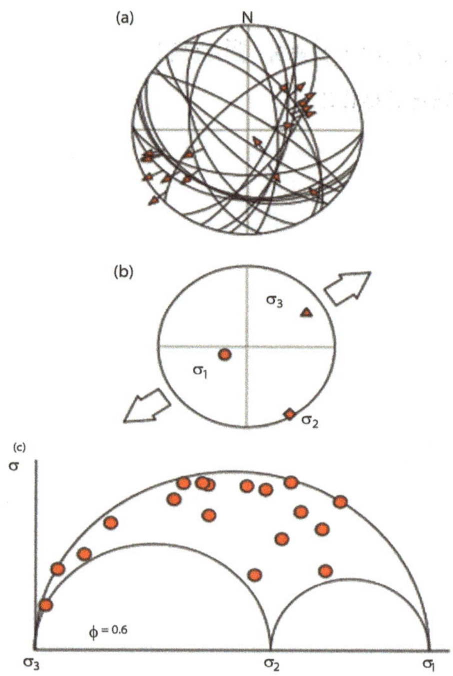

Figure 8.13. (a) Actual fault data, shown by great circles (fault surfaces) and arrows (lineations) on an equal-area lowerhemisphere plot. (b) Principal stresses found from stress inversion. (c) The data plotted in the dimensionless Mohr diagram, indicating the relative values of the principal stresses.

Source: https://elearning.uniroma1.it/pluginfile.php/624958/mod_resource/content/0/Haakon-Fossen-Structural_Geology_Part%201.pdf.

There can be data left over that don't fit in during the reduced stress tensor computation. To determine whether they collectively define a second tensor that might be connected to a different tectonic event, these can be examined independently. Using field criteria like cross-cutting relations and distinctive mineral phases on slip surfaces is an alternative—and typically safer—method of differentiating between subpopulations. Tangent-lineation diagrams are a geometric method of extracting stress from fault slip data. This is accomplished by plotting each fault plane and its lineation, creating an M-plane that contains the pole to each fault plane and the lineation, and drawing an arrow tangent to the M-plane at this pole, as illustrated in Figure 8.14.

The direction of the arrow reflects the movement of the footwall relative to the hanging wall. When differently oriented slip surfaces are plotted, such as the example presented in Figure 8.15, the resulting pattern may reveal both the orientations and the relative magnitudes (ϕ) of the principal stress axes (Figure 8.16). Software, such as FaultKin, can be used to plot the data collected in the field, but it is important that we understand what such programs do with our data.

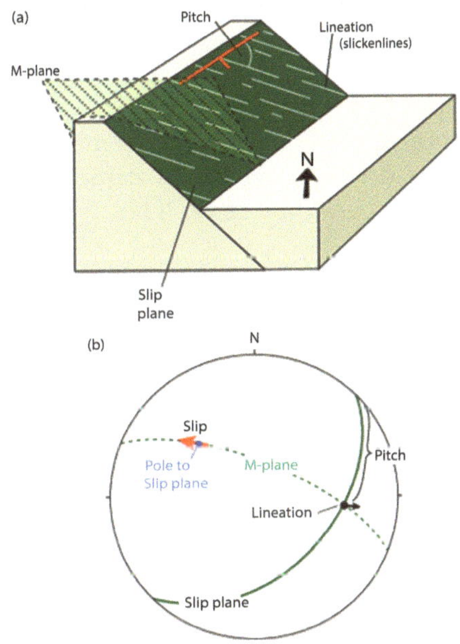

Figure 8.14. (a) Schematic illustration of pitch as measured on a fault surface. (b) The pitch plotted in a stereonet (lower hemisphere, equal area). This projection also shows how the tangent-lineation is found for a known slip plane (as shown in (a)) and its lineation (striation) by plotting a plane (the movement or M-plane) that contains the lineation and the pole to the slip plane. The tangent-lineation (red arrow) is the arrow drawn tangential to the M-plane at the pole to the slip plane. Its direction describes the movement of the footwall relative to the hanging wall; in the present case a normal movement, where the footwall moves up to the west.

Source: https://elearning.uniroma1.it/pluginfile.php/624958/mod_resource/content/0/Haakon-Fossen-Structural_Geology_Part%201.pdf.

It is obvious that in order to get a good result, a large number of data points from faults with different orientations are needed. Figure 8.15 provides an example, and it is evident from this plot that the data distribution is insufficiently consistent to constrain the relative stress magnitudes. This is a situation that occurs frequently when working with real data sets, particularly data sets derived from rocks that have polyphasally deformed rock structures. Paleostress analyses require caution for a number of reasons. One is that their success largely depends on our ability to locate fault populations that were created during the faulting history under a constant stress field. Another is that, within the study locality, the stress field must be uniform. We know that the stress field is likely to be perturbed by fault interaction and any mechanical layering, so this assumption is probably only an approximation. Paleostress analysis outputs must be interpreted in the context of these facts as well as other locality-specific factors.

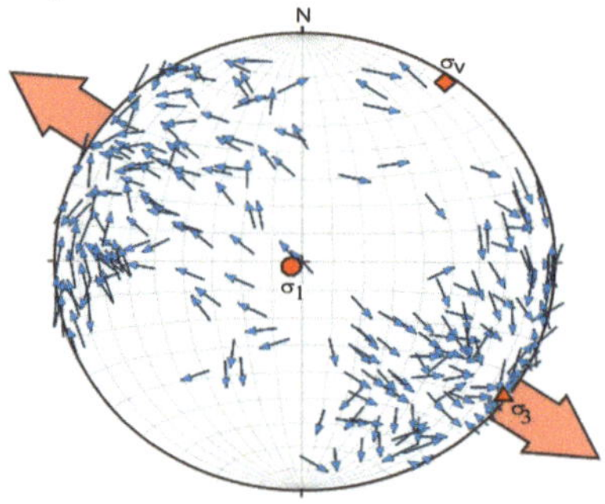

Figure 8.15. Tangent-lineation diagram for fault data from basement gneisses west of Bergen, Norway. σ_1 can be defined, but the ϕ-value, which in principle is found by comparing with the next figure, is more difficult to define in this case. This is quite common and is partly due to limited variation in fault orientations.

Source: https://elearning.uniroma1.it/pluginfile.php/624958/mod_resource/content/0/Haakon-Fossen-Structural_Geology_Part%201.pdf.

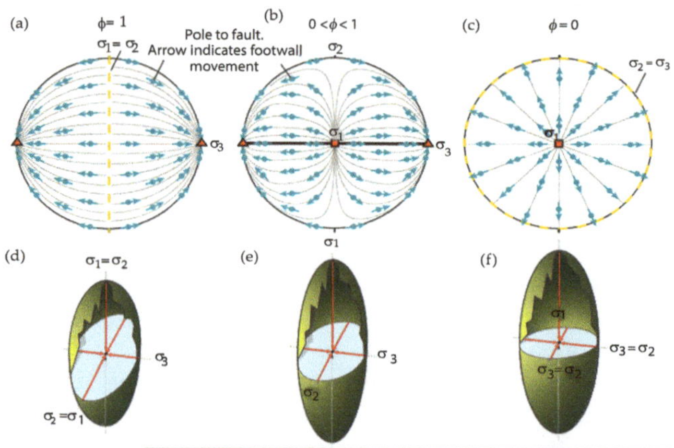

Figure 8.16. (a–c) Stereographic projection of poles to faults (points). Arrows indicate footwall versus hanging wall movements. The pattern emerges by considering faults with different orientations, and the assumption is made that the movement is parallel to the direction of maximum shear stress on each surface. The pattern depends on the ratio between the principal stresses, expressed by ϕ. By plotting field data in these diagrams one can obtain an estimate for the orientation of the principal stresses and f. The geometry of the strain ellipsoid is shown (d–ϕ). Based on Twiss and Moores (2007).

Source: https://elearning.uniroma1.it/pluginfile.php/624958/mod_resource/content/0/Haakon-Fossen-Structural_Geology_Part%201.pdf.

8.3. A KINEMATIC APPROACH TO FAULT SLIP DATA

The Wallace-Bott hypothesis is used by the paleostress method, which is based on the inversion of fault slip data to relate stress and slip. However, this method makes assumptions that glaringly oversimplify the true conditions in the crust. Because of this, some of us favor a strictly kinematic approach that treats strain or strain rate as opposed to stress. Plotting the fault planes and associated slip-related lineations in a stereoplot is the graphical representation of the kinematic method, as seen in Figure 8.17. P- and T-axes are determined for each fault plane based on the orientation and sense of slip, which is somewhat similar to focal mechanism solutions. 'These axes represent the contractional and extensional quadrants' symmetry axes. They are contained in the plane that is perpendicular to the fault plane and are perpendicular to each other. The lineation (slip direction) is also present in the fault plane. Moreover, P and T reach the fault plane at 450 degrees (Figure 8.17). Depending on the fault movement sense (normal, reverse, sinistral, or dextral), T and P's positions vary.

In the population, a pair of P- and T-axes are found for every fault. We obtain a distribution of P- and T-axes for fault populations with a distribution in fault orientations. Two orthogonal surfaces that are fitted to the plot should ideally be able to divide them apart. The line of intersection between the planes represents Y, the intermediate strain axis, while Z and X are situated in the center of the P- and T-fields, respectively (Figure 8.17c). This indicates the direction of the paleostrain axes; the magnitude can only be estimated by including details regarding the area and displacement of each fault in the analysis. Using a computer program, the kinematic method can be completed quickly. In most cases, the strain axes found using this method are very similar to the stress axes found by inverting slip data, as previously discussed. The kinematic approach also depends on a consistent strain field and accurate grouping of slip data produced at various stages of deformation.

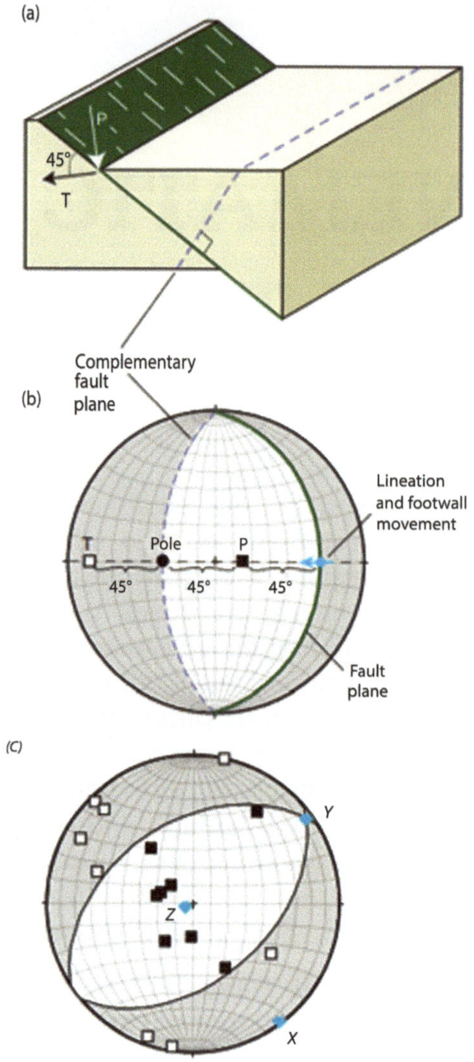

Figure 8.17. Kinematic analysis of fault data. (a) A simple normal fault. (b) Plot of the fault, the lineation and the sense of movement. The complementary shear plane oriented 90° to the fault is shown. The T- and P-axes bisect these two planes. (c) Data from 16 faults with different orientations. Each of the faults is plotted as illustrated in (b). Ideally, the P- and T-axes will plot within two sectors separated by two mutually orthogonal planes. If they do not, then the conditions are not fulfilled; they may for example have formed in two different stress fields. The few data are a minimum, but indicate a NW–SE stretching and vertical shortening, i.e., an extensional regime.

Source: https://elearning.uniroma1.it/pluginfile.php/624958/mod_resource/content/0/Haakon-Fossen-Structural_Geology_Part%201.pdf.

8.4. CONTRACTIONAL AND EXTENSIONAL STRUCTURES

Although shear fractures and faults are the focus of fault slip data analyses, contractional and extensional structures can also be helpful when analyzing stress or strain rates. In certain (mostly) brittlely deformed rocks, contractional structures—also referred to as anticracks—can be found as solution seams or stylolites. These structures typically indicate minor strains that rotate very little or not at all. Thus, it is reasonable to assume that the stress and strain axes are closely related. Low-strain extension structures like veins and joints operate similarly. While contractional structures tend to form perpendicular to σ_1, extensional fractures form perpendicular or at least at a high angle to σ_3. This makes joints and veins the structures that give the quickest and simplest connection between stress and strain. The combination of contraction and extension structures is particularly valuable, and when they occur in faulted rocks these structures can be used in or compared to the results of stress-inversion methods.

Non-rotated joints, veins and dikes are structures that immediately define the orientation of σ_3 at the time of initiation. A related kinematic method involves the reconstruction of walls displaced during dike intrusion. Magma overpressure during intrusion may cause fracturing of the surrounding rock, into which magma flows to form dikes. In the presence of a significant tectonic stress, the fractures and thus the dikes are oriented perpendicular to σ_3. In most cases, however, the rock contains preexisting fractures that will be filled with magma.

Fractures with an orientation perpendicular to σ_1 will preferentially open perpendicular to the walls, but if we want to find the orientation of σ_3 we have to look for points on each side of the dike that once were adjacent to each other. Such points can be corners where dikes step sideways (Figure 8.18), in which case displacement vectors can be constructed by connecting separated corners. Structures that were separated by dikes can also be used, such as veins, older dikes, fold hinges and steeply dipping layers. In these cases we observe strain (the local extension direction), but because of the small displacements usually involved it is reasonable to correlate the horizontal extension direction with σ_H.

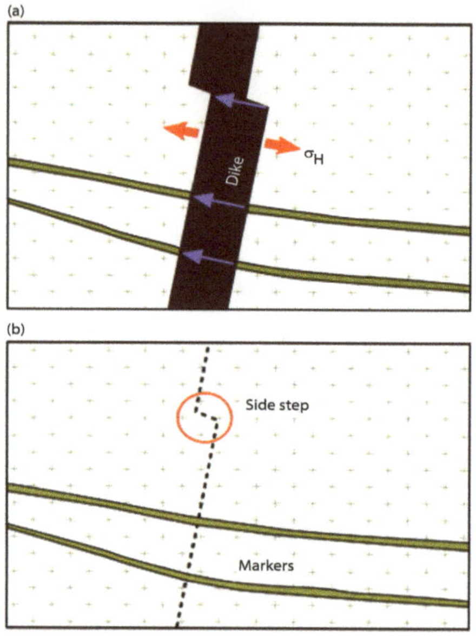

Figure 8.18. Reconstruction of a horizontal section (a) to the situation prior to dike intrusion (b). Arrows can be drawn between points that once were neighbors to find displacement vectors. For small strains these vectors approximate sH. From Fossen (1998).

Source: https://elearning.uniroma1.it/pluginfile.php/624958/mod_resource/content/0/Haakon-Fossen-Structural_Geology_Part%201.pdf.

Assuming that the vertical stress is a principal stress axis we can further assume that $\sigma_H = \sigma_3$. It is the local stress that is found, and it is not uncommon to find that dikes rotate in the vertical direction into a fringe-like geometry such as shown in Figure 8.19. Such geometries indicate that the σ_H changes direction in the vertical direction, similar to fringe zones around fractures

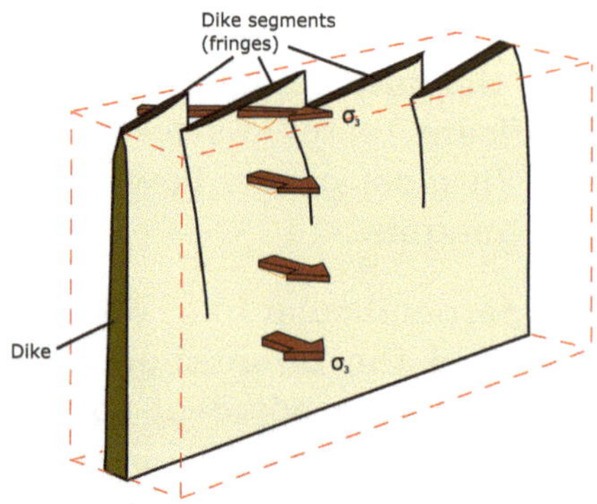

Figure 8.19. En-echelon dike system that is connected at depth – one of Anderson's popular interpretations published in 1951. Anderson interpreted this phenomenon as the result of an upward rotation of σ_3 towards the surface during magma intrusion.

Source: https://structuredatabase.wordpress.com/wp-content/uploads/2015/09/figure-3.png.

CASE STUDY

SAN ANDREAS FAULT, CALIFORNIA, USA

The San Andreas Fault is a transform fault, which is a type of geological fault where two tectonic plates slide past each other horizontally. In the case of the San Andreas Fault, it marks the boundary between two major tectonic plates: the Pacific Plate and the North American Plate. These plates are part of the Earth's lithosphere, and their movement is responsible for shaping California's geology and landscape.

Location

The San Andreas Fault extends approximately 800 miles (1,300 kilometers) through California. It stretches from the southern part of the state, near the Salton Sea, all the way up to the northern region, near Cape Mendocino. It traverses a diverse range of landscapes, including deserts, mountains, and coastal areas, making it a prominent geological feature in California.

Historical Significance

1. **Plate Tectonics Confirmation:** The San Andreas Fault played a pivotal role in confirming the theory of plate tectonics, which revolutionized the field of geology in the mid-20th century. It provided tangible evidence of the movement of Earth's lithospheric plates, as it clearly demonstrates the lateral movement of the Pacific and North American Plates.
2. **Large Earthquakes:** The San Andreas Fault is notorious for producing large and destructive earthquakes. Some of the most significant earthquakes in California's history, including the 1906 San Francisco earthquake, have been associated with this fault. These events have had a profound impact on the region's development and seismic hazard awareness.
3. **Seismic Hazard:** California is one of the most seismically active regions in the United States, and the San Andreas Fault is a primary source of this activity. Its presence has prompted extensive research and preparedness efforts to mitigate the potential impacts of future earthquakes. It has also influenced building codes and infrastructure development in the state.
4. **Geological Features:** The fault has created striking geological features, such as valleys, scarps, and offset streams, providing geologists with valuable insights into the history of the Earth's crustal movements. These features have contributed to the field of geomorphology.
5. **Tourism and Education**: The San Andreas Fault has become a tourist attraction and an educational resource. Visitors can explore areas where the fault is visible at the surface, such as the Carrizo Plain National Monument, and learn about the Earth's dynamic processes.

The San Andreas Fault is a significant geological feature located in California, where the Pacific Plate and North American Plate interact. Its historical significance includes confirming the theory of plate tectonics, generating large earthquakes, influencing seismic hazard preparedness, shaping geological features, and serving as a source of education and tourism.

CLASS ACTIVITY

ANALYZING REAL-WORLD FAULT SURFACES

Procedure:

Research

- Have students research real-world examples of fault surfaces, such as the San Andreas Fault, and look for images or case studies that highlight features like slickensides, fault gouge, and striations.

Comparative Analysis

- Compare the features observed in the classroom model with those found in real-world fault surfaces. Discuss similarities and differences.
- Use online resources (e.g., USGS, academic papers) to find detailed descriptions and images of fault surfaces.

Presentation

- Have students create a presentation or report summarizing their observations from both the hands-on model and their research. Include sketches, photos, and descriptions of fault surface features.

Questions for Students

1. How do the features of the modeled fault surface compare to those of real-world fault surfaces?
2. What can slickensides and striations tell us about the direction and nature of fault movement?
3. Why is understanding fault surface characteristics important for earthquake studies and hazard assessment?

SUMMARY

- Several kinematic criteria can help determine the sense of displacement on a slip surface. We should be able to make a sketch of the most important ones.
- Joints and veins are extension structures that initiate perpendicular to the smallest stress axis, and are therefore useful paleostress indicators.
- Conjugate faults are another useful set of structures that allows for a quick estimate of the stress field.
- Complex fault slip datasets may contain slip surfaces formed under two or more stress fields that must be separated to obtain useful results.
- Small fractures developed along a fault or slip surface may show geometric arrangements that carry information about the sense of slip on the fault.
- Models using analogue materials such as clay and sand have revealed a fairly consistent faulting sequence in experimental strike-slip fault zones. Fault zones usually begin with a set of relatively short R-shears arranged in en échelon arrays and coeval with minor, conjugate R'-shears.
- Folds associated to wrench fault systems are typically non-cylindrical, doubly plunging and relatively short with steeply dipping axial planes.
- Local strike-slip faults occur in the hanging wall blocks of low-angle faults and accommodate different amounts of displacement either on different parts of the fault or between the allochthonous and adjacent autochthonous rocks.
- Transform faults are strike-slip faults at plate boundaries, parallel to the direction of relative motion of the plates on either side. They include transfer faults along which plates slide past each other, but the kind of motion between plates is changed at the end of the fault.
- Fault observations used in paleostress analysis include the local strike and dip of the fault surface, the orientation of the lineation and the sense of movement.
- Conjugate fault sets are found in rocks that have experienced a single phase of brittle deformation. In many cases, fracture patterns show evidence of repeated activation under changing stress conditions and commonly more complicated arrangements than a simple conjugate pattern.

REVIEW QUESTIONS

1. Why must we be careful when interpreting lineations as displacement indicators?
2. What are the premises for successful paleostress analysis?
3. What is meant by the expression "slip inversion"?
4. What are conjugate faults, and what stress information do they give?
5. What does the Wallace–Bott hypothesis postulate?
6. What is the reduced stress tensor?

REFERENCES

1. Anderson, E. M. (1951). The Dynamics of Faulting. Edinburgh: Oliver & Boyd.
2. Angelier, J. (1994). Fault slip analysis and palaeostress reconstruction. In P. L. Hancock (Ed.), Continental Deformation.
3. Cashman, P. H., & Ellis, M. A. (1994). Fault interaction may generate multiple slip vectors on a single fault surface. Geology, 22, 1123–1126.
4. Dunne, W. M., & Hancock, P. L. (1994). Paleostress analysis of small-scale brittle structures. In P. L. Hancock (Ed.), Continental Deformation (pp. 101–120). Oxford: Pergamon Press.
5. Etchecopar, A., Vasseur, G., & Daignieres, M. (1981). An inverse problem in microtectonics for the determination of stress tensors from fault striation analysis. Journal of Structural Geology, 3, 51–65.
6. Fry, N. (2001). Stress space: Striated faults, deformation twins, and their constraints on paleostress. Journal of Structural Geology, 23, 1–9.
7. Harding, T. P. (1985). Seismic characteristics and identification of negative flower structures, positive flower structures, and positive structural inversion. Bulletin of the American Association of Petroleum Geologists, 69(4), 582–600.
8. Jolly, R. J. H., & Sanderson, D. J. (1995). Variation in the form and distribution of dykes in the Mull swarm, Scotland. Journal of Structural Geology, 17, 1543–1557.
9. Kearey, P., Klepeis, K. A., & Vine, F. J. (2009). Global Tectonics (3rd ed.). Oxford: Wiley-Blackwell.
10. Lisle, R. J., Orfie, T. O., Arlegui, L., Liesa, C., & Srivastava, D. C. (2006). Favoured states of palaeostress in the Earth's crust: Evidence from fault-slip data. Journal of Structural Geology, 28, 1051–1066.
11. Mann, P., Hempton, M. R., Bradley, D. C., & Burke, K. (1983). Development of pull-apart basins. Journal of Geology, 91(5), 529–554.
12. Marrett, R., & Allmendinger, R. W. (1990). Kinematic analysis of fault-slip data. Journal of Structural Geology, 12, 973–986.
13. Marrett, R., & Peacock, D. C. P. (1999). Strain and stress. Journal of Structural Geology, 21, 1057–1063.
14. Petit, J. P. (1987). Criteria for the sense of movement on fault surfaces in brittle rocks. Journal of Structural Geology, 9, 597–608.
15. Scholz, C. H. (1977). Transform fault systems of California and New Zealand: Similarities in their tectonic and seismic styles. Journal of the Geological Society of London, 133(3), 215–229.
16. Stein, R. S., Barka, A. A., & Dieterich, J. H. (1997). Progressive failure on the North Anatolian fault since 1939 by earthquake stress triggering. Geophysical Journal International, 128(3), 594–604.
17. Twiss, R. J., & Unruh, J. R. (1998). Analysis of fault slip inversions: Do they constrain stress or strain rate? Journal of Geophysical Research, 103(B6), 12,205–12,222.

18. Watterson, J. (1999). The future of failure: Stress or strain? Journal of Structural Geology, 21, 939–948.

19. Woodcock, N. H. (1986). The role of strike-slip fault systems at plate boundaries. Philosophical Transactions of the Royal Society of London A, 317, 13–29.

20. Woodcock, N. H., & Fischer, M. (1986). Strike-slip duplexes. Journal of Structural Geology, 8(7), 725–735.

CHAPTER 9

Salt Tectonics

LEARNING OBJECTIVES

After studying this chapter, you will be able to:

- Define salt tectonics
- Describe salt properties and rheology
- Understand salt diapirism, salt geometry and the flow of salt
- Discuss salt diapirism in the extensional regime
- Define salt diapirism in the extensional regime
- Know diapirism in the contractional regime
- Understand salt collapse by karstification
- Explain the concept of salt décollement

KEY TERMS FROM THIS CHAPTER

Contractional deformation	Diapirs
Differential loading	Extensional tectonics
Geologic modeling	Gravitational forces
Halokinetic cycles	Halotectonics
Poiseuille flow	Salt anticlines
Salt canopies	Salt tectonics
Sedimentary basins	Sedimentary facies

INTRODUCTION

The characteristics and behavior of salt as a rock differ greatly from those of most other rocks. This suggests that deformed sedimentary sequences containing salt layers take on unique structural styles that are frequently quite fascinating. Particular structures that are significant in a variety of contexts are salt ridges, pillows, diapirs, and even salt glaciers. Due to its propensity to act as a décollement, salt can control the structural expression and increase the areal extent of the deforming area even in situations where it is limited to a thin layer. Because many petroleum provinces contain salt layers or are deformed by salt tectonics, salt-related structures are of great importance to geologists working in both contractional and extensional tectonic regions. With examples from various locations throughout the world, this chapter offers an overview of salt structure geometries, processes, and tectonic settings with a focus on salt structures and salt tectonics.

9.1. SALT TECTONICS

The field of study known as "salt tectonics" examines the geologic deformation caused by salt, including how and why salt structures change, the three-dimensional shapes they produce, and how they interact with regional tectonics. The word "salt" is a handy way to describe rocks that are mostly made of evaporite minerals, particularly halite. Salt's primary geologic function in sedimentary basins is to lubricate deformation. Salt can also take on a variety of diapiric shapes as it rises through its overburden.

Throughout the world, salt tectonics has had a significant impact on over 130 sedimentary basins. In the shallow crust, salt can flex more easily than other rocks because of its weakness. Exploration for oil and gas and, to a lesser degree, the mining of certain minerals are highly interested in the effects of salt tectonics. Salt bodies are surrounded by copious hydrocarbon traps created by salt tectonics. Furthermore, large shrouds of salt form above hydrocarbon traps at deeper depths, influencing geophysical imaging of these exploration targets.

9.1.1. Passive Salt Structures

Salt structures can form solely due to gravitational instability from continued sedimentary loading, without any external tectonic influence. Pure halite has a density of 2160 kg/m^3, while initially deposited sediments have a lower density of 2000 kg/m^3. With loading and compaction, sediment density can increase to 2500 kg/m3, causing it to be denser than salt. As overlying layers become denser, the weak salt layer will deform into ridges and depressions due to Rayleigh-Taylor instability. Further sedimentation will accumulate in depressions, causing salt to move into ridges. Diapirs tend to form at ridge junctions, with salt movement along the ridges feeding their growth until salt supply is depleted. During later stages, the diapir's top remains near the surface, matching further burial with diapir rise, known as downbuilding. Examples of passive salt structures include the Schacht Asse II and Gorleben salt domes in Germany.

Not all salt layers that are buried beneath sedimentary overburdens give rise to such structures. This may be because the salt unit is interbedded with sedimentary layers that increase both its density and strength, or it may be the result of a relatively high strength overburden (Coward and Stewart, 1995).

9.1.2. Active Salt Structures

Salt structures will be more likely to form in areas of active tectonics. In the context of extensional tectonics, faulting will thin and weaken the overburden. Salt can rise into the cores of anticlines in an area impacted by thrust tectonics, as demonstrated by salt domes in the Zagros Mountains and El Cordo diapir (Coahuila fold-and-thrust belt, Northeastern Mexico), where the overburden layer has buckled.

Forceful diapirism is the process by which a salt body may be able to push through its overburden if the pressure inside it increases to a high enough level. Both active and passive salt movement may be present in many salt diapirs. It is possible for an active salt structure to break through its overburden and then proceed to develop as a completely passive salt diapir.

9.1.3. Reactive Salt Structures

Salt may still migrate into comparatively low pressure zones near forming folds and faults in situations where salt layers lack the prerequisites to form passive salt structures. Reactive structures are what these are known as.

9.1.4. Salt Detached Fault Systems

A distinctive collection of structures is created during extensional tectonics when one or more salt layers are present. Extensional faults originate in the middle of the crust and extend upward until they reach the salt layer. Because of the salt's weakness, the fault cannot spread. On the other hand, as the fault continues to shift, the salt's base is offset and the overburden layer bends. The stresses brought on by this bending will eventually be high enough to cause the overburden to fail. The initial thickness of the salt determines the kinds of structures that form. A system is considered unlinked when there is no clear spatial relationship between the faulting in the overburden and that beneath a very thick layer of salt. The overburden faults for intermediate salt thicknesses are referred to as soft-linked systems because they are spatially related to the deeper faults but offset from them, usually into the footwall. A hard-linked fault is created when the salt layer gets sufficiently thin that the fault that forms in the overburden is closely aligned with that beneath the salt and forms a continuous fault surface after only a small displacement.

The preferred detachment planes in thrust tectonic areas are salt layers. Variations in the thickness and, consequently, efficacy of the Hormuz salt, which dates from the late Neoproterozoic to the Early Cambrian, are thought to have had a fundamental control on the overall topography in the Zagros fold and thrust belt.

9.1.5. Salt Weld

The overburden and the sub-salt basement beneath it are effectively fused together when a salt layer becomes too thin to function as an effective detachment layer because of salt movement, dissolution, or removal by faulting. This is a crucial factor to take into account when modeling the migration of hydrocarbons since it could lead to the formation of new faults in the cover sequence. If the sides of a former diapir come into contact, salt welds can also form vertically.

9.2. SALT PROPERTIES AND "RHEOLOGY"

Salt differs significantly from the majority of other common rocks due to its unique physical and rheological characteristics. First of all, while pure halite can be found in salt layers in sedimentary sequences, anhydrite, gypsum, and clay minerals are frequently present as well. The density of pure halite is comparatively low at 2.160 g/cm^3, whereas impure salt has a slightly higher density. Because of this, salt is still denser than unlithified siliciclastic sediments but less dense than the majority of carbonate rocks.

9.2.1. Salt Properties

- Common constituent of sedimentary basins.
- Mechanically very weak.
- Low density (2.160 g/cm^2).
- High heat conductivity.
- Almost incompressible.
- Impermeable.
- Viscous, behaves like a fluid.
- Causes wide areas of deformation.
- Enhances horizontal detachments.
- Creates structural fluid traps.

The porosity (Φ) in this formula indicates the compaction. The average rock density for a siliciclastic sediment with 30% porosity, 2.7 g/cm^3 of average mineral density, and 1.04 g/cm^3 of salty pore water is roughly equal to 2.2 g/cm^3 of slightly impure salt. Therefore, for the overburden to become denser than the salt, a porosity value of less than or equal to thirty percent is needed. If the sediment is finely grained, this can happen at relatively shallow depths—600–700 m, maybe. For a sandy sequence, a density of 2-2 g/cm^3 might be reached at 1500–2000 m. Moreover, for a diapir to rise to the surface solely through buoyancy, the average density of the overburden needs to be higher than the average density of the underlying salt (Jackson, 1995).

This necessitates burial to a minimum depth of 1600 meters, but preferably closer to 3000 meters. In the Gulf of Mexico, this depth is approximately 2300 meters.

Even at surface conditions, salt deforms plastically during loading from a rheological perspective. Salt will only fracture when strain rates increase, as they do during mining operations, earthquakes, or hammer strikes. For the most part, salt flows as a viscoelastic medium in geologic settings. For this reason, salt mines are being used as disposal sites for waste.

Because of the low relaxation rates involved, the elastic component of salt deformation can be disregarded for the most part, and the salt can be thought of as behaving like a viscous material. More specifically, wet diffusion and dislocation creep are the two deformation mechanisms that predominate in salt deformation. When there is some water in the salt, the predominant deformation mechanism is wet diffusion, in which material dissolves and is carried along grain boundaries by a thin fluid film. Wet diffusion is typically faster and thus more significant in fine-grained salt than coarse-grained salt because it involves transport along grain boundaries and surface area increases with decreasing grain size. Wet diffusion is also aided by differential stresses and low strain rates. Dry salt is distinct because it lacks any liquids to carry matter. Unless the salt just fractures, in which case dislocation creep is the main mechanism of deformation (dry salt at or near surface conditions).

Since salt has a very low yield strength and flows easily, it is easy for either of these deformation mechanisms to be activated. Salt can be treated as a fluid in geologic modeling because of this and the fact that it is nearly incompressible in the majority of geologic conditions.

9.3. SALT DIAPIRISM, SALT GEOMETRY, AND THE FLOW OF SALT

9.3.1. Salt Structures

It has long been known that salt volumes in the crust can take on a wide range of geometric shapes, from more confined structures like salt stocks to longer structures like salt anticlines and salt pillows. As Figure 9.1 schematically illustrates, many of these structures go by different names even though they are generally referred to as diapirs. The Greek words dia (through) and peran (pierce) give rise to the term diapir, which is used in geology to characterize a body that gravitationally moves upward and intrudes into the overburden. Typically, the body is made of salt, magma, or water saturated mud or sand. Therefore, since they do not penetrate or pierce the overburden, certain structures—like salt pillows and salt anticlines—that merely bend and elevate the layers above are not considered diapirs in the strict sense of the word. But the majority of these structures show different phases that might eventually result in the creation of a real diapir (Kehle, 1988).

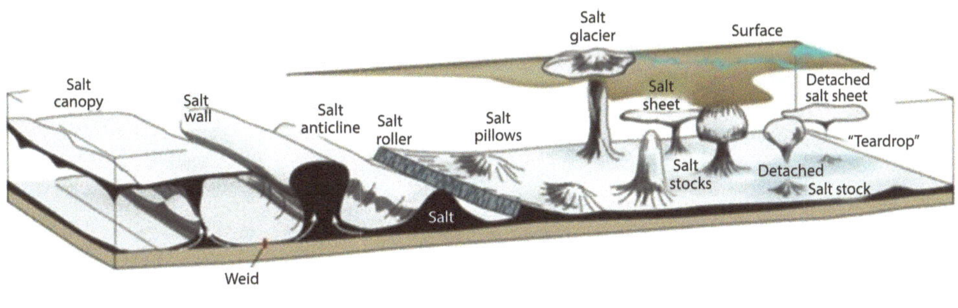

Figure 9.1. Different types of salt structures, their names and geometries. Maturity increases from the central part of the figure to the left and right.

Source: Haakon Fossen, Structural Geology Second Edition, ISBN 978-1-107-05764-7.

Salt expulsion, which is more accurate but less common, refers to the movement of salt from a salt layer into a salt structure. There are essentially two main categories or end-members of flow. One is known as a Poiseuille flow, which happens when salt seeps into a salt structure while a salt diapir or anticline is developing. In this instance, boundary drag—a phenomenon where viscous shear forces act along the salt's boundaries—restricts flow. Because of this effect, the salt moves more quickly through the middle of the salt layer than it does along the top and bottom. As a result, thin salt layers flow more slowly than thick ones, suggesting that when salt is thinned to a few tens of meters, the flow in a thick salt layer (tens of meters or more) slows down. If the salt becomes completely exhausted the boundary layers become attached to each other, and the contact is referred to as a salt weld, typically indicated on geologic and seismic sections by pairs of dots (Figure 9.2).

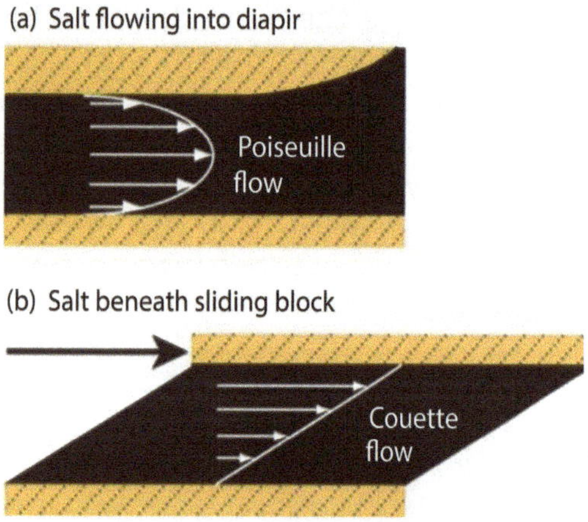

Figure 9.2. The two principal types of flow occurring in deforming salt layers. Arrows indicate velocity and the velocity profile is parabolic in (a) and linear in (b).

Source: Haakon Fossen, Structural Geology Second Edition, ISBN 978-1-107-05764-7.

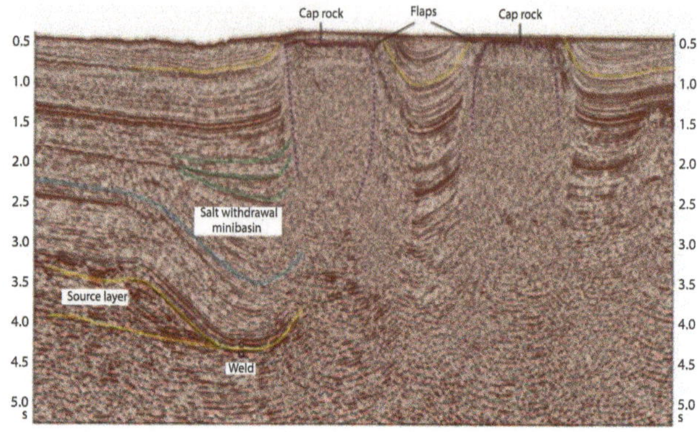

Figure 9.3. Salt structures imaged on a seismic 2-D line. The two structures have surfaced (now covered by thin Quaternary cover) and developed reflective caps. Deeper parts of the salt structures are obscured by seismic noise. The source layer has been interpreted based on a salt weld structure. The growth history is recorded in the sedimentary record: thickness changes near the diapirs start above the blue horizon, after deposition of a ~1.5–2 km sedimentary load. Note minibasin above the weld.

Source: Haakon Fossen, Structural Geology Second Edition, ISBN 978-1-107-05764-7.

There will always be some residual salt in salt welds; even if this salt is only a few centimeters thick, it will nonetheless constitute a weak area that is vulnerable to localized strain. A salt weld is significant because it stops salt from flowing laterally into nearby salt structure(s).

The other kind of flow is called Couette flow, and it happens when the overburden is translated in relation to the substrate. It simply involves shearing within the salt layer. Although the two types of flow may very well be superposed on top of one another, this type of flow is typical for salt layers functioning as décollements. A boundary effect similar to that seen in Poiseuille flow should ideally not exist in Couette flow.

9.3.2. Salt Diapir Geometries

Salt diapirs come in different shapes and sizes. They can appear longer in plan view, usually in the direction of graben axes or local faults. They might also have a circular form, as many of the Central Graben's diapirs do. Some show cross-sections of triangles with the salt thinning upward. This is typical in situations where faults in the overburden are connected to salt structures. Salt stocks are defined as salt diapirs with plug-like shapes. Salt walls are long, elongated salt structures that resemble stocks in cross-sections taken perpendicularly. Well-developed salt wall examples from the Moab area (Paradox Basin) in southeast Utah are depicted in Figure 9.4.

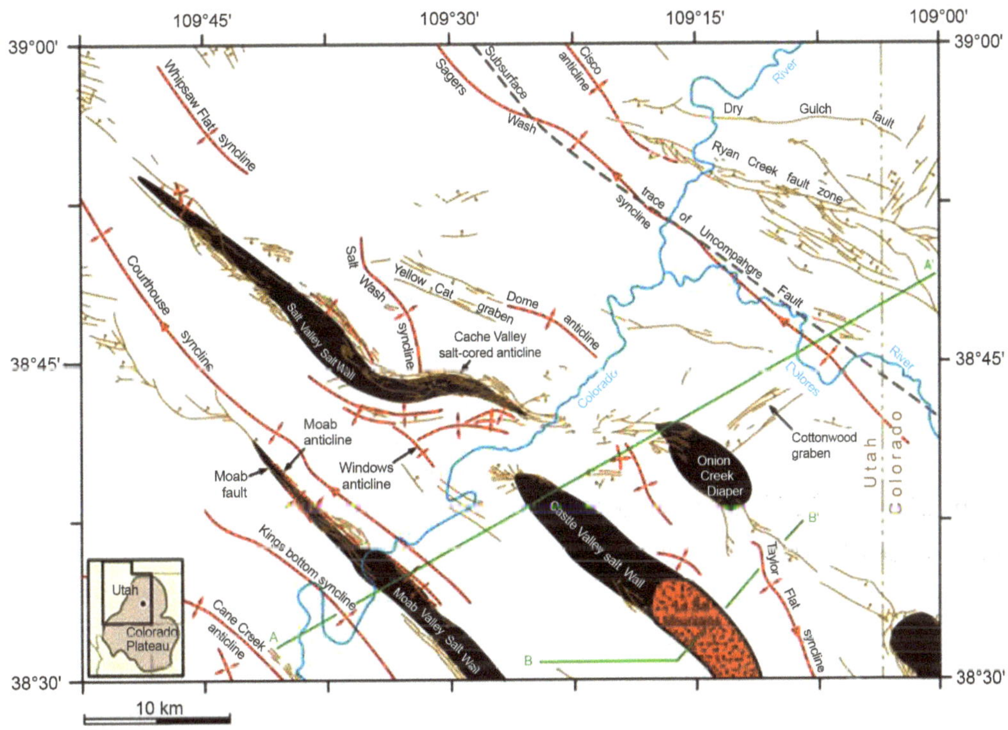

Figure 9.4. Salt structures in the Moab area of southeast Utah in map view and cross-section. Elongated salt walls dominate, but more circular diapirs also occur. The salt walls seem to follow the NW–SE trend of long-lived faults in the basement, while younger collapse related faults occur in the overburden.

Source: Haakon Fossen, Structural Geology Second Edition, ISBN 978-1-107-05764-7.

Numerous diapirs have a bulb, which is the broader upper portion, and a stem. In severe circumstances, the salt bulb may be entirely separated from its source layer due to the absence of the stem. A different name for isolated salt bulbs is teardrop diapirs. When salt moves laterally across younger sedimentary strata, it can produce asymmetrical geometries in these detached salt bodies. In this scenario, the salt is referred to as allochthonous. Diapirs can also form different kinds of salt stock canopies by flattening out and joining at one, or less frequently, multiple, stratigraphic levels.

Seismic data provides the majority of the information we currently know about the geometry of salt structures. Salt rarely creates seismic reflectors and is fairly homogeneous acoustically. As a result, areas that are "seismically transparent" and exhibit a distinctive noise pattern can be found on seismic lines that are relatively pure. The walls of the salt are frequently too steep to be seen on seismic lines, but the top of the salt is typically well imaged. Additionally, salt layers and structures absorb so much energy that it may be challenging to distinguish sub-salt reflectors. As a result, it might be challenging to differentiate between salt stocks and salt bulbs. It could even be challenging in some circumstances to tell impact structures from salt structures. However, seismic imaging is constantly developing, and with the help of contemporary 3-D surveys, images of salt can now be obtained that were not possible even a decade ago.

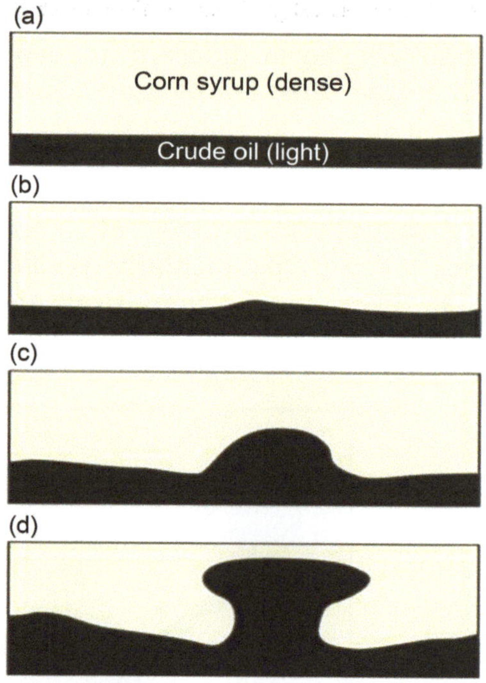

Figure 9.5. Experiment where a dense corn syrup is placed on top of a less dense oil.

Source: Haakon Fossen, Structural Geology Second Edition, ISBN 978-1-107-05764-7.

9.3.3. Modeling Salt Structures: The Fluid Approach

The ability of salt to flow even at temperatures below those found in geology was studied in the 1930s, as was the significance of the density contrast between the salt and the overburden. This resulted in a time when the salt and the denser sediments on top of it were both modeled as fluids. Lewis Lomax Nettleton showed in 1934 that it is comparatively simple to physically model this situation by adding a layer of corn oil on top of less dense crude oil. As a result of this unstable state, blobs of the light oil rose through the heavy oil, creating a diapir that resembled the so-called lava lamps, which have blobs of heated wax rising from the lamp's bottom. The dense fluid, called "sediment," in these fluid experiments will sink to the bottom, called "salt," which is less dense.

Since about 1960, a more advanced modeling tool called the centrifuge has been used. Hans Ramberg and his colleagues in his Uppsala, Sweden, laboratory have done more than anyone to explore the potential and push the limits of this tool. Using weak ductile material as analogues of rock, Ramberg and colleagues deformed their models in centrifuges at up to 2000g. Gravity therefore plays a more significant role in the centrifuge. Additionally, the experimental materials' densities and viscosities as well as the thicknesses of the stratigraphic layers (geometry) are planned to create a scaled-down representation of a real geologic setting while maintaining pertinent proportions and contrasts. Scaling is a crucial component of a physical experiment. It should be noted that realistic centrifuge modeling does not require that the densities and viscosities of the model layers match those of the geologic layers they are intended to represent. However, the model must match nature in terms of the density and viscosity contrasts between the different layers.

As it happened, the diapirs generated by the centrifuge experiments were strikingly similar to the salt diapirs found in the upper crust. With this kind of modeling, the progressive development of diapirs can be examined. Figure 9.6 provides an example of this evolution, showing how open domes or pillows give way to tall diapirs and then isolated or mushroom-shaped salt volumes that encroach on the overburden and eventually form salt canopies. The model can be stopped mid-run and additional sediment layers added. Salt eventually forms salt canopies at or near the surface. The centrifuge model has been successful in reproducing salt structures, such as mushroom-shaped diapirs and salt canopies, in the Dasht-e Kavir (great salt marsh) of Iran. This region's unusually high overburden ductility is caused by salt that has spread throughout some of these stratigraphic layers. Figure 9.8 provides a clear illustration of their ductile nature, which results in the formation of folds rather than faults and fractures.

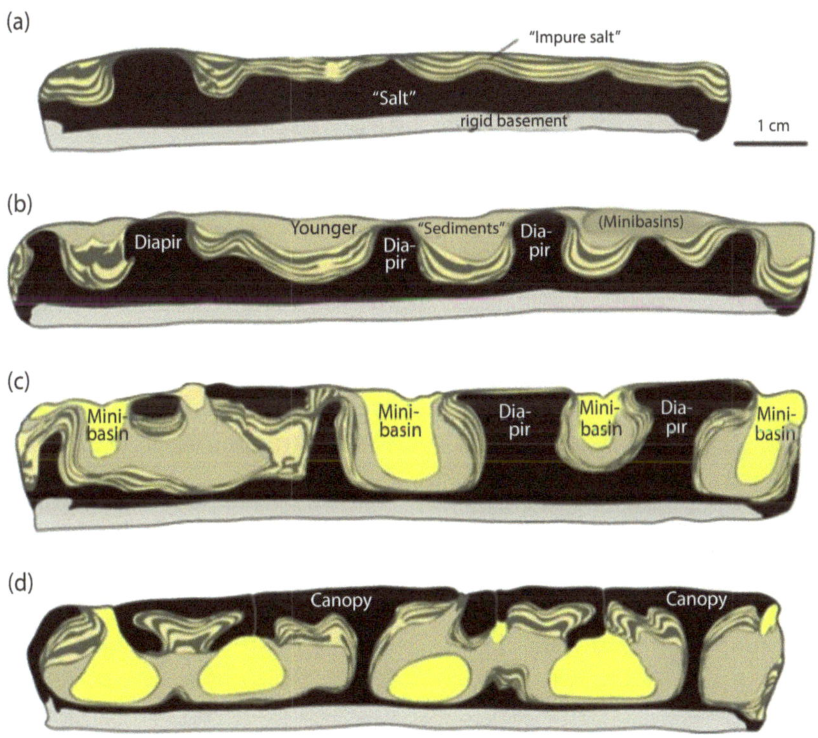

Figure 9.6. Sections through a scaled centrifuge model at different stages of development. The sections cannot be directly correlated, but give a good impression of the development of diapirs that end up forming a canopy.

Source: Haakon Fossen, Structural Geology Second Edition, ISBN 978-1-107-05764-7.

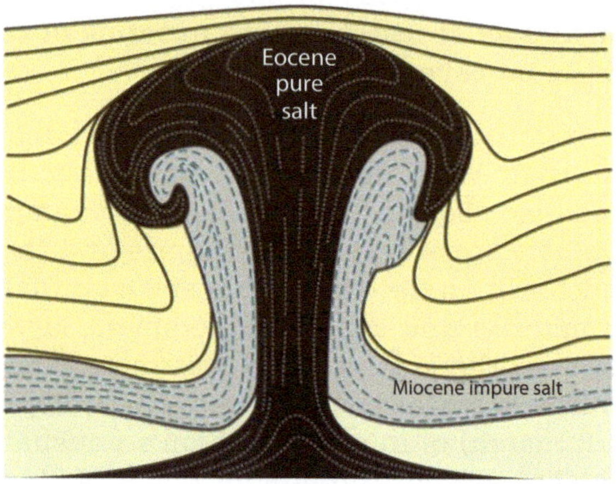

Figure 9.7. Typical geometry of mushroom-shaped diapirs found in the Dasht-e Kavir, Iran.

Source: Haakon Fossen, Structural Geology Second Edition, ISBN 978-1-107-05764-7.

Figure 9.8. Salt structures exposed in the Dasht-e Kavir, Iran. The circular shapes of the diapirs are visible. Note the folds in the overburden, which also contain some salt.

Source: Haakon Fossen, Structural Geology Second Edition, ISBN 978-1-107-05764-7.

The layers covering the salt will thin without breaking as the salt rises in the theoretical scenario where the overburden behaves like a perfect fluid, and the structure is not a diapir sensu stricto. Since these structures are rare, models should only include fluid properties for the mobile salt layer.

9.3.4. Modeling Salt Structures: The Brittle Approach

It is now widely acknowledged that simplification of the overburden by modeling it as a fluid is excessive. Shear strength is a property of actual rocks and sediments, not of fluids. While gravity is the primary force, a salt diapir must overcome the strength of its roof in order to begin. Thus, density contrast alone is insufficient to cause diapirism if the overburden is a strong, brittle rock (limestone or compacted siliciclastic sedimentary rock). Remember that a substantial density inversion requires a specific sedimentary thickness (approximately 2–3 km), indicating that the sedimentary strata covering salt have undergone lithification and consequent mechanical strengthening. Physical factors indicate that diapirism is highly unlikely to occur on its own without the help of tectonic faulting or fracturing. Conversely, traditional gravity-driven diapirism can proceed more readily once the roof is compromised and diapirism begins (Worrall & Snelson, 1989).

Because of this, simple deformation boxes have largely replaced the centrifuge model. In these boxes, sedimentation during diapirism can be controlled, gravitational forces can be added, and extension or contraction can be applied by adjusting the sidewalls. Over the past few decades, these models—which usually consist of viscous silicone gel overlaid by (brittle) sand—have been successfully used to reproduce a large number of seismically mappable salt structures in sedimentary basins. Significant experimental work has been conducted at the University of Texas at Austin, which produced the model depicted in Figure 9.9, as well as the University of Rennes (France).

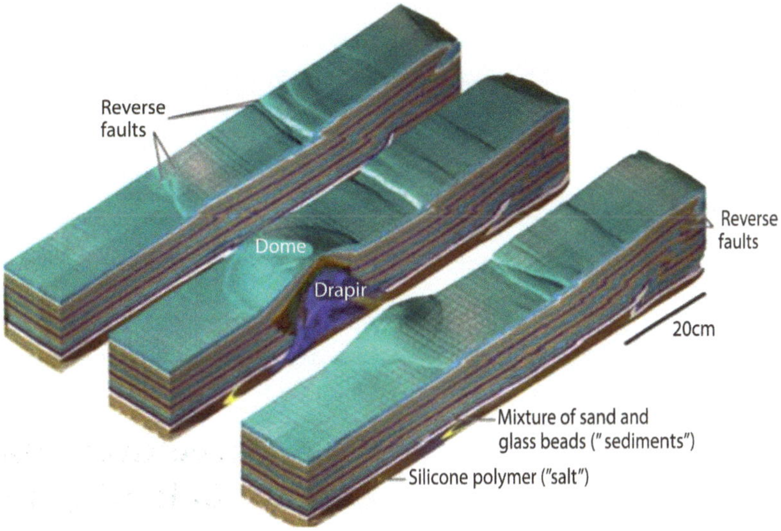

Figure 9.9. Example of physical experiment involving salt analog overlain by "sediments." This is one of several models that explore the result of contracting a sedimentary sequence with a preexisting salt diapir. During the experiment the diapir roof domed, and reverse faults developed along strike from the dome.

Source: Haakon Fossen, Structural Geology Second Edition, ISBN 978-1-107-05764-7.

9.3.5. Controls on Salt Flow

Despite the fact that salt covered in denser layers is gravitationally unstable, the salt won't move unless there is some sort of mechanical or gravitational anomaly. Differential loading refers to lateral variations in overburden thickness and/or density. A lot of people even view differences in strength as a form of differential loading. One of the main causes of salt movement is differential loading, which has to do with lateral variations in the stress state that is generated in the salt layer. We can discuss pressure differences because salt behaves like a viscous fluid, which salt cannot withstand over time. Therefore, the flow of salt toward low-pressure areas is initiated by differential loading. Because it is independent of density inversion, salt movement in response to differential loading differs significantly from buoyancy. This indicates that once new sediments are being deposited unevenly on top of the salt layer, differential loading is effective even at very shallow depths, causing salt to move.

Differential loading happens when salt begins to flow away from the area of maximum loading because there is some reason why that area is more heavily loaded (or eroded) than the rest of the area. This discrepancy may result from a lateral variation in the overburden thickness and/or slope, as shown in Figure 9.10a, or from variations in the lithology and rock density (local occurrence of lava, variations in the sedimentary facies, etc.). As shown in Figure 9.10b, a change in elevation can also cause salt flow.

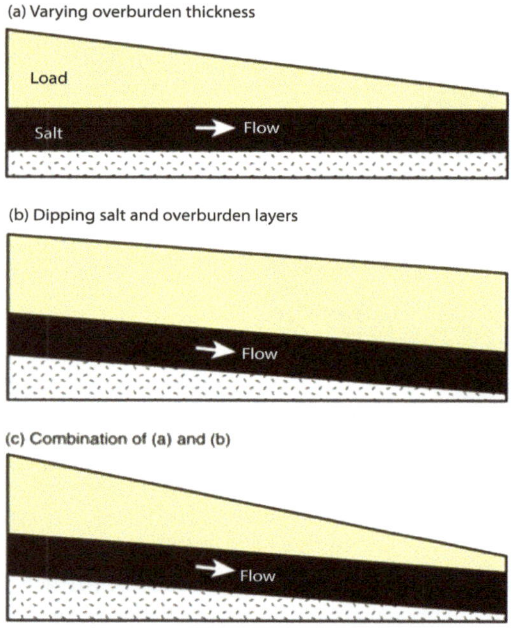

Figure 9.10. Situations where loading causes salt to flow. (a) Flow related to differential loading (e.g., a river delta), which causes differences in pressure. (b) Flow caused by elevation difference due to tilting. (c) Combination of the two, as commonly seen on continental margins. These simple models have restricted (pinned) ends so that the overburden cannot slide on the salt.

Source: Haakon Fossen, Structural Geology Second Edition, ISBN 978-1-107-05764-7.

Most salt structures are typically connected to folds and faults, and many (though not all) of these structures show that tectonic strain was a factor in the movement of salt. The weak salt will probably flow upward and shorten horizontally if a salt body is shortened, much like toothpaste squeezed out of an upright tube. When there is a regional extension, the salt expands laterally and thins vertically due to horizontal stretching or unloading. The flow of salt in this basic model is regulated by the boundaries on either side of the salt body; this boundary effect is known as displacement loading.

The phenomenon known as "thermal loading" describes how hot salt expands and becomes more buoyant. This may quicken the movement of salt toward the surface. Additionally, it has been proposed that convection—the rising and sinking of hot salt within the salt structure—may occur as a result of thermal loading. While this process might not be significant for most salt structures, it could account for some unusually coiled salt structures (also known as vortex structures) that are present in the Iranian Dasht-e Kavir.

9.3.6. Structures Above and Around Salt Diapirs

Just because of the friction between the flowing salt and its surroundings, salt flow may have an impact on the wall rock. Frictional drag folds that are created during the movement of salt may arise from this. These drag folds can develop in sedimentary layers that are poorly or completely consolidated, but in most other situations, the salt will be far weaker than the wall rock, negating the need for frictional drag to form.

However, layers frequently rotate (fold) around salt structures, and causes other than friction are discovered. For instance, because salt is nearly incompressible and nearby siliciclastic sediments compact, broad drag-like zones may form. Consequently, apparent drag along upward-thinning salt structures may result from such differential compaction. As will be covered in the next section, in other instances there is a forceful intrusion of salt into the overburden. Therefore, during diapirism, sedimentary layer flaps may be forcibly pushed upward. A third mechanism of folding is the resultant withdrawal of salt, i.e., as salt moves into the salt wall or salt diapir, the source layer thins. Localized subsidence of the overlying layers results from salt withdrawal, and a rim syncline or minibasin develops around the salt structure. Usually, this mechanism forms flanking minibasins with layers defining a downward-amplifying fold in conjunction with differential compaction.

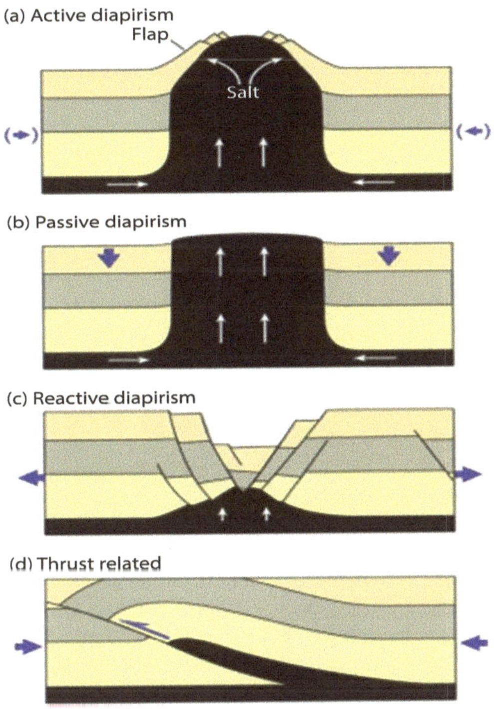

Figure 9.11. Main classes of diapirs and diapirism. (a) Active diapirism driven by density contrast and forceful ascent of salt (buoyancy), (b) passive evolution as sediments are deposited around the salt structure, (c) reactive formation in response to extension, and (d) salt structuring during contraction (thrusting).

Source: Haakon Fossen, Structural Geology Second Edition, ISBN 978-1-107-05764-7.

As shallow diapirs expand during fast sedimentation, a more intricate structure develops, as seen in Figure 9.12. Then, as sediments compact and the diapir rises, thin layers of sediments cover the diapir's roof and rim, but its surface relief increases. Eventually, the relief rises to a point where the salt either breaks through and becomes exposed or the scarp fails gravitationally. Before being buried under fresh sediments, the strata (flaps) along the diapir's rim are concurrently steepened and, in some cases, even overturned. As a result of this development repeating itself, stacking of so-called halokinetic cycles divided by nonconformities close to the salt structure may occur.

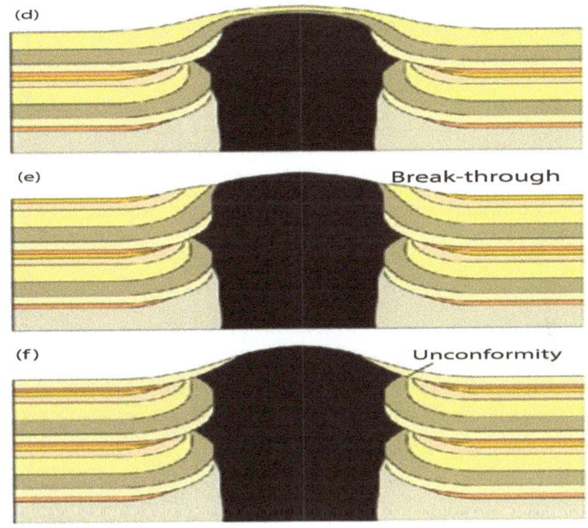

Figure 9.12. Formation of halokinetic sequences along a vertical salt structure. Occasionally the rising salt breaks through the roof (or the roof sediments gravitationally slide away) and bends the layers. An unconformity forms as the next sedimentary sequence starts depositing.

Source: Haakon Fossen, Structural Geology Second Edition, ISBN 978-1-107-05764-7.

Above salt structures that have not surfaced is typically a gentle dome structure. Once more, both salt growth and differential compaction may be connected to such a dome structure. Commonly, faults above the top of a salt diapir form as the layers bend and stretch during the salt movement. These faults are typically normal faults. Reverse faults only rarely develop above the salt diapir margins as a result of the salt forcibly jacking the roof. Either way, faults above salt structures typically mirror the geometry of the salt structure as seen from a map, meaning that faults with more linear traces develop along elliptical diapirs and salt walls, whereas circular diapirs and salt walls should ideally produce circular fault patterns. Nonetheless, the structural pattern above salt structures may be influenced by inherited fracture orientations or tectonic stress. Concentric faults, depicted by black faults in Figure 9.13 and frequently observed above circular salt diapirs, are frequently linked to the collapse of salt structures. Furthermore, some salt domes have been reported to have radial faults, which are the result of the underlying salt expanding. There have also been reports of concentric, radial, and other patterns of small-scale fractures forming above salt structures.

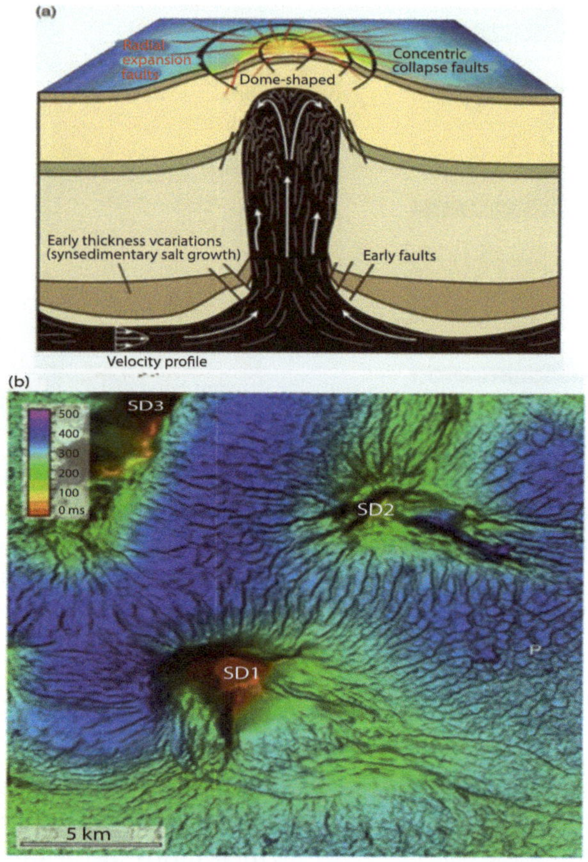

Figure 9.13. (a) Some common subsidiary structures associated with salt domes. (b) Illuminated structure map from seismic interpretation of the Eocene Top Brussels Marl in the Netherlands North Sea, color-coded for thickness to the Base North Sea Group (in time). Red color (no thickness) means that the salt pierced the Brussels Marl. Note radial structures (small faults) around salt domes (SD1–3), breaking up the regional polygonal fault pattern (P).

Source: Haakon Fossen, Structural Geology Second Edition, ISBN 978-1-107-05764-7.

There are notable strain gradients within the salt due to the decreasing flow velocity towards the surrounding rocks. While the flow is likely straightforward along the feeding salt layer, the patterns within the salt structure can be more intricate. This complexity is evident in the folded bedding observed in salt mines and domes, showcasing the non-linear flow within diapirs.

9.4. RISING DIAPIRS: PROCESSES

When a diapir begins to rise, it will eventually take on the shape of a diapir, which will be influenced by various factors such as the rate of sedimentation or erosion, tectonic regime, overburden strength, gravitational loading, salt temperature, salt layer thickness, salt extent (salt availability), and more. According to conventional diapirism models, diapirs can push through the overburden under the influence of displacement, thermal, or differential loading. Flaps of the roof are pushed aside during such active diapirism, which significantly rotates the layers along the upper portions of the salt walls upward. As previously stated, in most cases, buoyancy forces associated with gravity inversion are insufficient to start active diapirism. Diapirism can only be driven by buoyancy forces when a diapir is well established and the overburden thins.

When exposed or extremely shallow diapirs rise continuously at a rate that roughly corresponds with sedimentation, they are referred to as passive diapirism. Therefore, another mechanism must be used for the initiation of a diapiric structure. When a diapir structure is formed, salt is fed into it, causing the surrounding sediments to compact and sink into the source salt layer, which eventually thins. Downbuilding is the term used to describe this process, in which subsidence forms a minibasin around the salt diapir where new sedimentary strata accumulate. Tall salt diapirs may form if this process continues for a long time and if the source layer is sufficiently thick. When the rate of salt rise surpasses the rate of sedimentation, the diapir transforms into a structure that widens upward. On the other hand, the diapir narrows upward in the situation where the rise rate is lower than the sedimentation rate. Passive diapirs with steep walls are common, suggesting a balance between the rates of sedimentation and salt rise (flow rate). The diapir may become buried if the source layer is reduced or ceases to move salt.

As we have already mentioned, most salt structures cannot form unless the roof is weakened by faulting and fracturing. Such fracturing may involve very small strains. Tectonic faulting, however, can sometimes have a greater impact on the overburden, and salt fills in the gaps created by displaced fault blocks. Reactive diapirism is the term for the process wherein the salt rises in response to a significant tectonic strain, usually one of extension. Reactive diapirism is dependent on active extensional strain, as experiments have demonstrated that it stops once extension stops. Diapirism can also be caused by contractional deformation, since salt can be pushed into its overburden in a thrust fault's hanging wall. The majority of natural diapirs have components from two or all of these diapirism processes. A closer examination of salt tectonics in various tectonic regimes is necessary due to the significance of the interaction between faulting and salt diapirism.

9.5. SALT DIAPIRISM IN THE EXTENSIONAL REGIME

Passive continental margins and basins of continental extension are sites of numerous salt accumulations. Regional extensional tectonics affects many salt layers because these regions typically undergo extended or recurring periods of extension. In particular, the majority, if not all, of the salt diapir provinces were formed during regional extension, though contractional deformation may have subsequently altered them.

Extensional faulting and fracturing weaken the overburden, allowing reactive diapirism to initiate, as multiple physical experiments have demonstrated. As extension progresses, graben formation in the overburden usually produces a triangle-shaped diapir that rises and makes room for the salt. A local minibasin sits atop the graben structure above the salt at this point, though the depth of the basin is diminished by salt emplacement. With a single dominant master fault, the graben as depicted in Figure 9.14a can be asymmetrical in addition to being symmetric. A salt roller is a (non-equilateral) triangular-shaped volume of salt in the footwall created by such asymmetric structures. The salt on the fault's footwall side must thicken in order for a salt roller to continue evolving.

When the overburden above the salt diapir becomes thin enough and sufficiently weakened by extensional faulting, salt can begin to move in areas of faulting. If the total overburden above the source layer is, on average, denser than the salt, the salt can actively (buoyantly) force itself upward after an early stage of reactive salt diapirism. The surrounding area and the salt structure now have significantly different densities, and the compacting wall sediments' added weight presses the salt, forcing it to flow into the salt structure. At this point, active extension is not necessary because the loading may be sufficient to drive salt diapirism. The roof layers are stretched and domed during this passive stage, creating layers or flaps that dip sharply along the upper edges of the salt diapir.

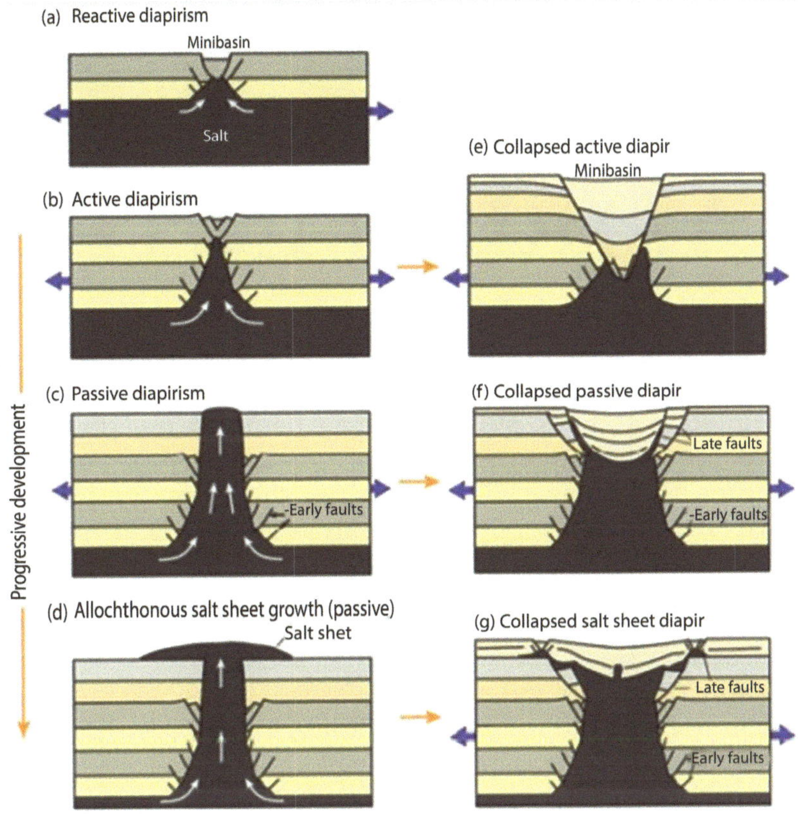

Figure 9.14. Illustration of a complete and idealized evolution of a surfacing salt diapir through stages of reactive (a), active (b), and passive (c, d) diapirism. At any stage the salt structure can collapse, giving rise to structures shown at the right-hand side (e–g).

Source: Haakon Fossen, Structural Geology Second Edition, ISBN 978-1-107-05764-7.

The salt can rise to the surface and function as a passive diapir if the source layer is sufficiently thick. Therefore, in an extensional setting, salt can flow through reactive, active, or passive diapirism during a diapir's rise. It is possible for the salt to form an allochthonous salt sheet that flows laterally at the surface, which is usually the sea floor.

9.6. DIAPIRISM IN THE CONTRACTIONAL REGIME

The sedimentary sequence above a salt layer may begin to buckle when the sequence beneath it is shortened due to regional contraction, whether tectonic or solely gravitational. The buckling facilitates the folding of the salt's overburden by forming low-pressure antiformal crests into which the salt flows. In this way, long salt anticlines form. If the process is synsedimentary, thinning or pinching out of layers toward the antiformal crests may result in significant variations in depositional thickness.

Since the overburden is still intact, the formation of salt anticlines during contractional folding of the overburden is not truly diapirism; instead, diapirism requires piercing of the roof. Thrust formation can cause piercing, as Figure 9.11d illustrates, but the resulting salt structure's geometry is highly asymmetrical and might not resemble traditional diapir structures at all.

The majority of large-scale diapir structures found in regions of regional contraction most likely originated during extension and were subsequently amplified or reworked during shortening, although buckling and reverse faulting are two ways that salt diapirs can begin during regional contraction. It is crucial to understand that salt diapirs that are already present are weak components that will preferentially deform if the area experiences horizontal shortening. When this happens, the salt structures narrow and salt is frequently squeezed laterally as well as upward. Depending on the salt content, structural development, and sedimentation, a variety of distinctive structures may emerge. Allowing contraction to form or reworking salt structures opens up a wider range of possible structural outcomes.

9.6.1. Teardrop Diapirs

The formation of teardrop diapirs, in which an initially upward-widening diapir is squeezed to the point where a salt weld forms in the middle of the structure, can occur from contractional amplification of pre-existing salt structures. The lower root structure is referred to as a pedestal, and the upper, isolated portion is called a teardrop. During this process, salt flows upward, often propelled by both the free surface at the top and the buoyancy component involved. Actually, the tectonic pressing or squeezing of the salt is enough to cause this process to happen; buoyancy is not needed for it to happen. Teardrop structures can also be found in extensional regimes, although they are most frequently observed in seismic data from regions where salt structures have shrunk.

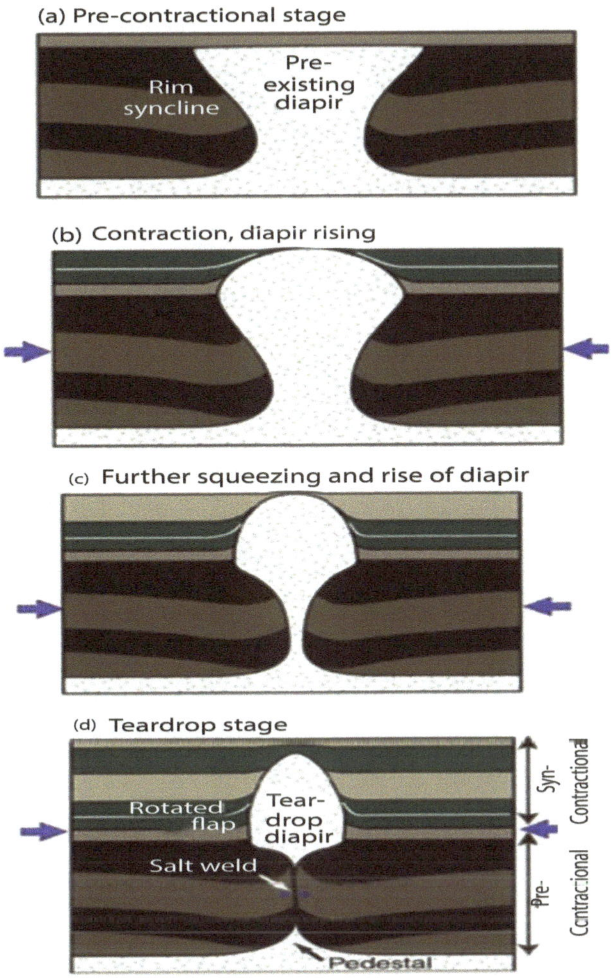

Figure 9.15. Schematic illustration of thc transformation of an hourglass-shaped diapir formed during extension (a) to a teardrop diapir (d) due to contraction.

Source: Haakon Fossen, Structural Geology Second Edition, ISBN 978-1-107-05764-7.

If a teardrop structure has an inclined salt weld, further shortening may cause the weld to activate as a reverse fault. The preexisting geometry and the rate of sedimentation determine the final geometry of a teardrop diapir and related structures. Furthermore, it matters whether the contraction is due solely to shear, as shown in Figure 9.15, or whether shear along the basal salt layer is also a factor.

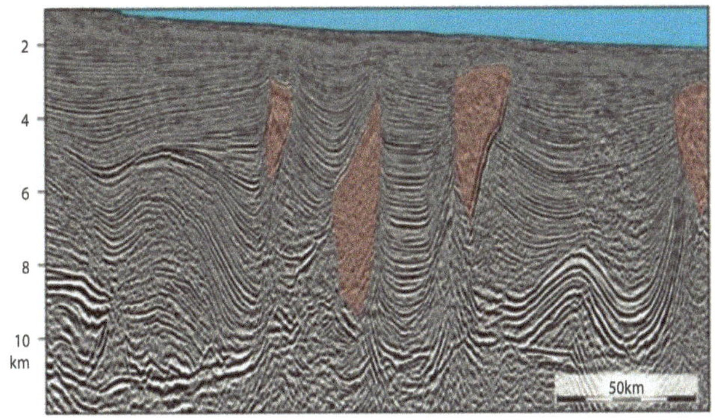

Figure 9.16. Depth-migrated seismic section from the submarine Mississippi Canyon, Gulf of Mexico, showing teardrop diapirs (detached salt stocks). The source salt layer is located at ~10–11 km depth (not interpreted). Folds formed as a result of salt withdrawal in an extensional setting.

Source: Haakon Fossen, Structural Geology Second Edition, ISBN 978-1-107-05764-7.

9.6.2. Salt Sheets

Occasionally, salt can rise to the surface (dry land or ocean floor) and extrude, resembling very slowly flowing lava erupting from a feeder volcano. While the structures of subaerial salt sheets have been understood for over a century, the majority of the thousands of salt sheets that have been mapped to date are submarine sheets that have been found in the last few decades through seismic imaging and drilling of rifts and passive margins. As demonstrated in Figure 9.14d, we have previously stated that this can happen in situations where the rate of sedimentation is outweighed by the vertical salt movement in a surfacing diapir (feeder), even in the absence of shortening. On the other hand, where a contractional squeeze is added, salt can more easily reach and cover the ocean floor or land surface and the structures become better developed.

Because it lies on top of younger, stratigraphically higher rock or sediment layers, resembling an allochthonous thrust nappe, salt that flows laterally as sheets on the surface or between stratigraphic layers is referred to as allochthonous. For this reason, salt sheets are also known as salt nappes.

The rate of sedimentation in relation to the rate of salt flow determines the extrusion of salt. The salt sheet follows the top of the upper bed in the absence of sedimentation, and upon burial, it will be limited to this stratigraphic or erosional level. This is the case with the subaerial salt structures found in southern Iran, which behave similarly to glaciers in that they flow downslope due to gravity. Folded markers in such salt glaciers show that differential flow within the salt is generally caused by basal irregularities and basal friction. The majority of the flow happens in the form of Couette flow close to the base, where the glacier's tongue rolls across the frontal ground. The flowing salt will, however, sporadically climb or ramp up-section away from the feeder at a rate that depends on the sedimentation rate versus the salt extrusion rate if sediments build up around the salt structure's margins. As a result, a sequence of ramps and flats may appear, with ramps being thought to represent periods of slower salt advance and faster sedimentation.

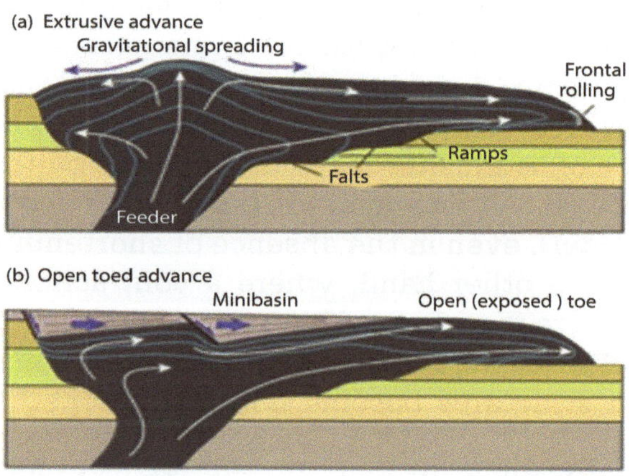

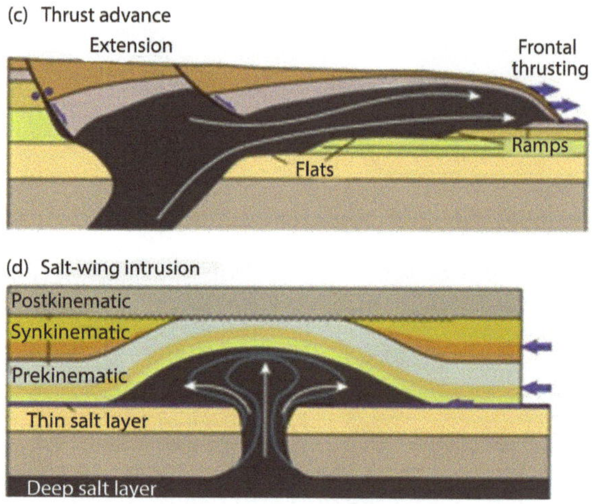

Figure 9.17. Schematic development of different allochthonous salt structures (salt sheets).

Source: Haakon Fossen, Structural Geology Second Edition, ISBN 978-1-107-05764-7.

Extrusive advance is the general term used to describe the process by which exposed salt sheets advance.

Sedimentation is a major factor in the formation of submarine salt sheets; numerous instances have been documented in which the frontal or toe portion of the salt sheet is exposed while the central or rear portion is covered in sediments. Open-toed advance is the mechanism by which such sheets advance by extrusion through the exposed toe. An open-toed advance emphasizes the asymmetry of the structure by causing lateral growth to occur in a preferred direction. Open-toed advance is caused by flow from sediments that build up on the top of the salt sheet and load the salt as the minibasin sinks into the salt, whereas extrusive advance is driven by salt flow up the feeder. Normal faults, as illustrated in Figure 9.17b, show that there is frictional drag between the salt and the minibasin's base and that the salt flows toward the toe. The minibasin will eventually be able to connect the feeder and the salt sheet by welding together at the base of the salt layer. In this scenario, the feeder's salt supply is cut off, and the isolated allochthonous salt continues to be extruded either by sedimentary loading or, in the event that the weld begins to act as a décollement (thrust), by tectonic translation.

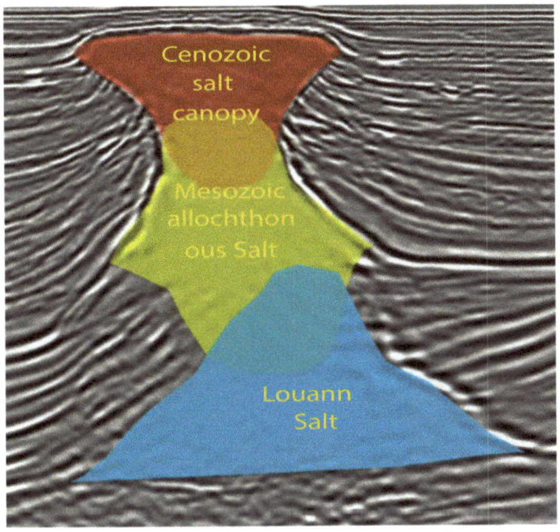

Figure 9.18. A typical stacking hourglass-shaped salt in the Central Mississippi Canyon area.

Source: https://csegrecorder.com/articles/view/salt-modeling-challenges-and-strategies mississippi-canyon-gulf-of-mexico.

Some fully submerged salt sheets move forward by thrusting, a process called thrust advance. In this instance, the toe zone contains a thrust. The presence of a weak layer at the bottom of the salt or in the uppermost portion of the underlying sediments is implied by tectonically driven thrusts where salt sheets are found in the hanging wall. This removes the frictional resistance against flow that arises during extrusive and open-toed advance. Rather than being related to changes in sedimentation rates, the ramps and flats seen along these thrusts are related to the thrust's propagation. It might be challenging to discern between the basal geometries created during extrusive and open-toed sheets on one side and thrusts on the other due to the striking similarity of the ramp-flat-ramp geometries.

A continuous unit, or salt canopy, that can cover a sizable area can form when multiple salt sheets form in the same area and get close enough to one another. With the formation of secondary diapirs that pierce even higher stratigraphic units, the canopy can function as any other salt layer once it has formed.

9.6.3. Salt Intrusions

The salt sheets mentioned above form at or close to the surface. At a deeper level, salt intrusion into a stratigraphic level is conceivable but uncommon. These formations, known as salt (wing) intrusions, are known from the southern North Sea, where a higher and relatively thin Triassic evaporite layer has been invaded by the main salt layer (Permian Zechstein salt). The resulting formations are similar to intrusions of magmatic laccolith. It seems that the formation of the salt depends on the existence of a weaker, higher evaporite layer into which the salt can intrude, as well as on regional contraction, which can pressurize the main salt and create décollement folds above the upper evaporite layer. Salt can flow into spaces created by décollement folding. Salt typically moves laterally, but when it invades an area, it moves up-section along faults.

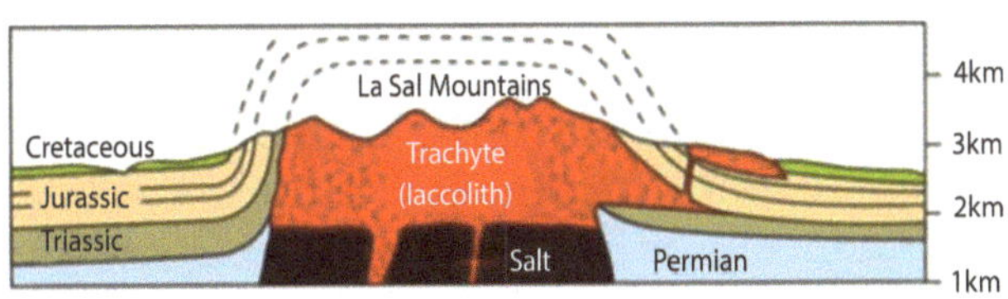

Figure 9.19. The rare occurrence of a salt diapir (salt wall) intruded by magma.

Source: Haakon Fossen, Structural Geology Second Edition, ISBN 978-1-107-05764-7.

9.7. SALT COLLAPSE BY KARSTIFICATION

Meteoric water has the potential to dissolve salt structures whose tops are at or near the surface. This can cause karstification, which is akin to the karst formation found in marbles and limestones, in addition to the formation of a cap rock of insoluble minerals and rock layers. Both situations have the potential for overlying rocks and sediments to collapse due to gravity as a result of karstification; however, because salt is more soluble (~360 g/l), there is a greater chance of this happening.

Long after the movement of salt stops, the collapse of karst structures can create fault systems above the salt structure. These fault systems can have simultaneous effects with crestal grabens and circular fault systems, which form during the extensional collapse of a salt structure during regional extension. It can be challenging to differentiate between the two types of fault structures. One of the many locations where salt dissolution has been suggested to have caused relatively recent collapse of salt walls is the Paradox Basin in southeast Utah. This is most likely due to reactivation of previously established faults at the crest. Salt dissolution is indicated by the presence of a cap rock, or residual rock, on top of the salt that is primarily composed of gypsum and clay. The Colorado Plateau was raised during the late Cenozoic and Quaternary, exposing these salt formations relatively recently. The collapse of salt structures' upper portions may be linked to subsurface dissolution, which could cause both the creation of new faults and the reactivation of preexisting ones.

9.8. SALT DÉCOLLEMENT

Whether in contractional environments like orogenic wedges (fold-and-thrust belts) or extensional environments like passive continental margins, salt plays a critical role as a mechanically weak décollement. Décollements can form in meter-thick or even thinner salt layers as long as the salt is laterally extensive, whereas salt structures like diapirs can only form in thick salt layers. The distribution and accommodation of strain can vary significantly when salt is present. As long as the salt layer is thick, salt's weak and viscous properties allow for the deformation of much larger regions. Additionally, it facilitates and frequently results in the decoupling of the substrate from the overburden. This indicates that the overburden's deformation pattern differs from the substrate's.

(Konstantinovskaya and Malavieille, 2011). One example of décollements related to salt is the folding of décollements. Because faults do not typically extend through salt layers unless the layer is very thin or the faults are very large, completely different fault patterns above and underneath the salt layer are also common.

Salt functions as décollements in orogenic wedges, dividing the more heavily imbricated and folded rock units from the less deformed or undeformed basement. Therefore, in collision zones like the foreland sections of the Alps and Pyrenees as well as the Iranian Zagros fold-and-thrust belt, salt encourages thin-skinned tectonics. Despite being transported tens or hundreds of kilometers, salt units in these settings localize tectonic strain and protect adjacent (overlying) rock layers from strain in this way. The above-lying wedge of deformed rocks becomes long and thin in cross-section as a result of the low friction of salt décollements; this allows it to extend farther into the foreland than it would have without a salt décollement. Therefore orogenic belts may become wider due to salt décollements.

In the brittle portion of the crust, where there is a significant rheologic and mechanical contrast between the salt and its surroundings, the role of salt in orogenic wedges is most noticeable. Thin limestone or phyllite units may have a similar function to that of salt at shallower depths in metamorphic depths, as other rock units weaken as plastic deformation mechanisms become more widely activated deeper down.

Gravity can also induce the formation of salt décollements. The most severe examples of salt décollement deformation can be found offshore Brazil and West Africa, as well as other passive continental margins. In these places, salt layers that dip into the ocean enable the overburden to slide due to gravity. Because of the salt's low strength, the dip—possibly only a few degrees—is sufficient but not particularly noticeable. Additionally, because the up-dip section is close to river delta systems, loading of the section occurs frequently. Similar to any gravity sliding process, the overall strain pattern is contraction in the front and extension in the back. Small-scale analogs can be observed in a variety of commonplace scenarios, such as snow covering a heated house sliding on its roof due to gravity. The heat from the house creates a thin layer of water that the snow slowly slides down-dip on, causing the lower part to fold and the upper part to expand.

9.8.1. Formation

Surface forces, aided by body forces (gravity sliding), "push" at convergent plate boundaries, causing décollements. Stepped thrusts, either over- or underthrusts, can form in strata with mechanically weak layers. These thrusts start at subduction zones and emerge deep into the foreland. Different lithologies of rock bodies result in different tectonic deformation characteristics. They may exhibit strong ductile deformation beneath the décollement surface and brittle behavior above it. Décollement horizons can form along planes of high pore pressure or at depths of up to 10 km. They are caused by high compressibility between different rock bodies.

The basal detachment of a fold-thrust belt's foreland portion usually lies at or close to the basement in a weak shale or evaporite. Allochthonous rocks are those above the décollement, while autochthonous rocks are those below. Material may be regarded as nappe if it is transported over a décollement of more than two kilometers. Thin-skinned tectonics refers to the faulting and folding that happens when a region's basal detachment occurs, but thick-skinned deformational regimes also experience décollements.

9.8.2. Compressional Setting

The décollement, which forms in the foreland basin of a subduction zone, is the lowest detachment in a fold-thrust belt. Other detachments, such as an imbricate fan of thrust faults and duplexes, as well as additional detachment horizons, can be found in a fold-thrust belt above the décollement. The layer directly above the décollement will undergo stronger deformation than other layers in compressional settings, while the layer below the décollement will experience weaker deformation.

9.8.2.1. Effect of Friction

Duplex formation is caused by décollements, whose geometry has a big impact on the thrust wedge's dynamics. The wedge's shape is determined by the amount of friction along the décollement; a low-angle slope indicates a low-friction décollement, while a higher-angle slope indicates a higher-friction basal detachment.

9.8.2.2. Types of Folding

At a décollement, two distinct folding patterns can happen. As part of the deformation that happens with a thrust fault, concentrated folding is characterized by uniform bed thickness throughout the fold and is invariably accompanied by detachment or a décollement. There is variation in bed thickness during disharmonic folding.

9.8.3. Extensional Setting

Décollements in extensional settings are accompanied by tectonic denudation and high cooling rates. They can form by several methods:

- According to the megalandslide model, normal faults will extend close to the initial fault source and shorten as they move farther away from it.
- The in situ model predicts numerous normal faults overlying one large décollement.
- According to the rooted, low angle normal fault model, the décollement is produced at depth when two thin rock sheets separate. Extensional faulting may be minimal or nonexistent close to the upper plate's thickest region, but as the upper plate thins, it becomes less coherent and may act like an extensional terrane with a thin outer layer.
- Normal faults with a high angle can result in décollements. A metamorphic core complex can be excavated thanks to uplift in a subsequent extension stage. Because of the high fault friction, a half graben forms but the orientation of stress remains unchanged. Subsequently, low effective friction caused by high pore pressure (Pp) forces ⊠1 to run parallel to the footwall fault. As it forms, a low-angle fault is prepared to function as a décollement. Normal faulting then thins the upper crust above the décollement. Novel high-angle faults facilitate crustal exhumation and regulate décollement propagation. At last, a significant and swift horizontal extension raises the ground both isostatically and isothermally. As an antiform moves toward shallower depths, a décollement forms.

9.8.4. Examples

9.8.4.1. Jura Décollement

Placed north of the Alps in the Jura Mountains, it was originally believed to be a folded décollement nappe. The nappe's thin skin was sheared off on deposits of Triassic evaporites that were a thousand meters thick. The Alpine orogenic wedge's most external limit is formed by the frontal basal detachment of the Jura fold-and-thrust belt, which has the youngest fold-and-thrust activity. The fold-and-thrust belt's Mesozoic and Cenozoic cover, as well as the nearby Molasse Basin, have been twisted over the weak basal décollement and moved at least 20 kilometers northwest.

9.8.4.2. Appalachian-Ouachita Décollement

The late Paleozoic fold-thrust belt of the Appalachian-Ouachita orogen, which is associated with lateral and vertical variations in rock lithologies, has a thin-skinned flat-and-ramp geometry along the southeastern edge of the North American craton. The surface of the décollement varies both along and across strikes. The décollement geometry preserves promontories and embayments in the late Precambrian–early Paleozoic rifted margin.

CASE STUDY

SALT DÉCOLLEMENT AND RIFT INHERITANCE CONTROLS ON CRUSTAL DEFORMATION IN OROGENS

A crustal-scale antiformal stack is a duplex of crustal thrust sheets with high overlap such that the roof thrust becomes folded into an anticlinal shape. Such a structure is observed in the Pyrenean Axial Zone and the Alpine Lepontine dome. Formation of such significant structures remains poorly understood, but requires high internal shortening in response to a subcritical wedge taper. This can be achieved by either increasing critical taper or lowering the surface slope. Proposed mechanisms include a lateral strength increase of the mid-crustal décollement, erosion of the internal orogen or sediment blanketing of the tip of the thick-skinned wedge. During orogenesis the distribution of shortening between pro- and retro-wedges may change significantly and can describe two clear phases: early symmetric inversion followed by asymmetric doubly-vergent collision.

Analogue and numerical models show that primary controlling factors of the lithospheric scale structural style of deformation in an orogen are rheology of the crust and mantle, strain weakening, convergence rate and inherited weakness zones. Noteworthy secondary factors are erosion and sedimentation, which remove mass from the internal orogen and add mass to the distal wedges, reducing the surface slope and thereby altering where shortening will be accommodated.

Décollement rheology is known to influence critical taper and the distribution of shortening in the overlying wedge. However, can a salt décollement also influence crustal deformation in the orogen as a whole? Efficient decoupling of the sedimentary cover might allow thin-skinned deformation to sufficiently change mass distribution at shallow depths to affect the crustal deformation in a manner similar to surface processes. Could a highly mobile cover trigger formation of an antiformal stack, and/or change distribution of crustal shortening?

Here, we build on previous work that investigates the role of rift inheritance and surface processes using lithosphere scale high-resolution 2D thermo-mechanical models. Critically, we use a highly mobile cover (viscous salt décollement layer) instead of the moderately mobile cover (frictional-plastic décollement layer) used by Erdős et al., To investigate the combined effect of rift inheritance and a highly mobile cover we present a novel measurement of the shortening distribution between the pro- and retro-wedges through time. We compare our results with the Pyrenees, an orogen with a crustal-scale antiformal stack and two-phase shortening distribution, and where a similar shortening distribution analysis was performed.

Methods

We use a modified 2D version of FANTOM, a high resolution (500 × 200 m in upper 25 km), thermo-mechanically coupled, arbitrary Lagrangian-Eulerian finite element code. Localization of deformation is incorporated through strain weakening of frictional-plastic materials. Surface processes were included with simple algorithms for elevation-dependent erosion and sedimentation up to 0 m base level.

Our models consist of a laterally homogeneous continental crust, comprising 3 km of pre-orogenic sediments underlain by a 1 km thick décollement layer, 21 km of upper crust and 10 km of strong lower crust (Figure 1). The lithospheric mantle extends to 120 km depth and sub-lithospheric mantle to 600 km depth. Convergence is imposed on both sides for a total convergence rate of 10 km/My. Initial localisation of deformation is seeded by a small weak seed at the top of the lower crust. Rift inheritance is generated by first extending for 50 km. Model 1 has a highly mobile cover with a weak viscous décollement layer representing salt with an effective viscosity of $n_{eff} = 10^{19}$ Pa.

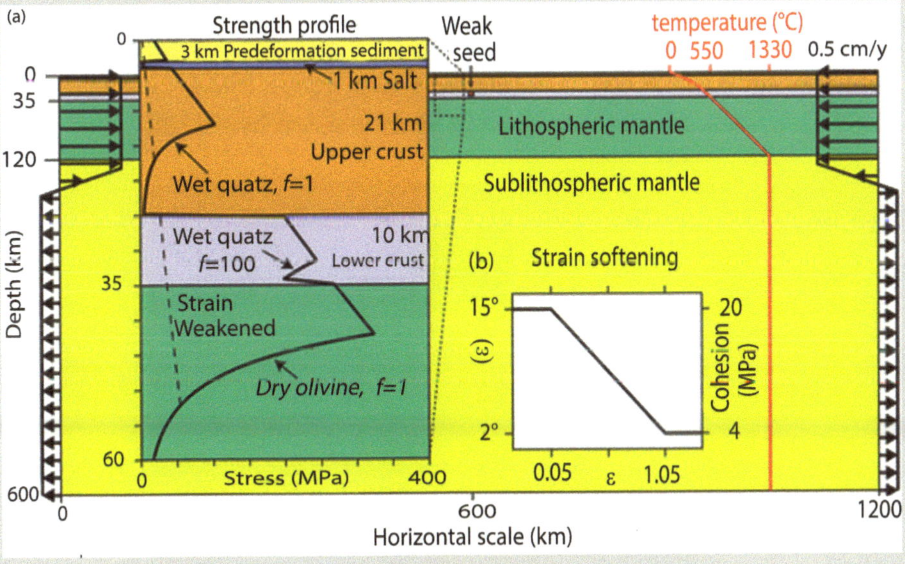

Figure 1. Model geometry showing layer thickness, initial strength profile indicating the strong coupling of the lower crust and lithospheric mantle, thermal profile and velocity boundary conditions.

Source: https://onlinelibrary.wiley.com/doi/full/10.1111/ter.12428

Results

Effects of salt, surface processes and extensional inheritance

Model 1 is characterized by a three-phase evolution. Phase 0 is the pre-orogenic rift phase, where at t = 5 My, 50 km of extension results in a narrow, roughly symmetrical rift bounded by two sets of conjugate frictional-plastic shear zones that are seeded in the lower crust, one through the upper crust and one through the lithospheric mantle. The pre-rift sedimentary cover has slid off the rift margins. No sediments were deposited in the rift.

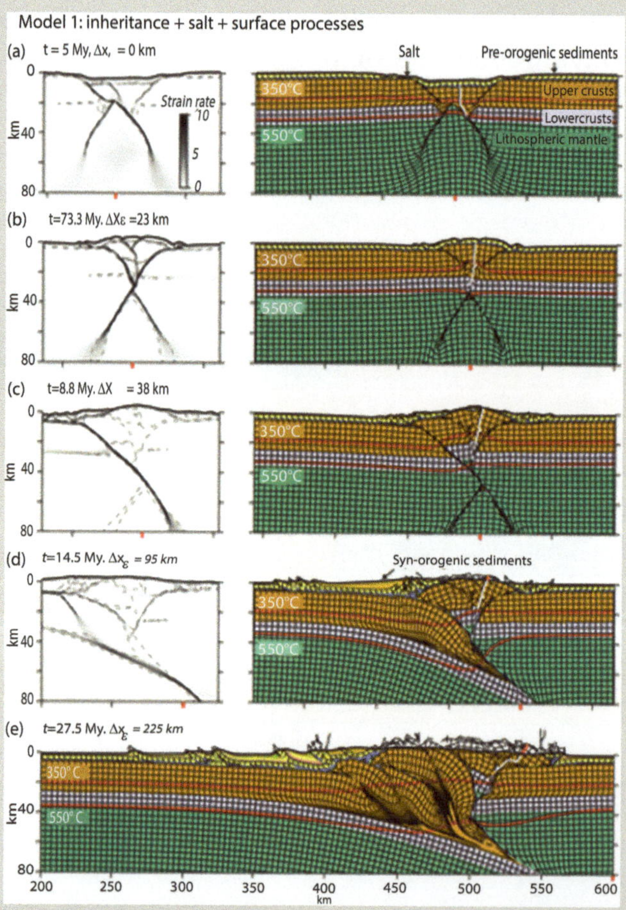

Figure 2. Evolution of Model 1, with a weak viscous salt décollement and extensional inheritance. The white line at the center is tracked to measure the shortening distribution, ignoring the red part above topography. Strain rate plots (left) show the square root of the second invariant of the deviatoric strain rate tensor. Grey dashed lines mark fully strain weakened areas. (a) Rift phase. (b) Symmetric inversion. (c) Onset of collision and asymmetry, retro-wedge abandoned. (d) Retro-wedge reactivated and dominant wide thin-skinned outer pro-wedge fold-and-thrust belt. (e) Final configuration. Well-developed retro-wedge, antiformal pro-wedge and wide thin-skinned outer pro-wedge fold-and-thrust belt.

Phase 1 initial lithospheric shortening leads to symmetric inversion of the rift zone. Inherited extensional shear zones are preferentially reactivated during the first 25 km of convergence, restoring crustal thickness, followed by uplift of a symmetric central block (Fig). The strain rate plot shows that both conjugate shear zones in the upper crust and upper mantle lithosphere are active simultaneously during this phase.

An asymmetric orogen develops during Phase 2. Strain localization on a single large-scale shear zone initiates asymmetric subduction of lower crust and mantle lithosphere and abandonment of other, previously active shear zones. Initial narrow thin-skinned fold-and-thrust belts not only root back into the deep shear zones, but also link to the gravitationally unstable cover that slides off the uplifting keystone block. At t = 14.5 My and 95 km of convergence, syn-orogenic sedimentation in a wedge-top basin promotes wide thin-skinned thrust sheets in the outer pro-wedge (Figure 2d). Below and at the same time, thick-skinned deformation propagates outward along the mid-crustal décollement due to progressive growth of the pro-wedge. As the crustal antiformal stack grows, over-steepening of its trailing edge reactivates the retro-wed

After 225 km of convergence (27.5. My; Figure 2e) the thin-skinned pro-wedge fold-and-thrust belt has efficiently propagated outward. The crustal antiformal stack continues to grow by frontal accretion of thrust sheets while the uplifting internal zone is continually erode.

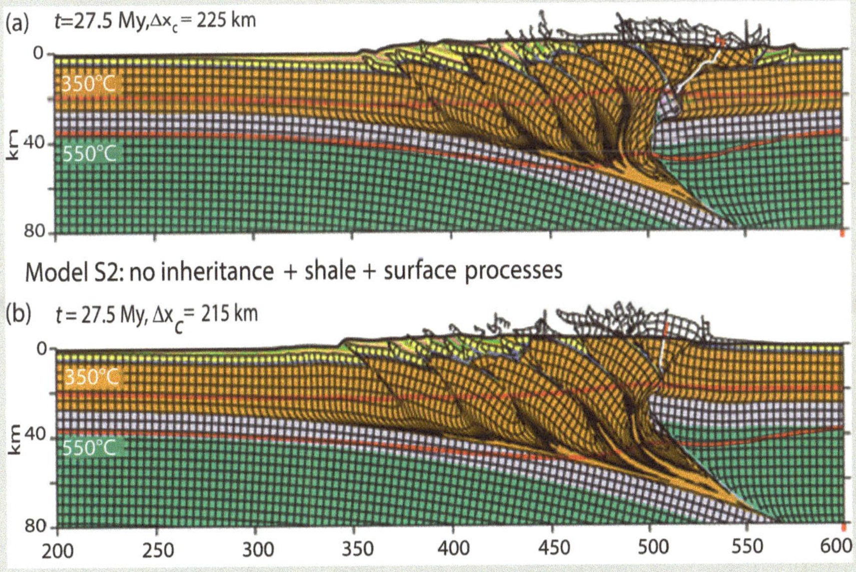

Figure 3. Compilation of supplementary models after 27.5. My. (a) Model S1, with a frictional shale décollement

Source: https://onlinelibrary.wiley.com/doi/full/10.1111/ter.12428

Shortening distribution

Quantitative analysis of the shortening distribution between the pro- and retro-wedges reveals that both models with rift inheritance show a very similar shortening distribution. Initial inversion during Phase 1 is roughly symmetric (Figure 4a). After Phase 1 the retro-wedge is briefly abandoned and then reactivated in Phase 2. During asymmetric orogenic growth in Phase 2, the shortening rate of the retro-wedge is only ~18%-20% of the overall convergence rate, with deformation focused into the pro-wedge. Overall, rift inheritance models accommodate more shortening in the retro-wedge than the model without extensional inheritance, stabilizing at around 18%-20% of total convergence after ~100 km (Figure 4c). In Model S2 (no inheritance) shortening is quickly concentrated almost exclusively into the pro-wedge and retro-wedge shortening stabilizes to around 6% of total convergence after only 50 km.

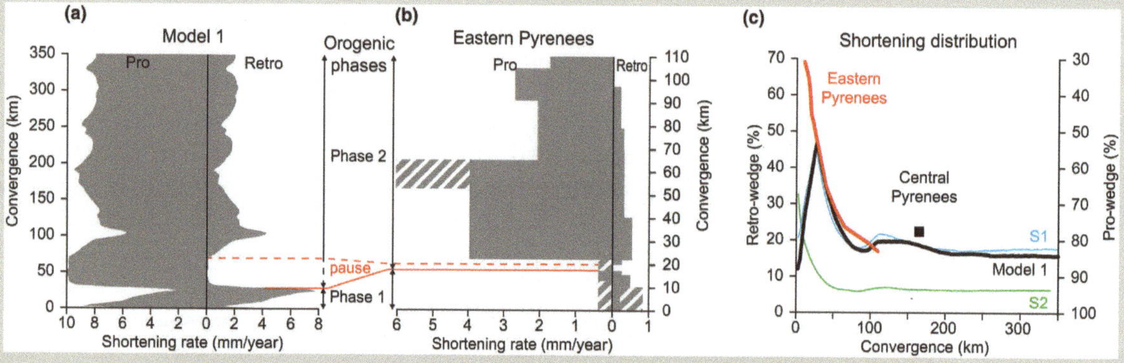

Figure 4. (a) Pro- and retro-wedge shortening rate of Model 1, plotted against total convergence. (b) Pro-and retro-wedge shortening rate of the Eastern Pyrenees, plotted against total convergence (note difference in scale). (c) Shortening distribution between the pro- and retro-wedges, plotted against total convergence.

Source: https://onlinelibrary.wiley.com/doi/full/10.1111/ter.12428.

Discussion

Model 1 shows that a highly mobile cover in combination with surface processes is sufficient to promote formation of a crustal antiformal stack. A weak salt décollement allows efficient decoupling of the cover. Together with the mid-crustal décollement, this forms a system of stacked, interacting décollements or nested critical wedges. The salt décollement cannot support the high surface slope necessary for the critical taper associated with the mid-crustal décollement, leading to gravitational collapse of the cover. Combined with surface processes that further reduce the surface slope, the thick-skinned pro-wedge taper is subcritical, promoting thick-skinned underthrusting in the core of the orogenic wedge. This, in turn, again generates gravitational collapse of cover on the uplifting crustal antiformal stack. This positive feedback loop promotes an antiformal stack flanked by a wide low taper thin-skinned fold-and-thrust belt in the pro-wedge. This structure emerges naturally in our model, not relying on prescribed weak zones and S-point or a hinterland that is even thicker than the antiformal stack itself to force formation of an antiformal stack. With a more competent frictional décollement, the fold-and-thrust belt cannot gravitationally collapse, thus avoiding the feedback loop that generates a crustal-scale antiformal stack. Thus, only a salt décollement combined with surface processes can promote formation of a crustal-scale antiformal stack. With less effective crustal strain weakening, this effect may not be sufficient to generate an antiformal stack.

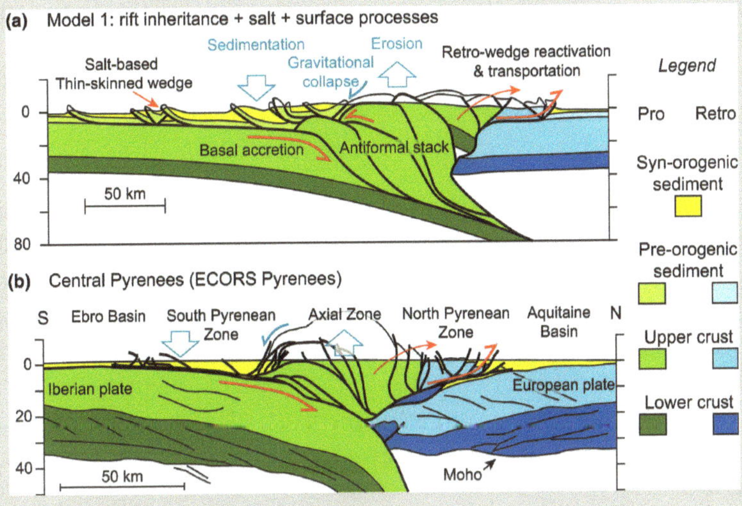

Figure 5. (a) Cartoon of Model 1 after 225 km of convergence showing factors (in blue) driving basal accretion and growth of an antiformal stack. The retro-wedge is reactivated and transported by over-steepened trailing edge pro wedge thrust sheets migrating onto the upper plate as the antiformal stack grows. (b) Cartoon of Central Pyrenees showing the same mechanism of shallow structure forcing underthrusting, growth of an antiformal stack, and reactivation of the retro-wedge.

Source: https://onlinelibrary.wiley.com/doi/full/10.1111/ter.12428.

Model 1 also shows that a two-phase shortening distribution is primarily controlled by rift inheritance. In Model 1 the early symmetric inversion (Phase 1) is caused by the pre-existing rift symmetry. In Phase 2 strain weakening and geometrical incompatibility of the two shear zones promotes asymmetric shear localization in the lithospheric mantle upon collision of the intact lower crust and lithospheric mantle. Without rift inheritance, asymmetric shortening is established from the moment the first lithospheric shear zone is formed, effectively skipping Phase 1 entirely.

In all our models, the onset of asymmetric shortening distribution in the upper crust (Phase 2) coincides perfectly with the onset of asymmetric subduction of the lithospheric mantle. All models with rift inheritance accommodate ~80% of shortening in the pro-wedge, regardless of thin-skinned décollement rheology. This suggests that it is the subduction asymmetry in the lithospheric mantle that controls shortening distribution in the upper crust. Thus, one could use the onset of asymmetric shortening to date the onset of lower lithosphere subduction.

Pyrenean Doubly-Vergent Orogen

Here, we compare the geometry and evolution of Model 1 to the Pyrenean orogen. The Pyrenees are often used as a type example of a mountain belt that resulted from inversion of a salt-rich, narrow, mantle-exhuming rift.

The crustal antiformal stack in the Pyrenean Axial Zone exhibits nearly complete overlap between thrust sheets. Model 1 reproduces this antiformal stack geometry. The same factors that promote an antiformal stack in Model 1 are found in the Pyrenees: Erosion efficiently removes mass from the central orogen; Sedimentation adds mass to the forelands; Thrusting of the cover above a salt décollement allows mass to migrate outward, particularly in the pro-wedge, to form a wide fold-and-thrust belt.

The evolution of the shortening distribution in the Eastern Pyrenees, based on a thorough field study and balanced cross-sections, fits remarkably well with that of our models that include rift inheritance, both qualitatively and quantitatively (Figure 4, note different scale of panels a and b). Phase 1 shortening in the Eastern Pyrenees is characterized by roughly symmetric rift inversion. Phase 2 is initiated simultaneously with the onset of asymmetric subduction of the lithospheric mantle, resulting in temporary abandonment of the retro-wedge. Over-steepening of the growing pro-wedge trailing edge allows the upper thrust sheets to migrate onto the upper plate, reactivating the retro-wedge albeit modestly.

The depth of the mid-crustal décollement in the reconstruction of Muñoz is significantly shallower than in our models. However, we did not set out to fully reproduce all details of the Pyrenees, and the depth of this décollement remains poorly constrained. Changing the depth of this décollement would probably not significantly change our main findings.

The combined effects of a highly mobile cover, orogen erosion and foreland sedimentation may also explain antiformal stack-like structures in the Central Alps and in the frontal zone of the Himalayas. (Mercier, Braun, & Beek, 2017).

Conclusions

Our numerical modelling results show that the presence of a weak viscous (salt) décollement can strongly control crustal deformation in a convergent orogen. The development of a crustal-scale antiformal stack is promoted by the combined effects of gravitational collapse of the cover above the salt décollement off the uplifting orogenic core, efficient erosion of the orogenic core and deposition of sediments in the foreland. Rift inheritance results in a two-phase orogenic evolution, with Phase 1 symmetric rift inversion and Phase 2 asymmetric main collision, triggered by asymmetric subduction of the lithospheric mantle. The onset of asymmetric shortening can be used to date the onset of subduction. Comparison to the Pyrenees shows that these controlling factors and resulting geometries are found in nature.

CLASS ACTIVITY

SEDIMENTARY ROCK IDENTIFICATION

Students use the simple tools of a geologist to identify the chemical and physical properties of four different types of sedimentary rocks. Students will use this information during a hike around Eagle Bluff to identify the two most common sedimentary rock. Students will also visit several karst features and learn how these rock layers dictate the shape of the landscape

SUMMARY

- Salt as a rock has properties and behavior that are very different from most other rocks. This implies that when sedimentary sequences containing salt layers are deformed, they develop their own characteristic and often very fascinating structural styles.
- Salt tectonics is the study of geologic deformation involving salt: how and why salt structures evolve, the three-dimensional forms that result, and their interactions with regional tectonics.
- When a salt layer becomes too thin to be an effective detachment layer, due to salt movement, dissolution or removal by faulting, the overburden and the underlying sub-salt basement become effectively welded together.
- Salt has physical and rheological properties that make it fundamentally different from most other common rocks.
- Salt volumes in the crust have for a long time been known to take on a variety of geometric shapes, from elongated structures known as salt anticlines and salt pillows to more localized structures such as salt stocks.
- The flow of salt from a salt layer into a salt structure is usually referred to as salt withdrawal or, more correctly but less commonly, salt expulsion.
- In the theoretical case where the overburden behaves like a perfect fluid, the layers overlying the salt will thin without fracturing as the salt rises, and the structure is not a diapir sensu stricto.
- In general, most salt structures are associated with faults and folds, and many (but not all) such structures indicate that tectonic strain played a role during salt movement.
- When a sedimentary sequence overlying a salt layer is shortened during regional contraction (tectonic or purely gravitational), the sequence above the salt may start to buckle.
- The formation of salt anticlines during contractional folding of the overburden is not really diapirism, since the overburden is still intact: diapirism requires piercing of the roof.
- Submarine salt sheets are usually influenced by sedimentation, and many cases have been found where the central or rear part of the salt sheet is covered by sediments, while the frontal part or toe remains exposed.

REVIEW QUESTIONS

1. Why does diapirism not initiate as a result of density inversion alone?
2. Differentiate between a salt dome and a salt diapir.
3. Discuss about schematic development of different allochthonous salt structures (salt sheets).

4. What is meant by reactive diapirism and in what tectonic regime(s) can it occur?
5. What is it about classic centrifuge models that in some way makes them unrealistic?
6. What is the difference between active and passive diapirism and how can we distinguish between them?

REFERENCES

1. Beauboeuf, R. T., & Friedman, S. J. (2000). High resolution seismic/sequence stratigraphic framework for the evolution of Pleistocene intra slope basins, western Gulf of Mexico: Depositional models and reservoir analogs. In Gulf Coast Section, SEPM, 20th Annual Research Conference, Deep-Water Reservoirs of the World, Dec. 3–6 (pp. 40–60).
2. Coward, M., & Stewart, S. (1995). Salt-influenced structures in the Mesozoic–Tertiary cover of the southern North Sea, U.K. In M. P. A. Jackson, D. G. Roberts, & S. Snelson (Eds.), Salt Tectonics: A Global Perspective. AAPG Memoir, vol. 65 (pp. 229–250).
3. Diegel, F. A., Karlo, J. F., Schuster, D. C., Shoup, R. C., & Tauvers, P. R. (1995). Cenozoic structural evolution and tectono-stratigraphic framework of the northern Gulf coast continental margin. In M. P. A. Jackson, D. G. Roberts, & S. Snelson (Eds.), Salt Tectonics: A Global Perspective. AAPG Memoir, vol. 65 (pp. 109–151).
4. Docherty, C., Sugrue, M., Willacy, C., & Strong, A. (1999). Sub-salt imagery and structure recognition using pre-SDM. World Oil, 220, 80–83.
5. Jackson, M. P. A. (1995). Retrospective salt tectonics. In M. P. A. Jackson, D. G. Roberts, & S. Snelson (Eds.), Salt Tectonics: A Global Perspective. AAPG Memoir, vol. 65 (pp. 1–28).
6. Kehle, R. O. (1988). The origin of salt structures. In B. C. Schreiber (Ed.), Evaporites and Hydrocarbons (pp. 345–404). New York: Columbia University Press.
7. Kockel, F. (1990). Morphology and genesis of northwest-German salt structures. In Proceedings of Symposium on Diapirism with Special Reference to Iran (pp. 225–248). Tehran University Governery of Hormozgan, Tehran, Iran.
8. Kockel, F. (1998). Salt problems in Northwest Germany and the German North Sea sector. In F. Kockel & R. Marschall (Eds.), Geology and Geophysics of Salt Structures. Journal of Seismic Exploration, vol. 7, 219–235.
9. Konstantinovskaya, E., & Malavieille, J. (2011). Thrust wedges with décollement: Levels and syntectonic erosion: A view from analog models. Tectonophysics, 502(3–4), 336–350.
10. Letouzey, J., Colletta, B., Vially, R., & Chermette, J. C. (1995). Evolution of salt-related structures in compressional settings. In M. P. A. Jackson, D. G. Roberts, & S. Snelson (Eds.), Salt Tectonics: A Global Perspective. AAPG Memoir, vol. 65 (pp. 41–60).
11. McGeary, D., & Plummer, C. C. (1994). Physical Geology: Earth Revealed. Dubuque: Wm. C. Brown Publishers. ISBN 0-697-12687-0.

12. Mercier, J., Braun, J., & van der Beek, P. (2017). Do along-strike tectonic variations in the Nepal Himalaya reflect different stages in the accretion cycle? Insights from numerical modelling. Earth and Planetary Science Letters, 472, 299–308.
13. Nilsen, K. T., Vendeville, B. C., & Johansen, J. T. (1995). Influence of regional tectonics on halokinesis in the Nordkapp Basin, Barents Sea. In M. P. A. Jackson, D. G. Roberts, & S. Snelson (Eds.), Salt Tectonics: A Global Perspective. AAPG Memoir, vol. 65 (pp. 413–436).
14. Prather, B. E. (2000). Calibration and visualization of depositional process models for above-grade slopes: A case study from the Gulf of Mexico. Marine and Petroleum Geology, 17, 619–638.
15. Rowan, M. G., Peel, F. J., & Vendeville, B. C. (2004). Gravity-driven fold belts on passive margins. In K. R. McClay (Ed.), Thrust Tectonics and Hydrocarbon Systems. AAPG Memoir, vol. 82 (pp. 157–182).
16. Schuster, D. C. (1995). Deformation of allochthonous salt and evolution of related salt-structural systems, eastern Louisiana Gulf Coast. In M. P. A. Jackson, D. G. Roberts, & S. Snelson (Eds.), Salt Tectonics: A Global Perspective. AAPG Memoir, vol. 65 (pp. 177–198).
17. Spiers, C. J., & Carter, N. L. (1998). Microphysics of rock salt flow in nature. In M. Aubertin & H. R. Hardy (Eds.), The Mechanical Behavior of Salt: Proceedings of the Fourth Conference Series on Rock and Soil Mechanics, vol. 22 (pp. 115–128). Clausthal-Zellerfeld: TTP Trans Tech Publications.
18. Talbot, C. J., & Alavi, M. (1996). The past of a future syntaxis across the Zagros. In G. I. Alsop, D. J. Blundell, & I. Davison (Eds.), Salt Tectonics. Geological Society London Special Publication, vol. 100 (pp. 89–110).
19. Worrall, D. M., & Snelson, S. (1989). Evolution of the northern Gulf of Mexico, with emphasis on Cenozoic growth faulting and the role of salt. In A. Bally & A. Palmer (Eds.), The Geology of North America—An Overview (pp. 97–138). Boulder, CO: Geological Society of America.

CHAPTER

10 Folds and Folding

LEARNING OBJECTIVES

After studying this chapter, you will be able to:

- Understand fold classification
- Identify causes of folding
- Focus on geometric description
- Explain folding mechanisms and processes
- Know fold interference patterns and refolded folds
- Describe folds in shear zones
- Explain folding at shallow crustal depths

KEY TERMS FROM THIS CHAPTER

Angle of inclination	Anticlines
Antiforms	Axial plane
Axis	Crustal deformation
Doubly plunging anticline	Ductile rocks
Earth's crust	Earth's surface
Fold classification	Fold style
Folding	Geometry

INTRODUCTION

In geology, folds are undulations or waves in the stratified rocks of Earth's crust. Originally formed from sediments deposited in flat horizontal sheets, stratified rocks can be found in various places where the strata have been warped. Some warping is so gentle that the inclination of the strata is barely visible, while in other cases, the warping is more pronounced causing the strata to be parallel or nearly flat. Folds come in a wide range of sizes, with some spanning several kilometers while others are just a few centimeters. Large folds often have their tops eroded away on the Earth's surface, revealing the cross sections of the inclined strata.

Folds are typically categorized based on the orientation of their axes and their appearance in cross sections perpendicular to the direction of the fold. The axial plane of a fold is the surface that divides the fold symmetrically. This plane can be vertical, horizontal, or at any intermediate angle. The axis of a fold is where the axial plane intersects with the strata of the fold. While some folds have horizontal or gently inclined axes, others may have steeply inclined or even vertical axes. The plunge of a fold is the angle of inclination of its axis measured from the horizontal. The flanks, limbs, or slopes of a fold are the portions between adjacent axes.

A convex upward fold is called an anticline, and a concave upward fold is called a syncline. A large syncline with minor folds superimposed is called a synclinorium, and a large anticline with minor folds superimposed is called an anticlinorium. A fold with a vertical axial plane is said to be symmetrical. An axial plane that is inclined is called an asymmetrical fold. The axial plane of an overturned fold, also known as an overfold, is inclined to the point where the strata on one limb are overturned. The axial plane of a recumbent fold is essentially horizontal. A fold is referred to as isoclinal when its two limbs are nearly parallel to the axial plane and, therefore, essentially parallel to one another.

A large number of folds are clearly linear, meaning that their length along the axis is many times greater than their width. Nonetheless, some folds have a roughly circular plan rather than a linear one. One such convex upward fold is a dome, in which the strata descend from a central point. A concave upward circular fold is called a basin. e. the layers incline inward toward the center.

The long linear folds that are typical of mountainous areas are thought to have been caused by compressional forces that were applied at right angles to the fold and parallel to the Earth's surface. According to some geologists, a lot of folds are caused by strata moving away from a region that has been vertically raised due to gravity. Weakly consolidated rocks may also be thrown into folds by the push of an advancing glacier, and gentle folds are created when sedimentary rocks are compacted over buried hills. Folds in nature are seldom the result of a single process, but rather of a confluence of processes.

10.1. FOLD CLASSIFICATION

Under stress, ductile rocks respond plastically by folding. Folding can happen even in the cool, relatively brittle shallow crust if the stress is applied gradually and gives the rock enough time to bend. Rocks in the shallow crust will behave like brittle solids and break if the stress is applied too quickly. Even in situations where stress and strain increase quickly, folding occurs more easily in deeper layers of the crust where the rocks are more ductile. There are two primary categories of folds: synclines, also known as down-folds, and anticlines, also known as up-folds.

10.1.1. Anticlines and Synclines

The two most fundamental kinds of folds are synclines and anticlines. Think of a rug that has had its sides pushed toward one another to create valleys and ridges; the valleys are folds that are "down," and the ridges are folds that are "up." When referring to geologic structures, upfolding folds are referred to as anticlines, and downfolding folds as synclines.

Block diagrams such as the ones below have the earth's horizontal surface, or the map view, at the top of the block. Cross-sections, or vertical cuts through the crust, make up the other two sides of the box that are visible. The stratified geologic formations, such as lava flows or sedimentary beds, that were initially horizontal are represented by the colored layers. To see the geologic structures' three dimensional forms, use the block diagrams. Remember that erosion has removed the upper portions of these structures, exposing their interiors in map view.

Figure 10.1. Anticline and syncline.

Source: https://courses.lumenlearning.com/geo/chapter/reading-folds/.

An anticline is represented on a map as parallel beds of the same kind of rock that slope away from the fold's center. The oldest beds in an anticline are those that were initially beneath the other beds, and they are located in the middle along the fold's axis. The center of the fold on the map is indicated by an imaginary line called the axis. A syncline can be seen as a series of parallel beds that dip toward the center when viewed in map view. The youngest beds in a syncline are located in the center, along the fold's axis. These beds were initially positioned on top of the other beds. In areas of the crust that are compressing, or where the crust is being forced together, anticlines and synclines form.

10.1.1.1. Plunging Anticlines and Synclines

A plunging anticline, also known as a plunging syncline, is a fold that, along its length, plunges into the earth due to a tilt in its axis away from the horizontal. The direction in which the fold's axis dips into the earth is known as the plunge direction.

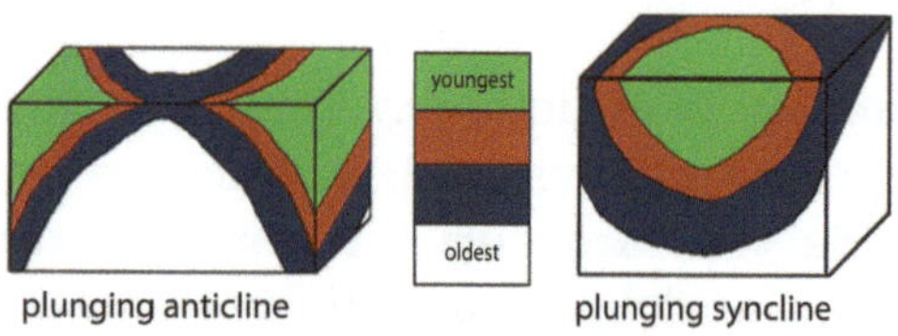

Figure 10.2. Plunging anticlines and synclines.

Source: https://courses.lumenlearning.com/geo/chapter/reading-folds/.

Anticlines that plunge create patterns on maps that point or close in the direction of the plunge, forming a U or V shape. An anticline and a cross-section taken at a straight angle to its axis are identical in appearance. A plunging syncline appears as a U- or V-shaped pattern on a map, opening in the direction of the plunge.

10.1.2. Anticlines

An anticline in structural geology is a type of fold that is convex upward and contains its oldest beds at its center. It should not be confused with an antiform, which is a general term for any fold that is convex upward. In cases where the age relationships between different layers are unclear, the term antiform should be utilized.

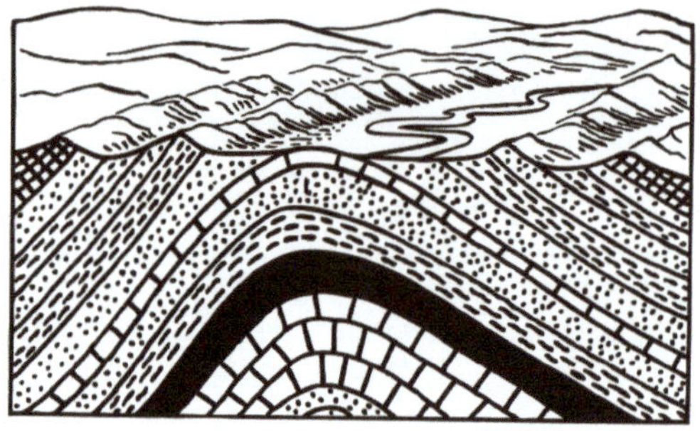

Figure 10.3. Diagram of an anticline.

Source: https://courses.lumenlearning.com/geo/chapter/reading-folds/.

The uplifted core of the fold is preferentially eroded to a deeper stratigraphic level relative to the topographically lower flanks, typically resulting in a sequence of rock layers that are progressively older toward the center of the fold on a geologic map. The layers descend from the fold's crest, or center.

Figure 10.4. Anticline with syncline visible at far right. Note the man standing before the formation, for scale.

Source: https://courses.lumenlearning.com/geo/chapter/reading-folds/.

Should an anticline fall (i.e., is inclined toward the surface of the Earth), the surface strata will create Vs that point downward. Although faulting can obfuscate and complicate their relationship, synclines typically flank anticlines. Because of the shortening that occurs along with orogenic mountain building, folds frequently form during crustal deformation. Anticlines, such as ramp and rollover anticlines, are frequently created by movement on non-planar faults during both shortening and extension.

10.1.2.1. Terminology

An antiform is any fold with an upwardly convex form. Anticlines are antiforms that have younger rocks progressively extending from their core.

An anticline is a geological formation that has a crest at the highest point on a particular layer of rock within a fold. The hinge in an anticline is where there is the most bending or curvature of the rock layer. An axis is an imaginary line that connects the hinges in different rock layers in a two dimensional cross-section of the anticline. In three dimensions, connecting these hinges creates an axial plane. In a symmetrical anticline, the axial plane aligns with the crest. In an asymmetrical anticline, the axial plane will be shifted towards the steeper side of the fold. An overturned anticline is an asymmetrical fold where one limb has been tilted beyond perpendicular, resulting in upside-down rock layers.

A dome is a structure that plunges in all directions to create an elongated or circular shape. Usually, a single major deformation event, diapirism from underlying magmatic intrusions, or the movement of upwardly mobile, mechanically ductile material like shale or rock salt (salt dome) result in the formation of domes. The Sahara's Richat Structure is thought to be a dome that erosion has exposed.

A doubly plunging anticline is formed when an anticline plunges at both ends. This structure may result from multiple deformations, the superposition of two sets of folds, or the geometry of the underlying detachment fault and the varying amount of displacement along its surface. The highest point on a doubly plunging anticline, or any geological structure, is known as the "culmination."

A pericline is an extended dome that formed as the sediments were being deposited.

An anticlinorium is a collection of anticlinal folds found within a larger regional anticline. Some well-known examples include the Purcell Anticlinorium in British Columbia from the Late Jurassic to Early Cretaceous period, the Blue Ridge anticlinorium in northern Virginia and Maryland within the Appalachian Mountains, and the Nittany Valley in central Pennsylvania.

10.1.2.2. Economic Significance

Oil and gas drilling is more common in double-plumming or faulted anticlines, culminations, and structural domes because of the low density of petroleum, which allows it to migrate buoyantly upward to the highest points of the fold until it is stopped by a low-permeability barrier like an impermeable stratum or fault zone. Shale, limestone, sandstone, and even salt domes are examples of low-permeability seals that hold hydrocarbons, or oil and gas, underground. As long as the stratum has low permeability, the type of stratum itself is irrelevant.

Figure 10.5. Structural trap: anticlinal fold.

Source: https://courses.lumenlearning.com/geo/chapter/reading-folds/.

Periclines serve as key hubs for the accumulation of hot, metal-rich formational brines, which can result in the formation of uranium, lead-zinc, and manto ore deposits, among other types of deposits. The preferred locations for the deposition of saddle-reef-style lode gold deposits are culminations in folded strata that are fractured by faults and shears.

10.1.3. Synclines

A syncline in structural geology is a fold that has younger layers nearer the structure's center. A large syncline with smaller folds superimposed is called a synclinorium (plural synclinoriums or synclinoria). Synclines are usually a fold that descends; they are called synformal synclines (i.e., a trough); however, when strata have been turned over and folded, antiformal synclines—that is, synclines that point upward or perched—can be discovered.

Figure 10.6. The Sideling Hill syncline as exposed in the Interstate 68 roadcut to the west of Hancock, Maryland, USA.

Source: https://courses.lumenlearning.com/geo/chapter/reading-folds/.

10.1.3.1. Characteristics

A sequence of progressively younger rock layers at the center of the fold, or hinge, and a reverse sequence of the same rock layers on the opposite side of the hinge are characteristics of synclines on a geologic map. The structure is a basin if the fold pattern is circular or elongated circular. Folds are usually the result of compression that occurs with orogenic mountain building during crustal deformation.

Notable Examples:

- Powder River Basin, Wyoming, USA.
- In western Maryland, USA, there is a road cut called Sideling Hill along Interstate 68, which exposes the Purslane Sandstone and the Rockwell Formation beneath it.

- Western Lake Superior, which occupies a basin created by the Midcontinent Rift System
- Saou, a commune in the Drôme department in southeastern France
- The Catlins, an area in the southeastern corner of the South Island of New Zealand

10.1.4. Basins and Domes

A basin is a depression in the strata (rock layers) that resembles a bowl. Similar to a syncline, a basin has a single point at its center rather than an axis. The youngest rock is located in the center, where all of the strata descend toward it. When viewed in map view, the strata surround the center point in the form of concentric circles, or a bull's eye pattern.

A stratum bulge is called a dome. Similar to an anticline, a dome has a single point at its center rather than an axis. The oldest rock is located in the center, and all of the strata dip away from it. When observed from a map view, the strata encircle the central point in the shape of concentric circles, resembling a bull's-eye pattern.

10.2. CAUSES OF FOLDING

Folds can be found in all rock types, on all scales, and at all crustal depths. There are numerous reasons why they occur.

10.2.1. Layer-Parallel Shortening

A sequence of layered rocks can undergo deformation in several ways, such as homogeneous shortening, reverse faulting, or folding, when it is shortened parallel to its layering. The mechanical layering's thickness and the differences in characteristics between the layers determine the response. These characteristics also affect the fold style, should the layering start to fold.

Figure 10.7. Box fold in La Herradura Formation, Morro Solar, Peru.

Source: By Miguel Vera León – https://www.flickr.com/photos/miguelveraleon/543121589/in/photolist-PZD8k-8Bypg5-NqSu3-uSJvBr-58ms3b-uUTmhK-uESM3j-uHrdnB-uqBmdh-P24kz-2jdQHB-2xbbL8-21NSWQ-21JvUR-21JuST-PZGQ2-5ubRb8-PZCPX-21NSjj-tLkg42-xmr9oE-tSmomn-tSkXMn-NrgBB-tzL2qy-tzL4cQ-tS1EcC-Nrf6H-uHbEc6-NHoxN-5cqjPL-5Mbkoa-5ugiko-53LcU2-2jeJWF-PZ6nL-QfeAM-53KSce-NreDV-2jdQBR-QdLGN-2jeJSk-2jeJV2-2jj8Yf-2jieEY-NHQfc-PZDDB-5uggvq-PZ5UC-NqTyh, CC BY 2.0, https://commons.wikimedia.org/w/index.php?curid=46012512.

The folding is controlled by isolated, thick, competent layers in a less competent matrix. These layers usually produce the traditional, rounded buckle folds that are made possible by matrix deformation. Kink bands, box folds, and chevron folds are typically produced in the case of regular alternations of layers with contrasting properties, such as sandstone-shale sequences.

10.2.2. Fault-Related Folding

Numerous folds are closely linked to faults; they are involved in their displacement, propagation, and ability to accommodate strains between adjacent faults.

10.2.2.1. Fault Bend Folding

A non-planar fault's displacement results in fault-bend folds. As displacement increases in non-vertical faults, the hanging wall flexes to accommodate the mismatch across the fault. Both thrust faulting and extensional faulting can produce fault bend folds. Listric faults create rollover anticlines in their hanging walls as a result. Ramp anticlines occur in thrusting when a thrust fault splits a section between two detachment levels. The folding is produced by displacement along this ramp with a greater angle.

10.2.2.2. Fault Propagation Folding

Tip-line folds, also known as fault propagation folds, are created when displacement happens on an already-existing fault without further propagation. This causes the overlying sequence to fold in both reverse and normal faults, frequently taking the shape of a monocline.

10.2.2.3. Detachment Folding

Detachment folds, usually in the form of a box-fold, may develop when a thrust fault keeps moving above a planar detachment without further fault propagation. These typically happen above a good detachment, like the one on middle Triassic evaporites in the Jura Mountains.

10.2.3. Folding in Shear Zones

In shear zones exhibiting characteristics of simple shear, minor asymmetric folds are often present, with the direction of overturning matching the overall shear direction. Some of these folds feature highly curved hinge-lines and are known as sheath folds. The folds in shear zones may be inherited, influenced by pre-shearing layering orientation, or formed as a result of instability in the shear flow.

Figure 10.8. Dextral sense shear folds in mylonites within a shear zone, Cap de Creus

Source: By Mikenorton – Own work, CC BY-SA 3.0, https://commons.wikimedia.org/w/index.php?curid=25156216

10.2.4. Folding in Sediments

Freshly deposited sediments typically exhibit low mechanical strength and are prone to remobilization prior to lithification, which can result in folding. These formations are known as synsedimentary structures, which are formed during sedimentation, to differentiate them from folds of tectonic origin.

- Folding of slumps: Folding of slumps, especially at their leading edges, is a common emplacement process that occurs in poorly consolidated sediments. Sequences of sedimentary rocks can be used to infer paleoslope directions based on the asymmetry of the slump folds.
- Dewatering: Seismic activity may cause rapid dewatering of sandy sediments, which can result in convoluted bedding.
- Compaction: Differential compaction over older structures, like fault blocks and reefs, can produce folds in a younger sequence.

10.2.5. Igneous Intrusion

Igneous intrusions have a tendency to erode the nearby country rock. When deformation occurs above high-level intrusions close to the Earth's surface, it typically takes the shape of folding, much like the upper surface of a laccolith.

10.2.6. Flow Folding

The term "competence" refers to the compliance of rock layers; an incompetent layer is relatively weak, whereas a competent layer or bed of rock can withstand an applied load without collapsing. Rock that exhibits fluid behavior, such as rock salt or any other extremely weak rock that is sufficiently buried, usually exhibits flow folding, also known as passive folding due to the lack of resistance. This means that the strata appear to have shifted undistortedly, taking on the shape that the surrounding, more rigid rocks have impressed upon them. The strata are merely indicators of the folding. Numerous igneous intrusions and glacier ice also have this folding characteristic.

10.3. GEOMETRIC DESCRIPTION

The limbs, the axial plane, and the fold axis are some of the components that define a fold's geometry. Depending on the type of fold, these components can have different orientations and shapes. The following are some typical fold geometric characteristics:

- Fold Axis: The fold axis is a line that indicates the fold's maximum curvature and is drawn along its crest. Depending on the type of fold, the fold axis' orientation can be anywhere from horizontal to vertical and inclined at different angles.
- Axial Plane: The fold is divided into two limbs by the hypothetical axial plane. The angle formed by the limbs is divided in half by the axial plane, which is perpendicular to the fold axis. Depending on the type of fold, the axial plane's orientation can change from horizontal to vertical and can be inclined at different angles.
- Limbs: The two sides of the fold that are divided by the axial plane are known as the limbs. The type of fold determines the various shapes and orientations of the limbs. In anticlines, the limbs typically dip away from the axis, whereas in synclines, they dip towards the axis.
- Fold Plunge: The angle formed by the fold axis and the horizontal plane is known as the fold plunge. A fold may plunge in one of three directions: vertically, horizontally, or at an angle.

Important details about the kind and degree of rock deformation can be learned from the geometry of folds. For instance, the direction and strength of tectonic forces acting on the rocks can be inferred from the orientation and shape of folds.

10.4. FOLDING: MECHANISMS AND PROCESSES

The most prevalent type of structure in the crust of the earth are folds. One of the best indicators of the deformation of a specific rock mass or landform throughout geological history is this fold. They take place at the macro, meso, and micro scales. Their folding pattern and geometry provide insight into the type of deformation. They are also crucial economically for mineral exploration, ore mining, and oil traps. Why folds form in the rock mass and how do they change over time? We must observe in the field in order to learn about the folds. We can learn about the folding history from this study, which is useful for comprehending the mechanism. The mechanism that produces the various fold types is called the mechanism.

10.4.1. Folding Mechanisms

Layer-parallel compression, uneven loading oblique to perpendicular to layers, or surface irregularities amplified during deformation flow can all cause folds. Folding mechanism types are numerous.

10.4.1.1. Bending

Bending is the application of forces at high angles to layers with or without competence contrasts. A layer that is bent is similar to a notebook that is filled in the middle and supported at both ends. When the load is positioned in the middle of Figure 10.9, the notebook bends downward.

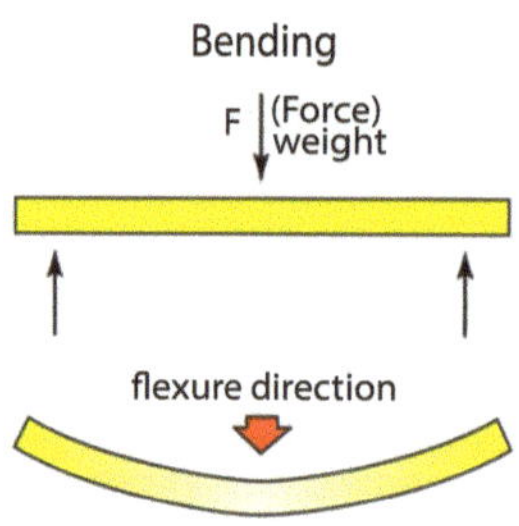

Figure 10.9. Folding mechanism by bending.

Source: https://www.uoanbar.edu.iq/eStoreImages/Bank/21079.pdf.

10.4.1.2. Buckling

A well-known active process for the formation of rounded folds in a competent layer is buckling (i.e., a layer with a low ductile flow rate) encased in an incompetent medium with a high ductile flow rate and enough viscosity contrast. This folding mechanism is replicated by gently pushing the two extremities of a paper sheet towards each other on a table. The sheet stays flat when the force is minimal. The force becomes abruptly curved as it is gradually increased. The development of a mechanical instability is the cause of this quick transition at a given force from a flat to a curved (buckled) form Figure 10.10.

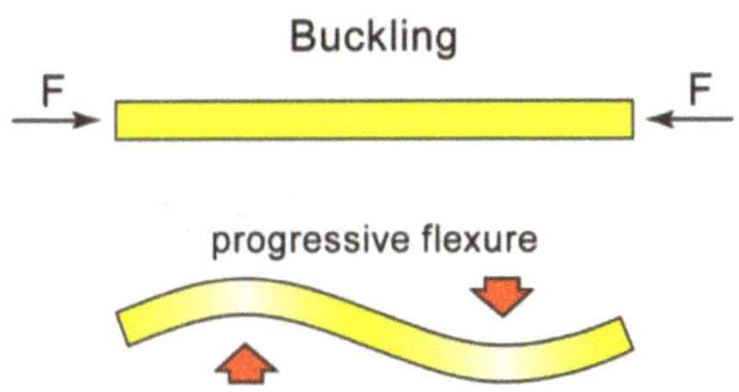

Figure 10.10. Folding mechanism by buckling.

Source: https://www.uoanbar.edu.iq/eStoreImages/Bank/21079.pdf.

10.4.1.2.1. Flexural-Slip and Flexural Flow

A multilayer can consist of alternating layers of sharply contrasting competence or a mound of competent layers divided by discontinuous surfaces. The mechanical effect is that the competent layers may readily slide in relation to one another on either side of the discontinuity or weak layer surface. When a stack deforms as a single layer, it is less possible for a fold to accommodate a greater flexure due to the shear "decoupling" of layers.

10.4.1.3. Flexural-Slip Mechanism

Discrete faulting, typically associated with bedding planes and accompanying folding, is described by the flexural-slip mechanism. Bending a book or stack of paper sheets is a classic simulation; increasing bending about the fold axis is accommodated by increasing slip between the book's pages or the pile's sheets. Because each sheet is the same thickness, they all fold in a parallel manner (i.e., surface layers stay parallel. Because layer-parallel stresses rise as limb rotation increases, slip plays a crucial role in folding (Figure 10.11).

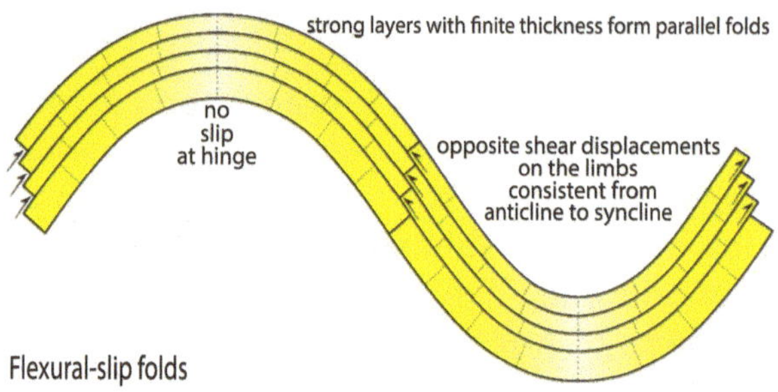

Figure 10.11. Folding mechanism by flexural slip.

Source: https://www.uoanbar.edu.iq/eStoreImages/Bank/21079.pdf.

10.4.1.4. Flexural Flow Mechanism

The bedding-parallel shear that is uniformly distributed within the ductile layer that is being folded between stiffer layers is described by the flexural flow mechanism. Similar to flexural slip, the axial plane exhibits an opposite bedding-parallel shear in the limbs.

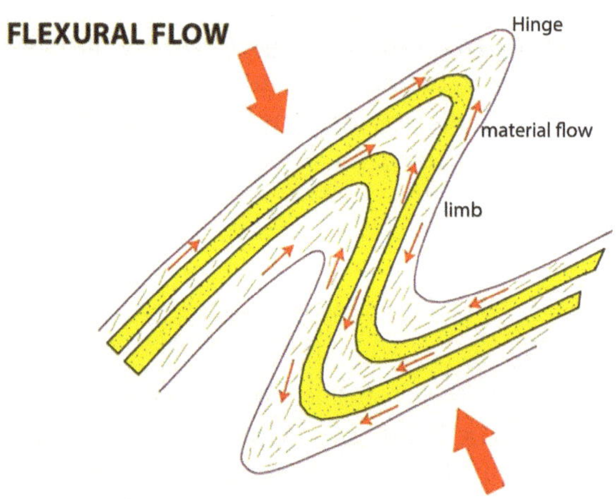

Figure 10.12. Folding mechanism by flexural flow.

Source: https://www.uoanbar.edu.iq/eStoreImages/Bank/21079.pdf.

The strain pattern caused by hingeward shear typically results in thicker hinges between limbs that have thinned, i.e., folds in flexural flow are largely comparable. Sometimes, weak layers that absorb bedding-parallel motion within larger parallel folds are subjected to flexural flow. In this instance, the incompetent layers experience flexural flow, while the stiff, active layers typically maintain their thickness throughout the deformation to create and regulate the overall shape of concentric and/or parallel folds. Figure 10.12 shows how ductile material exits the limbs and enters the hinges to preserve uniformity from bed to bed.

10.4.2. Folding Processes

The material enclosing the layers and body forces, which are both treated as viscous fluids in most models of fold formation, are ignored despite the fact that in reality these factors play a significant role in determining or changing the fold geometry. Curvature can take many forms in flat layers.

10.4.2.1. Rotation

Of course, folding causes some of the layers to rotate. The rock mass's compressive resistance decreases noticeably as a result of the stiff layers' constant orientational shift with respect to the shortening direction. This type of bulk weakening is brought about by a gradual alteration of the rock mass's internal geometry without a corresponding change in the material's characteristics (e.g., viscosity); structural softening is the term used to describe it. A rise in compressive resistance known as structural hardening occurs after structural softening. Instabilities must soften and harden in order to begin, grow, and eventually decay. It appears that any additional strain softening mechanism that might exist (e.g., a modification of the material's effective properties brought about by procedures like pressure solution, microfracturing, etc.) will also have an impact on the system's stability.

10.4.2.2. Lengthening

Fold hinges typically indicate the location where the folds started to enlarge. In order for any fold axis to plunge towards the two extremities of the hinge segment, this is typically done over a short distance. Lengthwise, though not always symmetrical, migration of the hinge tips is accompanied by additional limb rotation and the ensuing fold growth (amplification).

10.4.2.3. Mechanical Role of Layers: Active/Passive Folding

The displacement of material points, such as particles, occurs during any deformation. A passive particle only moves; it does not communicate with its neighbors. When an active particle interacts with its neighbors, the displacement of that particle is influenced by that of the neighbors.

In geology, the folding process and the strain pattern are influenced by the mechanical properties of compositional layering. Curvatures develop because, for instance, boundaries between layers with different strengths (viscosity) may slip or localize shear deformation. When deformation occurs at the layer scale, folding is active, and the deformation pattern is directly influenced by the strength difference between layers. On the other hand, passive folding occurs at the grain scale; layering only acts as a geometrical strain marker because layers lack significant competence contrasts and do not produce any stress acting across or parallel to the layer boundaries. The pattern of heterogeneous deformation is reflected in the shape of passive folds, which develop during heterogeneous flow. A stack of modeling paste layers is folded to test passive folding (consider oil traces or scum on slowly flowing water). One way to experiment with active folding is to bend a stack of cards that slide past one another.

10.5. FOLD INTERFERENCE PATTERNS AND REFOLDED FOLDS

Multiple folding episodes may cause areas to exhibit refolded folds in an astonishing range of orientations, resulting in a variety of fold interference patterns.

10.5.1. Classification of Fold Interference Patterns

We employ a classification (devised, like many other basic fold classifications, by J. Ramsay, 1967) to make sense of these. Only two generations of folds are covered by this classification; things become more complicated with three or more generations.

In each case we characterize a first generation of folds by:

- an axial surface
- a fold hinge that lies in that axial surface

We examine the effect of the second generation of folds on those two structures.

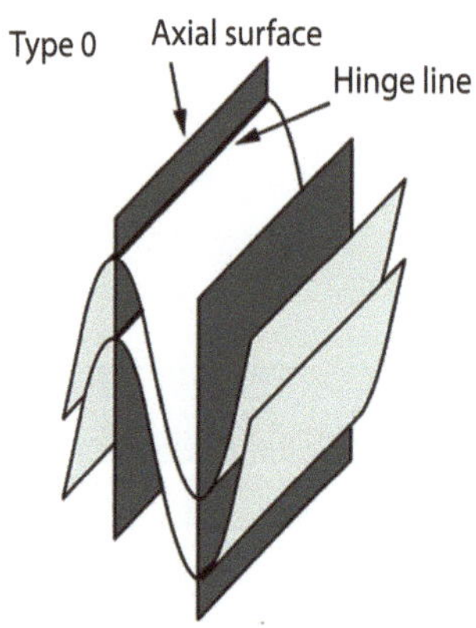

Figure 10.13. Type 0 fold interference pattern.

Source: https://geo.libretexts.org/Bookshelves/Geology/Geological_Structures_-_A_Practical_Introduction_(Waldron_and_Snyder)/01%3A_Topics/1.05%3A_Folds

In the second stage of deformation, the early folds are essentially just tightened if subsequent folding does not bend the axial surfaces or the hinges of the early folds. The fold pattern is known as "type 0," and geometry typically does not show that there have been two stages of deformation.

Type 1

Figure 10.14. Type 1 fold interference pattern.

Source: https://geo.libretexts.org/Bookshelves/Geology/Geological_Structures_-_A_Practical_Introduction_(Waldron_and_Snyder)/01%3A_Topics/1.05%3A_Folds

Later folding bends the early fold hinges but does not significantly bend the axial surfaces if later shortening is approximately parallel to the early fold hinges and the extension direction also lies in the early fold axial surfaces.

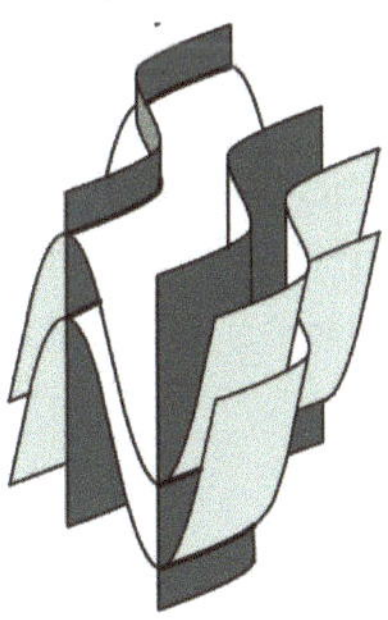

Type 2

Figure 10.15. Type 2 fold interference pattern.

Source: https://geo.libretexts.org/Bookshelves/Geology/Geological_Structures_-_A_Practical_Introduction_(Waldron_and_Snyder)/01%3A_Topics/1.05%3A_Folds

This usually results in oval outcrop patterns of layer traces on outcrop surfaces, as well as culminations and depressions on the early fold hinges. This pattern of fold interference, known as type 1, is characterized by basins and domes.

On the other hand, the early axial surfaces may also fold if the extension direction in the second deformation is at a high angle to them. In these conditions, complex shapes resembling bananas and mushrooms characterize outcrop patterns. This pattern of interference is referred to as type 2.

Type 3

Figure 10.16. Type 3 fold interference pattern.

Source: https://geo.libretexts.org/Bookshelves/Geology/Geological_Structures_-_A_Practical_Introduction_(Waldron_and_Snyder)/01%3A_Topics/1.05%3A_Folds

Thirdly, there's a situation where the extension direction is high angle to the early axial surfaces and the shortening direction is high angle to the early fold hinges. In these circumstances, the early axial surfaces are folded but the early hinges may stay roughly straight. The layering trace usually takes the form of a type 3 interference pattern, which is characterized by multiple zigzag shapes.

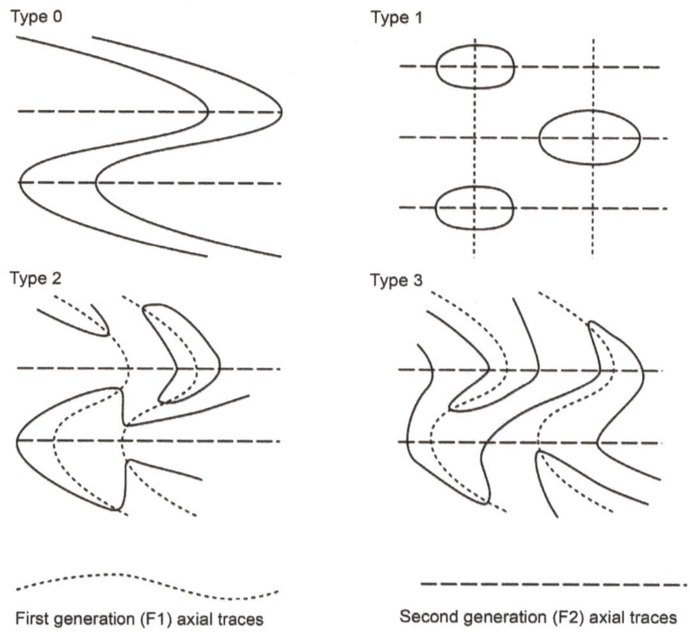

Figure 10.17. Interpretation of fold interference patterns.

Source: https://geo.libretexts.org/Bookshelves/Geology/Geological_Structures_-_A_Practical_Introduction_(Waldron_and_Snyder)/01%3A_Topics/1.05%3A_Folds

10.5.2. Analyze an Overprinted Fold Pattern

The most crucial first step in deciphering an overprinted fold pattern is to draw axial traces. The most recent set of axial traces is probably going to be roughly straight. Previous axial traces could be folded or not. (In general, type 1 interference patterns—in which there are no early axial surfaces folded—are the hardest to analyze; the folding order may not be able to be ascertained.)

The analysis of overprinted folds may also benefit from knowing the facing directions. For instance, downward-facing folds (antiformal synclines and synformal anticlines) are a reliable indicator of the presence of two or more generations of folds.

10.6. FOLDS IN SHEAR ZONES

A shear zone is a thin, severely deformed area in geology that results from the rock walls on either side of the zone sliding past one another. Shear zones occur in the upper mantle or crust of the Earth. The shear zone manifests as a fracture known as a fault in the brittle upper crust. The extreme pressure and temperature conditions in the lower crust and mantle cause the rock to become ductile. That is, like hot metal being worked by a blacksmith, the rock has the ability to gradually deform without breaking. The ductile rock in this area has gradually flowed to accommodate the relative motion of the rock walls on either side, creating a wider area known as the shear zone.

Shear zones are associated with a wide range of distinct rock types, each with their own distinctive structures, due to the fact that they can be found at varying depths.

Figure 10.18. Pegmatite dyke offset by a steeply dipping dextral shear zone, Cap de Creus.

Source: By Mikenorton – Own work, CC BY-SA 3.0, https://commons.wikimedia.org/w/index.php?curid=25156710

10.6.1. General Introduction

Rocks with a lower state of finite strain surround a zone of strong deformation, or high strain rate, which is known as a shear zone. A length to width ratio greater than 5:1 is one of its distinguishing features [Ramsay JG. (1987)]. The geological structures known as shear zones are a continuum that extends from brittle faults or shear zones to brittle–ductile or semibrittle, ductile–brittle, and finally ductile shear zones. Whereas the deformation in a ductile shear zone is dispersed over a larger zone, with the deformation state fluctuating continuously from wall to wall, the deformation in brittle shear zones is concentrated in a narrow fracture surface dividing the wall rocks. These geometric features can be combined in different proportions to form intermediate types of brittle–ductile (semibrittle) and ductile–brittle shear zones between these end-members.

The various deformation mechanisms that control the crust are reflected in the continuum present in the structural geometries of shear zones, i.e., the transition at or near the surface between brittle (fracturing) and ductile (flow) deformation, which occurs as depth increases. The ductile response to deformation is beginning to take hold as the material moves through the brittle-semibrittle transition. This transition happens across a certain depth range, known as the "alternating zone," where plastic flow and brittle fracturing coexist, rather than being restricted to a particular depth. The primary cause of this is that rocks typically have a heteromineral composition, meaning that different minerals react to applied stresses in different ways (e.g., quartz undergoes plastic reaction in response to stress much earlier than feldspars do). Thus, a different rheological response is determined by variations in lithology, grain size, and preexisting fabrics. The changeover depth is also influenced by additional, purely physical factors, such as:.

- geothermal gradient, i.e., ambient temperature.
- confinement pressure and fluid pressure.
- bulk strain rate.
- stress field orientation.

According to Scholz's model, the brittle-semibrittle transition begins at a depth of approximately 11 km and an ambient temperature of 300°C in a quartzo-feldspathic crust (using a geotherm from Southern California). After that, the underlying alternating zone reaches a depth of about 16 km and a temperature of roughly 360°C (Scholz, 2002). All that is present below a depth of about 16 km are ductile shear zones. The brittle domain, the schizosphere, is connected to the seismogenic zone, where earthquakes originate. The plastosphere is located beneath an alternating zone that acts as a mediator. True cataclasites begin to form in the seismogenic layer, which is located beneath an upper stability transition associated with an upper seismicity cutoff (typically located at a depth of 4-5 km). At a depth of 11 km, the seismogenic layer gives way to the alternating zone. Large earthquakes, however, have the ability to rupture deeply into the alternating zone, sometimes even reaching the plastosphere.

10.6.2. Rocks Produced in Shear Zones

The formation of distinctive fabrics and mineral assemblages reflecting the dominant pressure–temperature (pT) conditions, flow type, movement sense, and deformation history is caused by the deformations in shear zones. As a result, shear zones are crucial structures for understanding the past of a particular terrane.

A shear zone typically contains the following rock types, starting at the Earth's surface:

- uncohesive fault rocks. Examples being fault gouge, fault breccia, and foliated gouge.
- cohesive fault rocks like crush breccias and cataclasites (protocataclasite, cataclasite, and ultracataclasite).
- glassy pseudotachylites.

Both fault gouge and cataclasites are due to abrasive wear on brittle, seismogenic faults.

- foliated mylonites (phyllonites).
- striped gneiss.

When semibrittle behavior appears in the alternating zone with adhesive wear, manganites begin to form. Here, pseudotachylites are still present. Pseudotachylites vanish as one enters greenschist facies conditions, leaving only various mylonites. High-grade mylonites known as striped gneisses are found at the base of ductile shear zones.

10.6.3. Sense of Shear

Numerous microtectonic indicators and macroscopic structures can be used to infer the type of shear (dextral, sinistral, reverse, or normal) in a given shear zone.

10.6.3.1. Indicators

Slickensides, slicken fibers, and stretching- or mineral-lineations are the primary macroscopic indicators. They show which way the object is moving. The sense of shear can also be ascertained by using offset markers, such as displaced layering and dykes, or by bending or deflecting layering/ foliation into a shear zone.

Sheath folds and en echelon tension gash arrays, also known as extensional veins, which are indicative of ductile-brittle shear zones, can also be useful macroscopic shear-sense markers. The following structures make up microscopic indicators:

- asymmetric folds.
- foliations.
- imbrications.
- Crystallographic preferred orientation (CPO).
- mantled and winged porphyroclasts. Well-known examples are theta (Θ)-objects and phi (Φ)-porphyroclasts, as well as sigma (σ)- and delta (δ)-winged objects.
- mica fish (foliation fish).
- pressure shadows
- quarter structures.
- shear band cleavages.
- step-over sites.

10.6.4. Width of Shear Zones and Resulting Displacements

A single shear zone's width can be measured in kilometers as well as grains. Crustal scale shear zones, also known as megashears, can exhibit displacements ranging from tens to hundreds of kilometers, and they can even reach a width of 10 km. Faults, also known as brittle shear zones, typically enlarge as displacements increase and depth increases.

10.6.5. Strain Softening and Ductility

Since strain is localized in shear zones, strain softening of some kind is necessary for the impacted host material to deform more plastically. These phenomena have the potential to cause the softening:

- grain-size reductions.
- geometric softening.
- reaction softening.
- fluid-related softening.

The following deformation mechanisms (on a grain scale) must also be considered for a material to become more ductile (quasi-plastic) and undergo continuous deformation (flow) without fracturing:

- diffusion creep (various types).
- dislocation creep (various types).
- dynamic recrystallization
- pressure solution processes.
- grain-boundary sliding (superplasticity) and grain-boundary area reduction.

10.6.6. Occurrence and Examples of Shear Zones

All metamorphic facies have shear zones because of their deep penetration. The upper crust is essentially devoid of brittle shear zones. Ductile shear zones are limited to metamorphic terranes because they begin at greenschist facies conditions. The following geotectonic conditions can result in shear zones:

- transcurrent setting – steep to vertical:
  - strike-slip zones.
  - transform faults.
- compressive setting – low-angle:
  - recumbent fold nappes (at the base).
  - subduction zones.
  - thrust sheets (at the base).
- extensional setting – low-angle:
  - metamorphic core complex detachments.

Shear zones are independent of both geological age and type of rock. They typically form fractal-scaled, linked, anastomosing networks that, in their arrangement, reflect the underlying dominant sense of movement of the terrane at that particular time rather than being isolated in their occurrence. The North and South Armorican Shear Zones in Brittany, the North Anatolian Fault Zone in Turkey, and the Dead Sea Fault in Israel are a few notable instances of strike-slip shear zones. The San Andreas Fault in California and the Alpine Fault in New Zealand are examples of transform-type shear zones. The Moine Thrust in northwest Scotland is an example of a thrust-type shear zone. The Japan Median Tectonic Line is an illustration of a subduction zone. Shear zones associated with separation faults are located in southeast California, e.g., the Detachment Fault of Whipple Mountain. The Brazilian Borborema Shear Zone is an illustration of a large anastomosing shear zone.

10.6.7. Importance

Shear zones are significant weak points in the Earth's crust that can occasionally reach the upper mantle, which makes them important. These characteristics often exhibit evidence of multiple overprinting stages of activity and can have very long lifespans. They can carry materials either upwards or downwards, with water circulating dissolved ions being the most significant example. This may even re-fertilize mantle material and cause metasomatism in the host rocks. Important gold deposits in Precambrian terranes are one example of how economically viable mineralizations can be found in shear zones.

10.7. FOLDING AT SHALLOW CRUSTAL DEPTHS

Folding is a ductile deformation process that maintains layer continuity at the observational scale, in contrast to faulting and fracturing. Folding is supported at the microscale by crystal-plastic and diffusion processes in the primarily plastic middle and lower crust. Folding in the upper, primarily brittle crust is primarily supported by brittle deformation mechanisms, i.e., with the exception of carbonaceous layers, where twinning and wet solution (pressure solution) are also significant, and salt layers, where crystal-plastic processes take place. However, viscosity contrasts—or lack thereof—between the involved layers determine fold geometry rather than the deformation mechanism itself (Hudleston and Treagus, 2010). For instance, room temperature active folding experiments are usually conducted on frictionally deforming layers of sand and clay. These experiments yield folds with geometric properties that are directly analogous to naturally occurring folds with entirely distinct microscale deformation mechanisms that form at high P-T conditions deep within the crust. Elasticity contributes to buckling as well, but temperature-controlled deformation mechanisms and elasticity do not directly correlate with one another. This also applies to passive folding, which necessitates little to no viscosity contrast between the layers. Therefore, folds formed in any material at any crustal level are subject to the same laws and principles. Nevertheless, there are features of the folding of rocks and sediments in the upper crust that have significant ramifications. These include the estimation of strain, the exploration and exploitation of hydrocarbons, the sequestration of CO_2, and the quantification of fold evolution through the use of syn-kinematic sedimentary sequences. We'll concentrate on these traits in this section.

The low temperature and pressure conditions of the upper crust allow for the formation of the full geometric range of fold types, but two general characteristics of upper crustal folds can be identified from observations. Firstly, passive folding and bending are especially frequent. This is especially true for easily bent and sheared sedimentary rocks and soft sediments. Second, faulting and upper crustal folding are frequently closely associated. Indeed, faults are a defining characteristic of upper crustal deformation; they allow for highly localized deformation that is easily reflected in their walls, especially in lithologies and soft sediments. Faults and folds together represent brittle-ductile behavior; ductile deformation is defined as that which maintains layer continuity at the observational scale, whereas brittle deformation breaks up layers and other markers. We will go into more detail about fault-related folds below due to their significance for petroleum geology and foreland orogenic deformation.

10.7.1. Detachment Folding

At different scales in the uppermost crust, folding of more competent layers occurs above a weak detachment or décollement, without a ramp fault coring the uplift (Figure 10.19a). They are typically found in fold-and-thrust belts that are both deep underwater and subaerial, like the Jura Mountains (Figure 10.19b), the Canadian Rocky Mountains, the Appalachian fold belt, the western Gulf of Mexico (e.g., the Zagros fold-and-thrust belt (Yarbuh and Contreras, 2015), and have been crucial to the kinematics of thin-skinned tectonics. Additionally, they can form as gravity-driven structures above brittle layers, like the margins of continental downslope regions. Usually composed of salt or overpressured shale, the décollement is layered with more capable sandstone or limestone. These capable layers produce buckle folds that adhere to the general buckling laws in terms of wavelength, amplitude, and other geometric characteristics.

They exhibit lift-off, multi-detachment, symmetric, asymmetric, and disharmonic styles and can have concentric, chevron, or box shapes. But the folds abruptly disappear in the direction of the underlying detachment. This detachment, which is similar to the folds created by pushing a carpet on a floor, divides the less-or undeformed footwall from the shortened hanging wall. If there is enough ductile material in the detachment to allow it to pass through the core of the growing anticline, then tight and disharmonic folds may evolve in the core of the detachment folds. Shortening is typically caused by secondary faults in the antiform as a result of local space issues if there is insufficient material to fill the fold's core, as shown in Figure 10.19b.

Evaluation of the dominant folding mechanism can also be done using fold geometries at shallow crustal depths. Schmalholz et al., (2002) identified three distinct folding mechanisms based on the material parameters that controlled the folding: (i) folding that was matrix-controlled (regulated by the viscosity ratio between the layer and matrix), (ii) folding that was detachment-controlled (regulated by the thickness of the weak layer beneath a strong layer), and (iii) folding that was gravity-controlled (regulated by the gravity to viscous stress ratio). A phase diagram based on the ratios of wavelength to total thickness and matrix thickness to layer thickness was also displayed by them. This diagram makes it possible to determine the predominant folding mechanism just by looking at the fold geometry.

10.7.2. Fault-Bend Folding

The forced folding of hanging-wall layers as they pass over a fault bend or kink with flat-ramp-flat geometry is known as fault-bend folding (i.e., a trajectory of a staircase). Rich (1934) identified these folds for the first time in the southern Appalachian fold-and-thrust belt's Pine Mountain. When thrusting occurs, the bend is usually connected to a ramp because the detachment fault connects two weak detachment levels by transferring displacement to a higher stratigraphic level. The limbs of fault-bend folds are typically symmetrical. A pair of anticlines and synclines is the result of movement of the hangingwall layer over the fault bend. Folding happens in two circumstances. The fault-bend fold's backlimb is where the hanging-wall layers first bend upward along the ramp with continuous limb dips before bending back to horizontal at the upper detachment level. On the other hand, fault-bend folds passively move their forelimb along the upper flat. Since the layering is parallel to the fault, bedding interfaces are prone to slippage, which causes the folds to become flexural-slip folds. The first geometric models of fault-bend folds were based on the tensile slip and flow, which imply conservation of area and bed-length. Based on well and outcrop data, these simplified, kink-like models made geometric construction simple.

The fault parallel flow model is a related model in which hanging-wall particles travel parallel to the fault at the same speed (Ziesch et al., 2014). The bed thickness varies in this model. Layer thickness variations and kink-like geometries are also implied by the vertical or inclined shear models, in which particles only travel along vertical or continuously inclined paths. The majority of fault-bend folds are curved rather than kink-like, and the more recent trishear model has been employed, along with accurate measurement techniques and fold geometry classification. It might also indicate that some of the underlying assumptions are false or erroneous, or that some of the schemes only accurately capture the fold geometry in specific regions of fold structures. However, different fold geometries in various deformation zones and tectonic settings, geometrical and/or kinematic interactions of fold structures, different rheology, pressure, and temperature conditions can also contribute to such variations in results and the accuracy of classification schemes.

The simplified and idealized fold forms are necessary. Researchers who confine themselves to 2D idealized schemes distance themselves from the intricacy of natural fold structures and instead reinforce rather than challenge these idealized models. In a similar vein, 2D fold profiles have their uses, but since folds are by nature 3D structures, they can vary in terms of length scale, amplitude, wavelength, arc-length, closure direction, fracture patterns, etc. so that it is preferable to treat them in three dimensions because the majority of them exhibit non-cylindrical geometries, nonplanar axial surfaces, and related cleavage. However, obtaining sufficient information from natural macroscopic and regional folds to accurately assess their geometry, kinematics, and mechanics is typically challenging. To put it another way, based solely on field observations, geometric descriptions of folds such as amplitude-wavelength ratio, thickness variations, and interlimb angles only reveal a limited amount of information about the type and mechanism of deformation. Future research in this area should concentrate on combining field-based observations, subsurface data sets, and three-dimensional numerical modeling of folds in various model configurations (the number of layers, thickness of the layers, type of perturbation, and amplitude of the perturbation within the layer) (Figure 10.19).

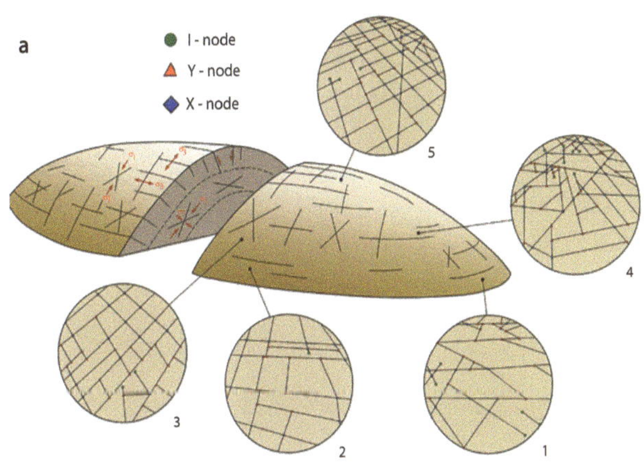

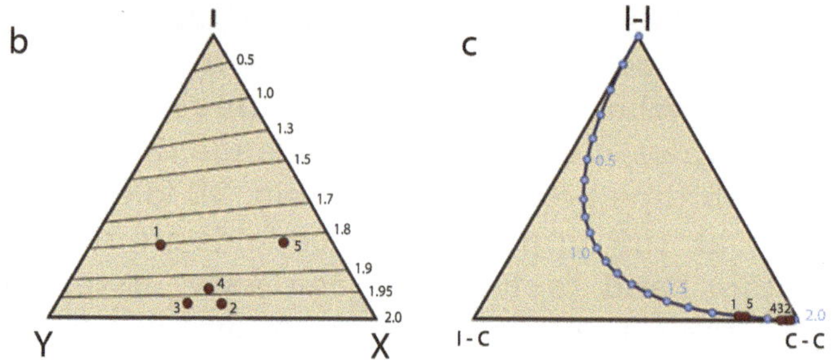

Sample Number	N_I	N_Y	N_X	N_B	C_B	I - I	I - C	C - C
1	6	18	11	52	1.8	1	3	37
2	1	11	12	41	1.97	0	1	30
3	2	20	14	59	1.96	0	2	51
4	7	36	34	126	1.94	0	7	69
5	3	26	44	128	1.82	0	3	123

Figure 10.19. (a) Schematic illustration of fold-related fracture networks on the Maleh-Kuh periclinal fold structure, showing how the orientations and topology of fracture networks vary depending on the structural position. Circular trace maps of five fracture networks exhibit different node and branch types. The diameter of each circle is 2 m. (b and c) Ternary diagrams of the different node and branch types with dark purple dots representing the topology of fracture network traces in (a). The contours represent the average number of connections per branch (CB) (i.e., node triangle) (b) and distribution of I-I (isolated), I-C (partly connected) and C-C (doubly connected) branches (i.e., branch triangle) (c), along with the table of computed data, and how they change relative to the topology. NI, NY, NX and NB, respectively, are the number of Inode, the number of Y-node, the number of X-node, and the number of total branches t.

Source: Seyed Tohid Nabavi, Haakon Fossen, "Fold geometry and folding – a review," Earth-Science Reviews 222 (2021) 103812, https://doi.org/10.1016/j.earscirev.2021.103812

Geometrical techniques for classifying folds should prioritize practical approaches, as geologists need to be able to use them in the field to identify and characterize various fold types with ease. It could be difficult due to the abundance of schemes and techniques used for fold classification and characterization. However, every scheme has benefits, and in order to select the best scheme for a given data set or area, one must be aware of the various schemes.

CLASS ACTIVITY

COMPARING SOIL PROFILES

You will notice different soil layers if you examine the earth in a soil pit or on a roadside cut. We refer to these strata as soil horizons. A soil profile is the configuration of these horizons in a soil. In order to categorize and interpret the soil for diverse purposes, soil scientists—also known as pedologists—observe and characterize soil profiles and soil horizons.

Many easily observable soil characteristics, including color, texture, structure, and thickness, vary amongst soil horizons. Some attributes are not as noticeable. Specific laboratory testing is needed for properties like consistency, reaction, and mineral and chemical content. Soil horizon types are defined by all of these characteristics.

To identify the master horizons, soil scientists use the capital letters O, A, B, C, and E. To distinguish between these horizons, they use lowercase letters. The three main horizons found in most soils are the substratum (C), the subsoil (B), and the surface horizon (A). An organic horizon (O) may be present on the surface of some soils, but it can also be buried. Eluviation, or the substantial loss of minerals from subsurface horizons, is referred to as the master horizon, or E. R is used for hard bedrock, which is not soil.

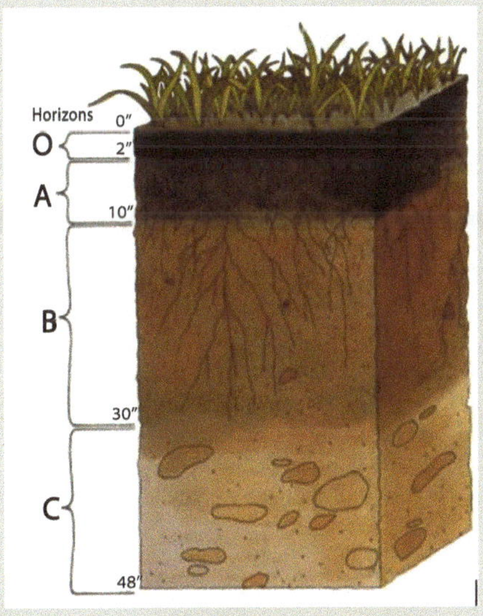

Figure 1. A Soil Profile.

Source: https://www.nrcs.usda.gov/resources/education-and-teaching-materials/a-soil-profile.

1. You can either draw your own design on a 3" x 5" note card or print the soil profile cards (above) onto cardstock paper.
2. Divide the cards into pieces.
3. Put a little piece of carpet tape on the card. Double-sided tape rolls are available in different widths. You only need one-inch of tape.
4. To indicate the depth of this soil, place soil from the surface horizon after pulling back some of the sticky tape at the top.
5. Using the same method, peel back the tape for each successive layer one at a time.
6. To understand more about the relationships between plants and soil, you can also gather some surface vegetation to keep with your type of soil.

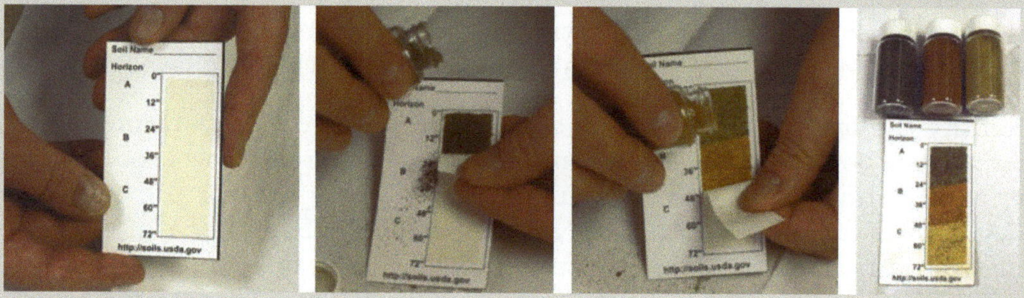

Figure 2. Various types for soils found in different locations.

Source: https://www.nrcs.usda.gov/resources/education-and-teaching-materials/a-soil-profile.

By changing the A horizon's depth or displaying different types of soil for various locations, you can illustrate the severity of erosion. Utilizing the tape on a larger card or poster, displays of multiple soils can be created on a drawing of a hill slope or other landscape.

SUMMARY

- Ductile rocks behave plastically and become folded in response to stress. Even in the shallow crust where rocks are cool and relatively brittle, folding can occur if the stress is slow and steady and gives the rock enough time to gradually bend.
- The most basic types of folds are anticlines and synclines. Imagine a rug, the sides of which have been pushed toward each other forming ridges and valleys – the ridges are "up" folds and the valleys are "down" folds. In terms of geologic structures, the up folds are called anticlines and the down folds are called synclines.
- In structural geology, an anticline is a fold that is convex up and has its oldest beds at its core. The term is not to be confused with antiform, which is a purely descriptive term for any fold that is convex up. Therefore if age relationships between various strata are unknown, the term antiform should be used.
- Synclines are typically a downward fold, termed a synformal syncline (i.e., a trough); but synclines that point upwards, or perched, can be found when strata have been overturned and folded (an antiformal syncline).
- A basin is a bowl-like depression in the strata (layers of rock). A basin is similar to a syncline, but instead of an axis it has a single point at the center.
- A dome is a bulge in strata. A dome is similar to an anticline, but instead of an axis it has a single point at the center. The strata all dip away from the center point and the oldest rock is at the center. In map view, the strata form concentric circles – a bull's eye pattern – around the center point.
- The geometry of a fold is defined by several elements, including the fold axis, the axial plane, and the limbs. These elements can vary in orientation and shape depending on the type of fold.
- Folds can result from layer parallel compression, uneven loading oblique to perpendicular to layers, or from amplification of surface irregularities during deformation flow.
- Most models of fold formation ignore body forces and the effect of the material enclosing the layers (both are treated as viscous fluids), which in practice has a very important role in determining or modifying the fold geometry.
- Areas affected by multiple episodes of folding may display refolded folds in a bewildering variety of orientations, producing a variety of fold interference patterns.
- When trying to understand an overprinted fold pattern, the most important first step is to draw axial traces. In general, the latest set of axial traces is likely to be approximately straight. Earlier axial traces may or may not be folded.
- A shear zone is a thin zone within the Earth's crust or upper mantle that has been strongly deformed, due to the walls of rock on either side of the zone slipping past each other. In the upper crust, where rock is brittle, the shear zone takes the form of a fracture called a fault. In the lower crust and mantle, the extreme conditions of pressure and temperature make the rock ductile.

REVIEW QUESTIONS

1. What is a folded structure?
2. What are the geometric aspects of fold?
3. What is the folding mechanism?
4. What are the processes of folding?
5. What are the different types of folds?
6. What is a fold interference pattern?
7. What is fold zone?

REFERENCES

1. Billings, M. P. (1990). Structural Geology. Prentice Hall of India Pvt Ltd. ISBN-0-87692-059-8, p. 606.
2. Fossen, H. (2010). Structural Geology. Cambridge University Press. ISBN 978-0-521-51664-8.
3. Hudleston, P. J., & Treagus, S. H. (2010). Information from folds: A review. Journal of Structural Geology, 32, 2042–2071. https://doi.org/10.1016/j.jsg.2010.08.011.
4. Ramsay, J. G. (1967). Folding and Fracturing of Rocks. San Francisco: McGraw Hill. ISBN: 0070511705, 9780070511705.
5. Ramsay, J. G. (1987). The Techniques of Modern Structural Geology. Volume 2: Folds and Fractures. Academic Press. ISBN 0-12-576902-4
6. Rich, J. L. (1934). Mechanics of Low-Angle Overthrust Faulting as Illustrated by Cumberland Thrust Block, Virginia, Kentucky, and Tennessee. American Association of Petroleum Geology Bulletin, 18(12), 1584–1596. https://doi.org/10.1306/3D932C94-16B1-11D7-8645000102C1865D.
7. Schmalholz, S. M., Podladchikov, Y. Y., & Burg, J. P. (2002). Control of folding by gravity and matrix thickness: Implications for large-scale folding. Journal of Geophysical Research: Solid Earth, 107(B1). https://doi.org/10.1029/2001JB000355.
8. Scholz, C. H. (2002). The Mechanics of Earthquakes and Faulting. Cambridge University Press. ISBN 0-521-65540-4.
9. Yarbuh, I., & Contreras, J. (2015). The interplay between deformation, erosion and sedimentation in the deep-water Mexican Ridge foldbelt, western Gulf of Mexico basin. Basin Research, 29, 446–464. https://doi.org/10.1111/bre.12157.
10. Ziesch, J., Tanner, D. C., & Krawczyk, C. M. (2014). Strain associated with the fault-parallel flow algorithm during kinematic fault displacement. Mathematical Geosciences, 46, 59–73. https://doi.org/10.1007/s11004-013-9464-3.

Index

E

FSC
www.fsc.org
MIX
Papier | Fördert gute Waldnutzung
FSC® C083411

Kolibri 360 GmbH
Ferdinand-Jühlke-Straße 7
99095 Erfurt, Deutschland
produktsicherheit@kolibri360.de